Springer Proceedings in Physics 63

Springer Proceedings in Physics

Managing Editor: H. K. V. Lotsch

44 *Optical Fiber Sensors*
Editors: H. J. Arditty, J. P. Dakin, and R. Th. Kersten

45 *Computer Simulation Studies in Condensed Matter Physics II: New Directions*
Editors: D. P. Landau, K. K. Mon, and H.-B. Schüttler

46 *Cellular Automata and Modeling of Complex Physical Systems*
Editors: P. Manneville, N. Boccara, G. Y. Vichniac, and R. Bidaux

47 *Number Theory and Physics*
Editors: J.-M. Luck, P. Moussa, and M. Waldschmidt

48 *Many-Atom Interactions in Solids*
Editors: R .M. Nieminen, M. J. Puska, and M. J. Manninen

49 *Ultrafast Phenomena in Spectroscopy*
Editors: E. Klose and B. Wilhelmi

50 *Magnetic Properties of Low-Dimensional Systems II: New Developments*
Editors: L. M. Falicov, F. Mejía-Lira, and J. L. Morán-López

51 *The Physics and Chemistry of Organic Superconductors*
Editors: G. Saito and S. Kagoshima

52 *Dynamics and Patterns in Complex Fluids: New Aspects of the Physics – Chemistry Interface*
Editors: A. Onuki and K. Kawasaki

53 *Computer Simulation Studies in Condensed Matter Physics III*
Editors: D. P. Landau, K. K. Mon, and H.-B. Schüttler

54 *Polycrystalline Semiconductors II*
Editors: J. H. Werner and H. P. Strunk

55 *Nonlinear Dynamics and Quantum Phenomena in Optical Systems*
Editors: R. Vilaseca and R. Corbalán

56 *Amorphous and Crystalline Silicon Carbide III, and Other Group IV – IV Materials*
Editors: G. L. Harris, M. G. Spencer, and C. Y.-W. Yang

57 *Evolutionary Trends in the Physical Sciences*
Editors: M. Suzuki and R. Kubo

58 *New Trends in Nuclear Collective Dynamics*
Editors: Y. Abe, H. Horiuchi, and K. Matsuyanagi

59 *Exotic Atoms in Condensed Matter*
Editors: G. Benedek and H. Schneuwly

60 *The Physics and Chemistry of Oxide Superconductors*
Editors: Y. Iye and H. Yasuoka

61 *Surface X-Ray and Neutron Scattering*
Editors: H. Zabel and I. K. Robinson

62 *Surface Science: Lectures on Basic Concepts and Applications*
Editors: F. A. Ponce and M. Cardona

63 *Coherent Raman Spectroscopy: Recent Advances*
Editors: G. Marowsky and V. V. Smirnov

64 *Superconducting Devices and Their Applications*
Editors: H. Koch and H. Lübbig

Volumes 1–43 are listed on the back inside cover

G. Marowsky V. V. Smirnov (Eds.)

Coherent Raman Spectroscopy

Recent Advances

Proceedings of the International Symposium on Coherent Raman Spectroscopy, Samarkand, USSR, September 18–20, 1990

With 196 Figures

Springer-Verlag
Berlin Heidelberg New York
London Paris Tokyo
Hong Kong Barcelona
Budapest

Prof. Dr. Gerd Marowsky
Abteilung Laserphysik, Max-Planck-Institut für Biophysikalische Chemie, Am Fassberg, W-3400 Göttingen, Fed. Rep. of Germany

Prof. Dr. Valery V. Smirnov
General Physics Institute, Academy of Sciences, Vavilov Street 38, SU-117942 Moscow, USSR

Symposium Chairmen

Academician Prof. A. M. Prokhorov
Academician Prof. A. K. Atakhodjaev
Prof. V. V. Smirnov

ISOCRS Organizers

Academy of Sciences of the USSR
General Physics Institute of the USSR Academy of Sciences
Samarkand State University, Uzbekistan Soviet Republic, USSR

ISOCRS Sponsors

General Physics Institute of the USSR Academy of Sciences
Moscow State University
Institute of Physics of the Belorussian Academy of Sciences
Coherent Physics GmbH, FRG
Quantel, France

ISBN-13:978-3-642-77196-5 eISBN-13:978-3-642-77194-1
DOI:10.1007/978-3-642-77194-1

Softcover reprint of the hardcover 1st edition 1992

Typesetting: Camera ready by authors

54/3140-543210 – Printed on acid-free paper

Preface

Progress made during the last few years in the field of nonlinear optics and quantum electronics has significantly increased our understanding of the interaction between light and matter and led to the development of new spectroscopic techniques. Of great importance are the methods based upon coherent Raman scattering processes, such as CARS (Coherent Anti-Stokes Raman Scattering), RIKES (Raman Induced Kerr Effect Spectroscopy), and SRS (Stimulated Raman Scattering). In the past, scientific results obtained with these Raman techniques were presented at a variety of conferences dealing with Raman or molecular spectroscopy.

In 1990 an international symposium on coherent Raman spectroscopy was organized in Samarkand (Uzbekistan, USSR) and scientists from many disciplines came together to discuss their common interest in coherent scattering spectroscopy methods, techniques, and applications. The symposium provided an informal atmosphere in which approximately 100 leading scientists from 13 countries could discuss the fundamentals and applied problems of the various coherent Raman scattering processes.

These proceedings reflect the state of the art in this field. In particular, they provide an overview of the various highly efficient coherent Raman techniques and devices that make available novel information about the structure of energy levels, collisional dynamics of atoms and molecules, and processes of internal molecular energy transformation. In addition, these techniques allow the creation of practical local nonperturbing diagnostic methods for the determination of gas parameters such as chemical composition, temperature, density, velocity, and the energy distribution between the internal degrees of freedom. The contributions to this book report the latest theoretical and experimental results in the field of coherent Raman techniques, grouped under the following headings: New Techniques and Methods, High-Resolution Spectroscopy, Studies of Nonstationary Processes, Selected Applications.

We would like to thank all participants and contributors to the symposium. Especially, we would like to thank the members of the steering committee for their efforts and enthusiasm, without which the success of this meeting would not have been possible. In addition, we express our appreciation to our sponsors for their financial support. The editors owe special thanks to the Academy of Sciences of the USSR and the Deutsche Forschungsgemeinschaft, who enabled a scientific exchange programme in the field of CARS spectroscopy, facilitating the participation of several German scientists in the Samarkand symposium. In addition the editors thank Dr. H. Lotsch and D. Hollis of Springer-Verlag for their continuous interest in and support of this scientific "joint venture" between East and West, using these terms as they were understood at the beginning of the 1990s.

Göttingen,
Moscow,
June 1991

G. Marowsky
V.V. Smirnov

Contents

Part I

New Techniques and Methods

Infrared Resonant CARS in CH_3F

V.A. Batanov, V.S. Petriv, A.O. Radkevich, A.L. Telyatnikov, and A.Yu. Volkov

Institute of Physics and Technology, USSR Academy of Sciences, Krasikova 25A, SU-117218 Moscow, USSR

The model developed in /1-4/ describing the behavior of multi-photon processes in fully resonant molecular media allows us to describe four-photon parametric processes CARS and CSRS as well. It was shown in /5/ that fully resonant CARS and CSRS processes are of great interest in the field of investigation of active media. At least a 4-level fully resonant scheme should be used for adequate description of spectroscopic features of these resonant CARS and CSRS processes.

In the present paper we discuss the specific features of resonant infrared CARS and CSRS processes for the example of the active media of far infrared (FIR) lasers. A distinguishing feature of our model is the absence of any limitations on the intensities of the electric fields involved and on the frequency detunings from resonance.

Unlike the model discussed in /1/, we discuss here a four-photon IR parametric process (Fig.1), which describes degenerate four-wave mixing of three IR fields, which relate to corresponding rovibrational transitions between the levels of a symmetric top molecule (in this particular case we consider CH_3F).

The polarization of an active medium at the interacting fields' frequencies is defined by a system of steady-state density-matrix equations:

$$i\hbar(\omega_\gamma-\omega_{ij})\rho_{ij}=[H,\rho]_{ij}-i\hbar\frac{\rho_{ij}}{\tau_{ij}}\;; \qquad \frac{\rho_{ii}-\rho^e_{ii}}{\tau_{ii}}=\frac{[H,\rho]_{ii}}{i\hbar}\,. \tag{1}$$

The perturbation Hamiltonian is given by

$$H=-(\mu_{13}\varepsilon_L)-(\mu_{24}\varepsilon_L)-(\mu_{23}\varepsilon_S)-(\mu_{14}\varepsilon_A), \tag{2}$$

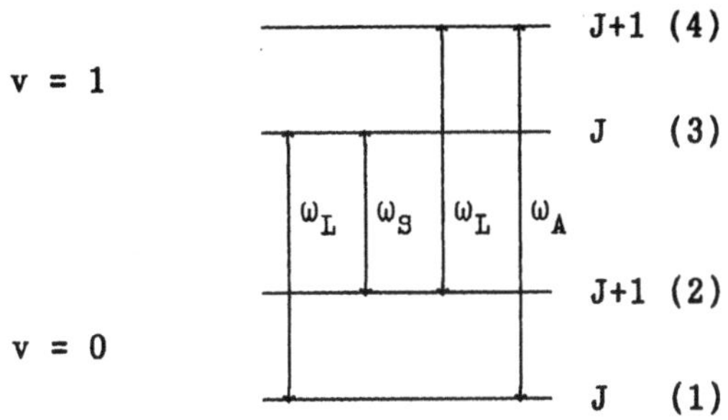

Fig.1. Resonant four-level system for CARS and CSRS processes.

Springer Proceedings in Physics, Volume 63 Coherent Raman Spectroscopy
Eds.: G. Marowsky and V.V. Smirnov

where ac fields and corresponding polarizations are given by

$$\varepsilon_\gamma=\frac{1}{2}\cdot[A_\gamma\cdot\exp(i\omega_\gamma t-ik_\gamma x+i\psi_\gamma) + c.c.], \quad (3)$$

$$\rho_{ij}=P_{ij}\cdot\exp(i\omega_\gamma t-ik_\gamma x+i\psi_\gamma), \quad (4)$$

where P_{ij} are complex amplitudes of polarizations; ω_{ij} are the eigen frequencies; τ_{ij} the rotational lifetimes; μ_{ij} the transition dipole moments; A_γ, ψ_γ, ω_γ the real amplitudes, phases and frequencies of ac electric fields, respectively; $k_\gamma=c/\omega_\gamma$ is the wavenumber; c the speed of light; and γ=S,A,L. We suppose all lifetimes to be equal $\tau_{ij}=\tau_{ii}=\tau$. As in /1/, we discuss the case of all field polarizations being collinear.

Density matrix equations can be written as a system of linear equations /1/:

$$\begin{aligned}
&L_{12}P_{12}=-B_{13}P_{32}-B_{14}P_{42}\varphi+B_{32}P_{13}+B_{42}P_{14}\varphi\\
&L_{13}P_{13}=B_{13}\Delta r_{31}-B_{14}P_{43}\varphi +B_{23}P_{12}\\
&L_{14}P_{14}=B_{14}\Delta r_{14}-B_{13}P_{34}\varphi^*+B_{24}P_{12}\varphi^*\\
&L_{23}P_{23}=B_{23}\Delta r_{23}-B_{24}P_{43}+B_{13}P_{21}\\
&L_{24}P_{24}=B_{24}\Delta r_{42}-B_{23}P_{34}+B_{14}P_{21}\varphi^*\\
&L_{34}P_{34}=-B_{31}P_{14}\varphi -B_{32}P_{24}+B_{14}P_{31}\varphi+B_{24}P_{32},
\end{aligned} \quad (5)$$

where $P_{ji} = P^*_{ij}$, (*) means a complex conjugate,

$$\begin{aligned}
&L_{ij} = i - \tau\cdot(\omega_\gamma-\omega_{ij})=i+\tau\cdot\delta_{ij}, \quad B_{ij} = \mu_{ij}A_\gamma\tau/2\hbar,\\
&\quad \Delta r_{ij}=\rho_{ii}-\rho_{jj}, \quad \Delta=\tau\cdot(\omega_{34}-\omega_{12}),\\
&\rho_{11}=\rho^e_{11} + 2B_{13}\mathrm{Im}(P_{13}) + 2B_{14}\mathrm{Im}(P_{14})\\
&\rho_{22}=\rho^e_{22} + 2B_{23}\mathrm{Im}(P_{23}) + 2B_{24}\mathrm{Im}(P_{24})\\
&\rho_{33}=\rho^e_{33} - 2B_{13}\mathrm{Im}(P_{13}) - 2B_{23}\mathrm{Im}(P_{23})\\
&\rho_{44}=\rho^e_{44} - 2B_{14}\mathrm{Im}(P_{14}) - 2B_{24}\mathrm{Im}(P_{24}).\\
&\varphi =\exp(i\theta), \ \theta=2\psi_L-\psi_P-\psi_S.
\end{aligned} \quad (6)$$

In a real molecular medium anti-Stokes gain can be calculated as follows:

$$\tilde{\alpha}_A=\sum_{K,M}\alpha_A(K,M), \quad (7)$$

$$\alpha_A = \alpha_A(K,M) = \frac{4\pi\omega_A}{E_{14}}\cdot\mathrm{Im}(\mu_{14}P_{14}) \quad (8)$$

The analysis of (7) is complicated, but the analysis of (8) for $\alpha_A=\alpha_A(K,M)$ enables one to reveal the qualitative behavior of real tuning curves of CARS.

The solution of (5) with the assumption of $B_{14}\ll1$ gives rise to the analytical expression for an amplitude of polarization at the anti-Stokes frequency:

$$P_{14}=B_L^2B_S\varphi^*[\beta_1\gamma_2+\beta_2\gamma_1-(\beta_1+\beta_2)\gamma_3]/\Gamma. \quad (9)$$

We designate $B_{23}=B_S$, $B_{14}=B_A$, $B_{13}\approx B_{24}=B_L$ and

$$\beta_1=\Delta r_{32}/L_{23}^*-\Delta r_{31}/L_{13},\quad \beta_2=\Delta r_{42}/L_{24}-\Delta r_{32}/L_{23}^*,$$

$$\gamma_1=L_{24}(L_{12}L_{23}^*L_{13}L_{14}+B_L^2L_{13}(L_{14}-L_{23}^*)-B_S^2L_{14}L_{23}^*),$$

$$\gamma_2=L_{13}(L_{14}L_{24}L_{23}^*L_{34}+B_L^2\ L_{24}(L_{14}-L_{23}^*)-B_S^2\ L_{14}L_{23}^*),$$

$$\gamma_3=2B_L^2\ L_{13}L_{24}(L_{14}-L_{23}^*),\quad \Gamma=\Gamma_1+\Gamma_2,$$

$$\Gamma_1=L_{14}L_{23}^*(L_{12}L_{13}-B_S^2\ )(L_{34}L_{24}-B_S^2\ ),$$

$$\Gamma_2=B_L^2\ (L_{14}-L_{23}^*)[(L_{12}+L_{34})L_{13}L_{24}-B_S^2\ (L_{13}+L_{24})].$$

The expression (9) allows us to investigate the tuning curves for various values of B_L and B_S.

Let us discuss two marginal cases of pump intensities: 1) $B_L \gg B_S$, $B_S<1$; and 2) $B_L \ll B_S$, $B_L<1$. Tuning curves of the CARS process correspond to the ridges of α_A, that is, to the zeros of the denominator Γ, supposing all L_{ij} to be real (so called sharp line limit). In the first case ($B_L \gg B_S$) the solution for the variables $v=X$ and $u=Y+\Delta/2$ (where $X=\delta_{14}\tau$, $Y=\delta_{34}\tau$) is

$$v = u \pm u\cdot\sqrt{1 - B_L^2\ /[u^2-\Delta^2/4]}\ . \tag{10}$$

In the second case ($B_L \ll B_S$) one can acquire the solution for the following variables $v=X$ and $u=X-Y-\Delta/2$:

$$v = 0,\ v = u \pm \Delta/2 + B_S^2\ /[u \mp \Delta/2]. \tag{11}$$

These curves are given in Fig.2b,c in the coordinates of the detunings of laser $\delta_L=X-Y+\Delta/2$ and anti-Stokes $\delta_S=X$ fields. One can

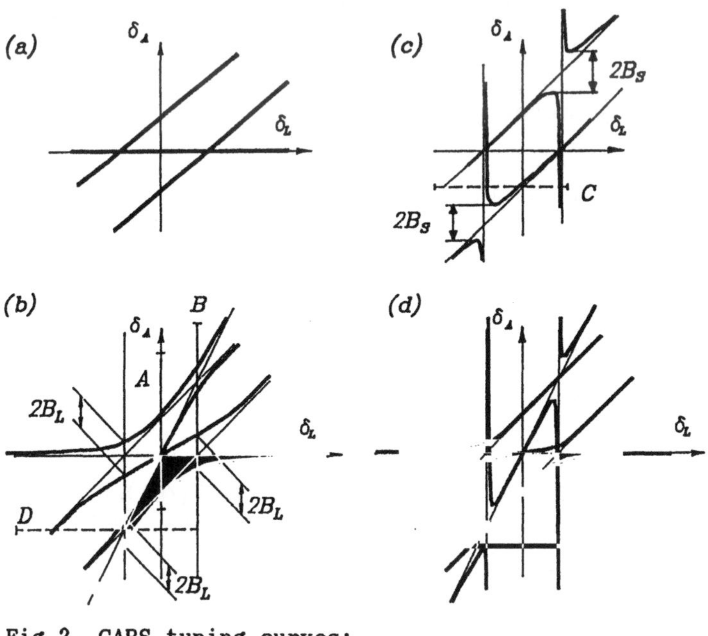

Fig.2. CARS tuning curves:
a)$B_S<1$, $B_L<1$; b)$B_L \gg 1$, $B_S<1$; c)$B_S \gg 1$, $B_L<1$; d)$B_L \cong B_S=B \gg 1$.

see that the behavior of the curves in Fig.2b,c,d differs significantly from the tuning curves for weak Stokes and laser fields (discussed, for example, in /5/), given in Fig.2a. These tuning curves were also numerically obtained from equation (9) for the intermediate case of $B_S \approx B_L \gg 1$ and are presented in Fig.2d.

Let us discuss the nature of the appearance and propagation of parametric waves. Wave propagation depends to a great extent on the phase factor $\exp(i\theta(z))$, which varies significantly along the wave direction, in short, this factor contains the inseparable effects of phase and wave synchronism, the so-called phase-locking effect.

As can be seen from (9), media polarization $\tilde{P}_A$ can be expressed in the following way:

$$\tilde{P}_A = \sum_{K,M} \mu_{14} P_A(K,M) = A_A \exp(i\psi_A)\exp(i\theta),$$

where A_A and ψ_A are the real amplitude and phase which do not depend on θ. Expression (9) is valid in the weak anti-Stokes field limit. It can also be derived from the system of equations (5)-(6) that the expression for P_γ in the case of arbitrary field intensities of all the fields is

$$\tilde{P}_\gamma = \sum_{K,M} \mu_{1j} P_\gamma(K,M) = C_\gamma \exp(i\psi_\gamma)\exp(\pm i\theta) + D_\gamma \exp(i\nu_\gamma), \tag{12}$$

where C_γ and D_γ, ψ_γ and ν_γ are real amplitudes and phases which do not depend on θ ('-' for γ=A,S, and '+' for γ=L in the second exponent). Wave propagation of the fields involved (at the initial part of the wave path, where one can suppose laser and Stokes field intensities to be constant) is given by the system of differential equations for slowly-varying phases and amplitudes:

$$dE_\gamma/dz = G_\gamma \mathrm{Im}(\tilde{P}_\gamma), \tag{13}$$

$$d\theta/dz = 2G_S \mathrm{Re}(\tilde{P}_S)/E_S + 2G_L \mathrm{Re}(\tilde{P}_L)/E_L + 2G_A \mathrm{Re}(\tilde{P}_A)/E_A, \tag{14}$$

where $G_\gamma = 4\pi\omega_\gamma \mu_\gamma / c$. When the anti-Stokes field is much weaker than the others, then $d\theta/dz$ is defined mainly by the polarization at the anti-Stokes field $\tilde{P}_A$, and the first (parametric) part of (12) dominates. The system in this case be expressed in the following form:

$$d(dE_A/dz)/dz = æ_1 E_A (d\theta/dz), \tag{15a}$$

$$d^2\theta/dz^2 = -æ_2 (dE_A/dz)/E_A. \tag{15b}$$

Utilizing the steadiness conditions for θ:

$$d\theta/dz = 0, \quad (16a) \qquad d^2\theta/dz^2 < 0 \tag{16b}$$

we acquire from (15a,16a) that $d^2E_A/dz^2 = 0$, so $\tilde{\alpha}_A = (dE_A/dz)/E_A$ has a maximum because of the parametric nature of α_A and the fact that $\alpha_A > 0$, derived from (15b) and (16b). Thus the effect of phase locking clearly appears in the case of anti-Stokes field intensity to be much weaker than the Stokes and laser fields.

The problem of the phase behavior cannot be solved in a general case, so we will discuss some specific cases:

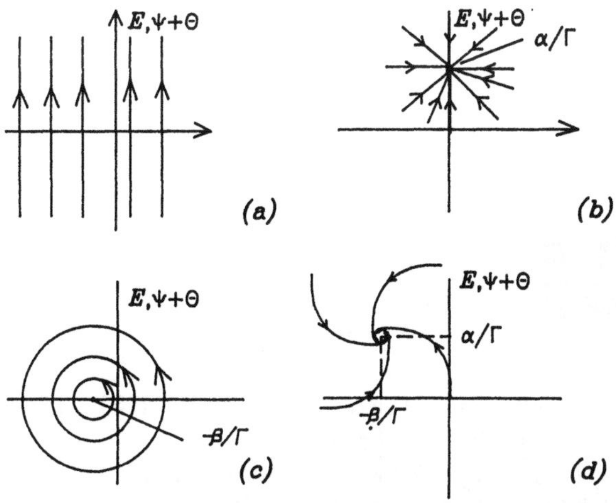

Fig.3. Phase paths for resonant CARS in polar coordinates (E) and $(\psi+\theta)$. The conditions for (a) - (d) are explained in the text.

1. $E_A=E\ll 1$ (Parametric case). Then the equation (13-14) are

$$dE/dz=\Gamma\sin(\psi+\theta),\qquad d\theta/dz=\Gamma\cos(\psi+\theta)/E \tag{17}$$

where $\Gamma=G_A\cdot P_A$ is a real positive value.
The solution for phase paths (E vs. $\psi+\theta$) are $E\cdot\cos(\psi+\theta)=C=E_0\cdot\cos(\psi+\theta_0)$ (C is an arbitrary constant). The paths are given in Fig.3a. The dependence of E and θ vs. z is given by

$$\begin{aligned} E(z) &= \sqrt{E_0^2\cos^2(\psi+\theta_0)+(\Gamma z+E_0\sin(\psi+\theta))^2}, \\ \theta(z) &= -\psi+\mathrm{arctg}(\Gamma z/E-\mathrm{tg}(\psi+\theta_0)). \end{aligned} \tag{18}$$

When $E_0\to 0$ we have $E(z)=\Gamma z$, $\theta(z)=\pi/2-\psi$, the case of instantaneous phase locking.

2. The case when one- or two-photon absorption process is added to the parametric amplification. The equations for phase paths are

$$dE/dz=\Gamma\sin(\psi+\theta)-\alpha E,\quad d\theta/dz=\Gamma\cos(\psi+\theta)/E \tag{19}$$

where α is the absorption factor. Thus we have the equation

$$dE/d\theta=E\cdot(\Gamma\sin(\psi+\theta)-\alpha E)/(\Gamma\cos(\psi+\theta))$$

which gives rise to the solution :

$$E\cdot\sin(\psi+\theta)=C\cdot E\cdot\cos(\psi+\theta)+\Gamma/\alpha \tag{20}$$

These phase paths are given in Fig.3b. In this case $E\to\Gamma/\alpha$, $\theta\to\pi/2-\psi$.

3. The case of parametric gain and phase independent real part of polarizations (off-resonance case) is described by the equations

$$dE/dz=\Gamma\sin(\psi+\theta),\qquad d\theta/dz=\Gamma\cos(\psi+\theta)/E+\beta \tag{21}$$

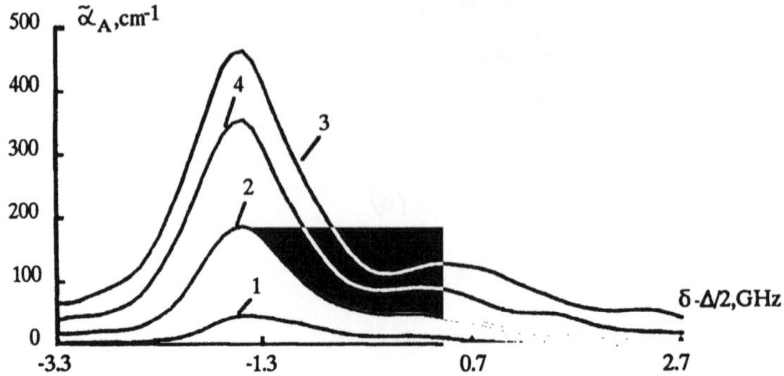

Fig.4. Gain spectra of anti-Stokes field for CH_3F R(3) line pumping for section A, Fig.2b ($\delta_L=0$, $B_L \gg B_S$).
1) $I_L=5\cdot10^5 W/cm^2$, $I_S=5\cdot10^4 W/cm^2$, $I_A=10^{-3} W/cm^2$, p= 5 torr.
2) $I_L=2\cdot10^6 W/cm^2$, $I_S=2\cdot10^5 W/cm^2$, $I_A=10^{-3} W/cm^2$, p=10 torr.
3) $I_L=5\cdot10^6 W/cm^2$, $I_S=5\cdot10^5 W/cm^2$, $I_A=10^{-3} W/cm^2$, p=15 torr.
4) $I_L=1\cdot10^7 W/cm^2$, $I_S=1\cdot10^6 W/cm^2$, $I_A=10^{-3} W/cm^2$, p=20 torr.

Phase paths are the circles (Fig.3c)

$$(E\cdot\cos(\psi+\theta)+\Gamma/\beta)^2+(E\cdot\sin(\psi+\theta))^2=(\Gamma/\beta)^2+C, \tag{22}$$

that is, the system is oscillating round the steady state position without phase locking. This case can not exist in reality because we can never neglect the absorption.

4. The case when the combination of two above cases (19) and (21) is realized:

$$dE/dz=\Gamma\sin(\psi+\theta)-\alpha E, \qquad d\theta/dz=\Gamma\cos(\psi+\theta)/E+\beta. \tag{23}$$

This general case of the equation for phase paths cannot be solved analytically. The paths are shown in Fig.3d. It has one special point: $E=\Gamma/(\alpha^2+\beta^2)$, $\theta=-\psi+\mathrm{arctg}(-\alpha/\beta)$, which is a stable focus; the rate of convergence is defined by the ratio α/β. In the case of $\beta\to0$ we have the second, and for $\alpha\to0$ the third case discussed above.

In a real situation of long optical paths, taking account of change of pump fields leads to the movement of the focus towards the origin. Thus the phase locking conditions are realized for the weak anti-Stokes field for all the cases discussed. This allowed us to analyze the behavior of CARS spectra using α for the condition $\theta=\theta_{opt}$. One can also see (Fig.3d) that the value of the stationary phase can differ from the optimal one.

As an example we made a numerical simulation of $\tilde{\alpha}_A$ for the R(3) line of CH_3F (Δ=2.7 GHz). Fig.4 presents the calculated spectra corresponding to the section designated as A in Fig.2b for different values of laser and Stokes intensities ($B_L \gg B_S$), for a weak anti-Stokes signal. This spectrum has maxima of different magnitudes, which correspond to the tuning curve in Fig.2b. Analogous spectra for the section B, Fig.2b, are given in Fig.5. The spectra for the second case ($B_L \ll B_S$) are given in Fig.6 (section C in Fig.2c).

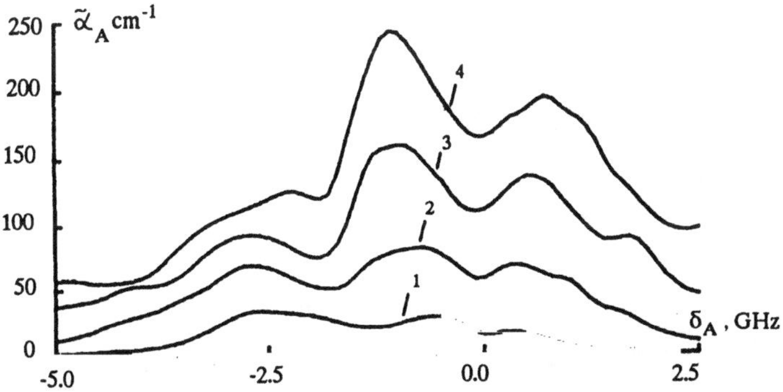

Fig.5. Gain spectra of anti-Stokes field for CH_3F R(3) line pumping for section B, Fig.2b ($\delta_L=\Delta/2$, $B_L \gg B_S$).

1) $I_L=5\cdot10^5 W/cm^2$, $I_S=5\cdot10^4 W/cm^2$, $I_A=10^{-3} W/cm^2$, p= 5 torr.
2) $I_L=2\cdot10^6 W/cm^2$, $I_S=2\cdot10^5 W/cm^2$, $I_A=10^{-3} W/cm^2$, p=10 torr.
3) $I_L=5\cdot10^6 W/cm^2$, $I_S=5\cdot10^5 W/cm^2$, $I_A=10^{-3} W/cm^2$, p=15 torr.
4) $I_L=1\cdot10^7 W/cm^2$, $I_S=1\cdot10^6 W/cm^2$, $I_A=10^{-3} W/cm^2$, p=20 torr.

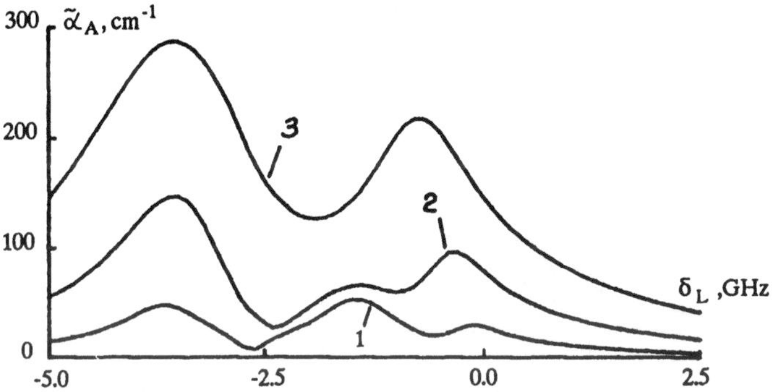

Fig.6. Gain spectra of anti-Stokes field for CH_3F R(3) line pumping for section C, Fig.2c ($\delta_A=\Delta/2$, $B_L \gg B_S$).

1) $I_S=2\cdot10^6 W/cm^2$, $I_L=2\cdot10^5 W/cm^2$, $I_A=10^{-3} W/cm^2$, p= 10 torr.
2) $I_S=5\cdot10^6 W/cm^2$, $I_L=5\cdot10^5 W/cm^2$, $I_A=10^{-3} W/cm^2$, p=15 torr.
3) $I_S=1\cdot10^7 W/cm^2$, $I_L=1\cdot10^6 W/cm^2$, $I_A=10^{-3} W/cm^2$, p=20 torr.

Fig.7 demonstrates a saturation of CARS process by the anti-Stokes field. Fig.7 corresponds to the section B in Fig.2b. The gain saturates rapidly because the anti-Stokes field is resonant in this case. The calculated spectra for section D in Fig.2b, which are presented in Fig.8, corresponding to large offsets of the anti-Stokes field from resonance give evidence that α_A is not saturated up to anti-Stokes intensities of 30 kW/cm^2.

The combined solution of (5),(6),(9) allows us to describe wave propagation in the CH_3F medium. Numerical calculations which describe wave propagation in active media show that for the appropriate conditions the output anti-Stokes energy can be up to 0.5-1% of the initial energy of the strongest field among laser and Stokes ones for moderate pump intensities of 3-10 MW/cm^2.

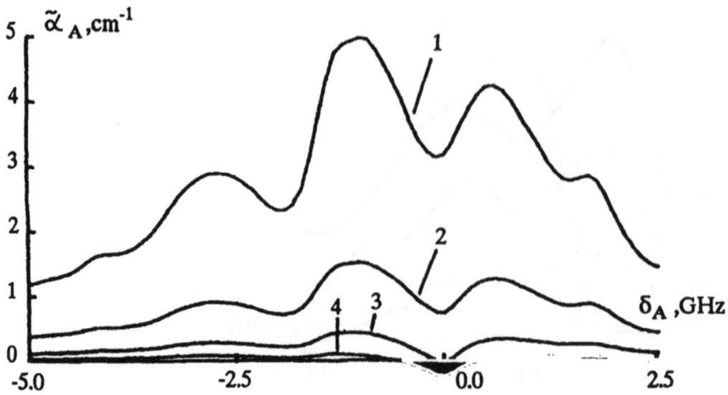

Fig.7. Saturated gain spectra of anti-Stokes field for CH_3F R(3) line pumping for section B, Fig.2b.
$I_S=5\cdot10^6 W/cm^2$, $I_L=5\cdot10^5 W/cm^2$, p= 15 torr.
1) I_A=1 W/cm^2, 2) I_A=10 W/cm^2, 3) I_A=100 W/cm^2, 4) I_A=1 kW/cm^2.

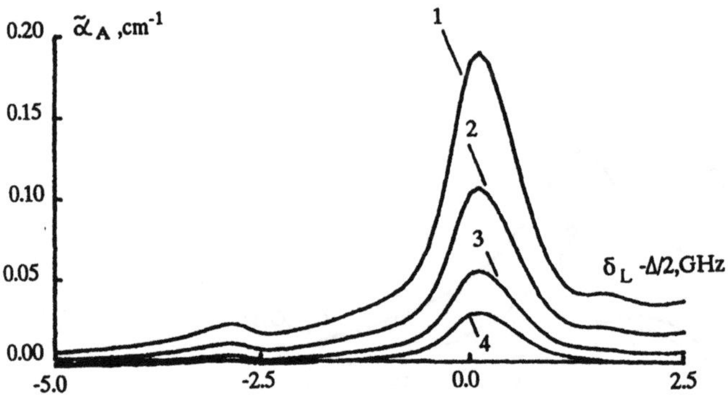

Fig.8. Saturated gain spectra of anti-Stokes field for CH_3F R(3) line pumping for section D, Fig.2b.
$I_S=5\cdot10^6 W/cm^2$, $I_L=5\cdot10^5 W/cm^2$, p= 15 torr.
1) I_A=1 kW/cm^2, 2) I_A=3 kW/cm^2, 3) I_A=10 kW/cm^2, 4) I_A=30 kW/cm^2.

These results show the rather high efficiency of CARS (and CSRS) processes in CH_3F lasers.

Thus the figures discussed above show that the anti-Stokes gain spectra change significantly with the change of the laser and Stokes fields. The saturation by these fields changes these spectra in a quite complicated manner, changing the number of gain peaks and also their positions and heights.

In the present paper, on the basis of an example of resonant infrared CARS in CH_3F gas, the possibility of modeling the spectral behavior of coherent parametric radiation in a fully resonant scheme for different pump intensities and medium densities was shown. We suppose this result to be of interest for the utilization of CARS as a method of active media diagnostics.

References.

1. Batanov V.A., Radkevich A.O., Telyatnikov A.L., Volkov A.Yu., Bakos J.S. Parametric processes in optically pumped FIR lasers. Int. J. of IR & MM Waves. v.9, p.761, (1988).
2. Batanov V.A., Radkevich A.O., Telyatnikov A.L., Volkov A.Yu. Emission spectra of the tunable Raman CH_3F optically pumped FIR laser. Int. J. of IR & MM Waves. v.11, p.31, (1990).
3. Batanov V.A.,Fleurov V.B., Kuzmin K.Yu., Radkevich A.O., Telyatnikov A.L., Timofeev S.V., Volkov A.Yu., Bakos J.S. Degenerate four-photon parametric interactions (DFPI) in optically pumped CH_3F laser. Int. J. of IR & MM Waves. v.11, p.443, (1990).
4. Batanov V.A.,Petriv V.S., Radkevich A.O., Telyatnikov A.L., Volkov A.Yu. Tunable Raman and parametric pulsed optically pumped FIR lasers. Int. Conf. on IR & MM Waves. Orlando, USA, (1990).
5. Attal-Tretout B., Berlemont P., Taran J.P. Tree-color CARS spectroscopy of OH radical at triple resonance. Molecular Phys. v.70, p.1, (1990).

Hydrogen CARS Spectra Influenced by High Laser Intensities

R. Bombach, B. Hemmerling, and W. Hubschmid

Paul Scherrer Institute, CH-5232 Villigen PSI, Switzerland

Molecular hydrogen CARS spectra are strongly distorted under the influence of high laser intensities. The underlying effects, mainly saturation and dynamic Stark shift are analysed by comparing the measured spectra with simulations on the basis of a two level model of H_2.

1 Introduction

One of the most popular applications of the coherent anti- Stokes Raman scattering (CARS) technique is its use as a *non-intrusive* method for determining temperature and species concentration [1]. However, there are some effects which call into question the *non-intrusive* nature of this method. At extremely high laser power, dielectric breakdown severely influences the probed medium. At lower laser power saturation of the Raman transition and the dynamic Stark effect limit the quantitative application of CARS as a diagnostic tool. The last two effects have been investigated on Q-branch transitions of nitrogen and hydrogen by Péalat et al. [2] and Bombach et al. [3], respectively and on Q-branch as well as pure rotational transitions of hydrogen by Lucht et al. [4]. The aim of this paper is to extend the studies of saturation and AC-Stark effect. A standard scanning CARS instrument (resolution$\approx$0.1 cm^{-1}) proved to be sufficient to resolve these effects. Besides the saturation splitting and broadening of the lines, and the line shift due to the dynamic Stark effect, the interference with neighbouring lines has to be taken into account. The importance of the different contributions was determined by synthesizing the spectrum on the basis of a system of coupled differential equations (optical Bloch equations).

2 Experimental

The experimental setup consists of a standard three-dimensional BOXCARS arrangement (see Fig. 1). The output of a frequency-doubled Nd:YAG laser (QUANTEL YG 581-10 with intracavity etalon) operating at 10 Hz was split into two parts. 60 to 80 mJ were used as CARS pump beams, whereas about 150 mJ were used to pump a tuneable dye laser working with a mixture of DCM and Pyridine 1 dissolved in methanol. The use of a two stage dye laser resulted in good beam quality but in poor efficiency, resulting in a Stokes laser output of only 12 mJ. To achieve the desired high intensities for easy observation of the anticipated effects a lens with unusually tight focusing (f=63 mm) was used to direct the three beams into the measuring cell. As the intensities reached 1000 GW/cm^2 - well above dielectric breakdown in atmospheric hydrogen -

Springer Proceedings in Physics, Volume 63 **Coherent Raman Spectroscopy**
Eds.: G. Marowsky and V.V. Smirnov

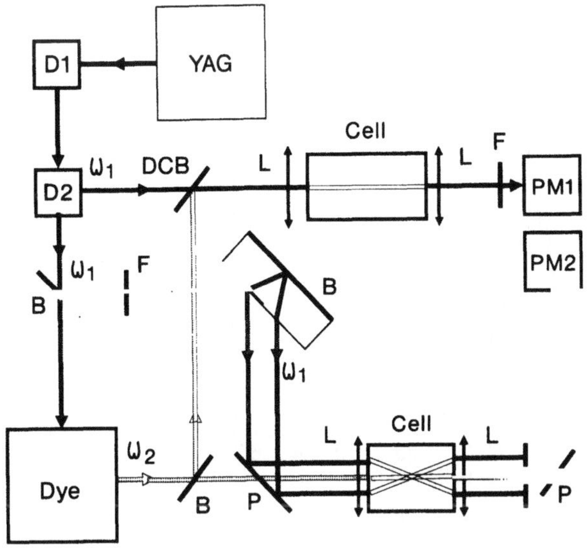

Fig. 1 Three-dimensional BOXCARS experimental setup (D1, D2 frequency doubler, B beam splitter, P reflecting prisms, F filter, L lens, PM photomultiplier, DCB dichroitic beamsplitter)

the pressure was reduced to 300 mbar. Further increases in intensity by tighter focusing is not possible with a singlet lens due to spherical aberration. Close attention was paid to spatial filtering of the anti-Stokes radiation because at or near breakdown intensities strong anti-Stokes emission in the direction of the two pump beams was observed.

For wavelength calibration a second cell (l=400 mm) using a collinear CARS configuration was added. Within this cell, weak focusing (f=200 mm) and low laser energies (8 mJ pump beam provided by a second doubler, and 0.4 mJ Stokes beam) were used to obtain a calibration signal of 0.1 cm^{-1} FWHM. The width of the signal is dominated by the bandwidth of the YAG and dye lasers (each 0.08 cm^{-1}). Other contributions to the linewidth are the Doppler effect (0.04 cm^{-1} at room temperature), Dicke narrowing (−0.01 cm^{-1} at 300 mbar) and pressure broadening ($6 \cdot 10^{-4}$ cm^{-1} at 300 mbar) [5]. However, these effects remain unobservable with the given instrumental resolution. Even with strong focusing and with high laser energies no saturation effects were visible when using a collinear arrangement. The signal contributions from regions of lower laser intensity by far surpass the contributions from the focal volume. Thus, the second cell enabled the accurate observation of very small shifts and provided a scan width calibration. Both of the CARS signals were recorded simultaneously on a shot to shot basis and were subsequently stored in a personal computer.

3 Measurements and Results

Figure 2 shows a part of the CARS spectrum featuring the most striking consequences of the high laser intensities used. Line shifts, broadenings and distortions are clearly visible. The Q(1) line is asymmetrically split into two components. The position of the

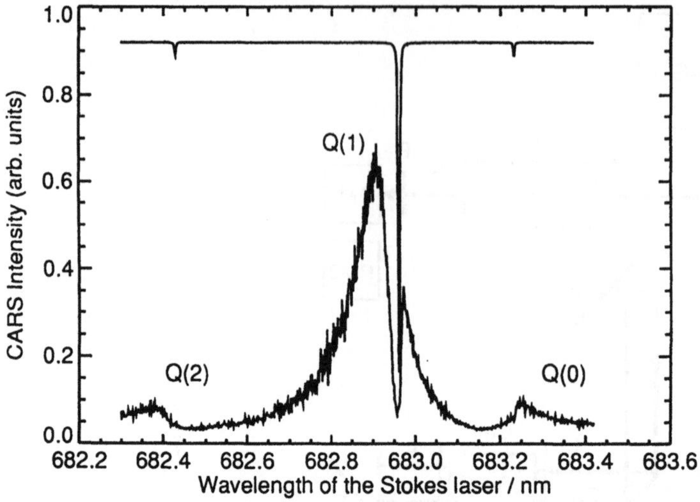

Fig. 2 Q-branch CARS spectrum of H_2 ($v = 0 \rightarrow v' = 1$) at high laser intensities (lower trace). The upper trace (shown upside down) represents the reference spectrum. The hydrogen pressure in both cells was 300 mbar.

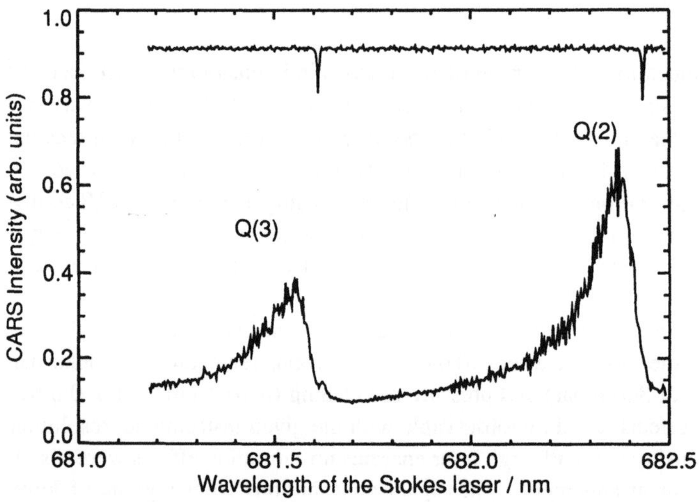

Fig. 3 Q-branch CARS spectrum of H_2 (same conditions as in Fig. 2).

dip agrees fairly well with the original position of the line. The stronger component is shifted towards lower Raman frequency. The Q(0) and the Q(2) lines appear to be pushed away from the Q(1) line and there is no dip visible. The appearance of Q-lines at higher rotational quantum numbers is quite similar to the Q(2) line. Figure 3 shows a part of the Q-branch CARS spectrum representing the Q(2) line together with the Q(3) line. Another point worth mentioning is the difference in noise amplitude on the wings of the Q(1)-line. The noise is more pronounced at the side towards lower

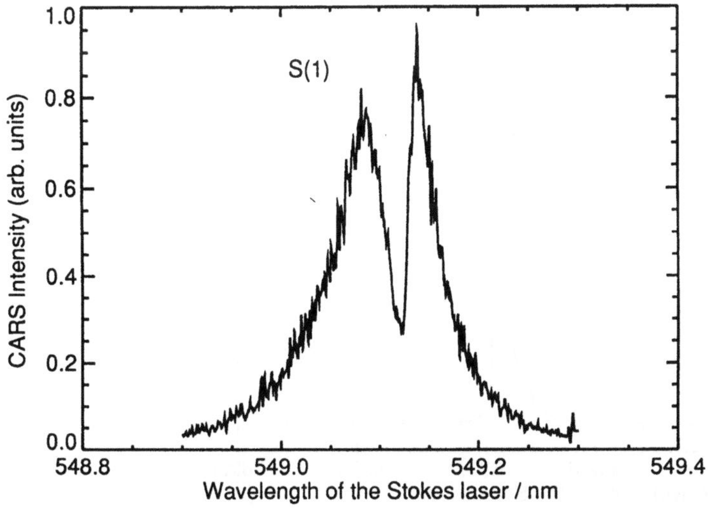

Fig. 4 Pure rotational CARS spectrum of H_2 ($v = 0 \rightarrow v' = 0$) at high laser intensities.

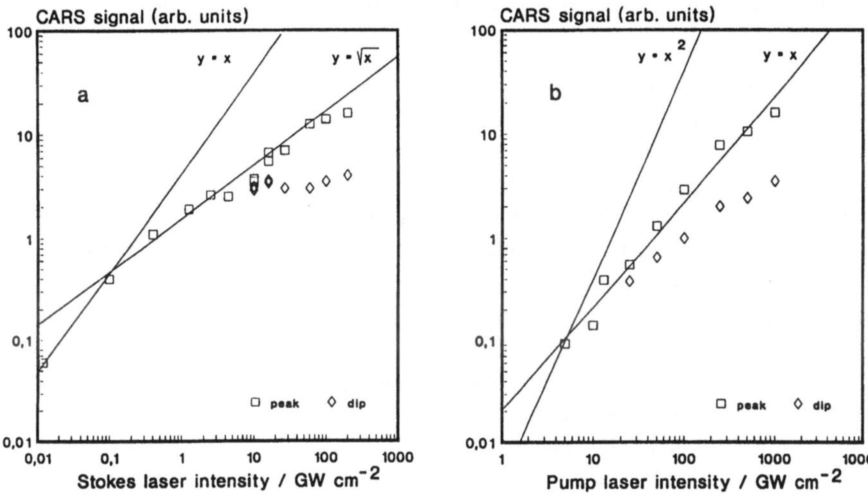

Fig. 5 Dependence of the H_2 Q(1) CARS signal on a) Stokes laser intensity and b) pump laser intensity. In a) the pump laser intensity was $\approx$850 GW/cm^2 whereas in b) the Stokes laser intensity was kept at $\approx$200 GW/cm^2.

Raman frequency. The pure rotational transition S(1) depicted in Fig. 4 exhibits a more symmetrical splitting than the Q(1) line. Furthermore, the noise amplitude on both wings is quite similar.

The pressure dependence of the CARS line shape was investigated together with its dependence on the intensities of the pump and Stokes beams, respectively. Figure 5(a) shows the anti-Stokes intensity as a function of Stokes laser energy at a constant pump energy of 60 mJ ($\approx$850 GW/cm^2). Except for the lowest energies measured, a significant nonlinearity is apparent. This behaviour is anticipated since the $I_p I_s$ product lies in the

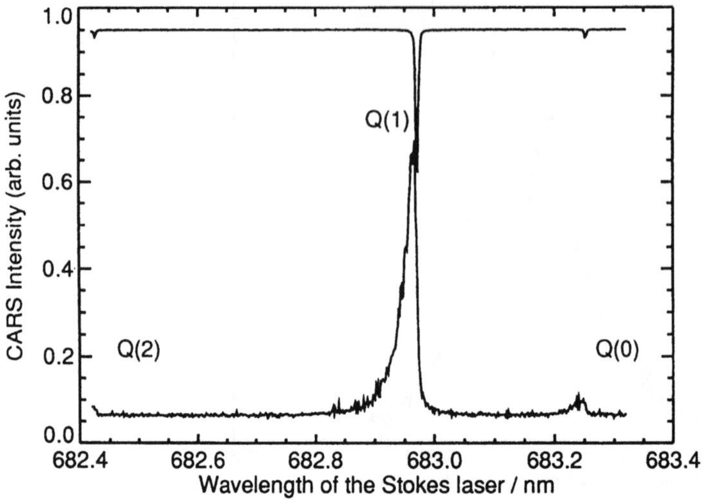

Fig. 6 Line shape of the Q(1) ($v = 0 \rightarrow v' = 1$) line of hydrogen at very low Stokes laser intensity. The original Stokes beam (≈200 GW/cm^2) was attenuated by a 10^{-4} filter while the pump laser intensity was kept at ≈850 GW/cm^2.

order of magnitude where the onset of saturation for nitrogen has been reported [6]. Even at Stokes energies of about 1 μJ (≈20 MW/cm^2, well below the energy densities used in broadband CARS) line shape distortions were found (see Fig. 6). A splitting into two peaks occurs quite suddenly at Stokes energies of 0.5 mJ (≈9 GW/cm^2). Saturation effects are also apparent from the more or less linear dependence of the anti-Stokes signal upon the pump beam energy rather than the expected quadratic relationship [see Fig. 5(b)]. This is considered to be strong evidence for saturation by stimulated Raman pumping. The observed line splitting disappears at pump beam energies of about 0.6 mJ (≈8 GW/cm^2) at full Stokes energy (≈200GW/cm^2).

Due to these distorted lineforms a straightforward determination of the Stark shift is not possible. This becomes particularly apparent when looking at the shape and the position of the Q(0) and Q(2) lines in Fig. 2.

Spectra were recorded in the pressure range of 3 to 300 mbar at Stokes energies of 13 mJ (≈230 GW/cm^2) and pump energies of 70 mJ (≈1000 GW/cm^2). No visible change in the line shape could be observed. Contributions from Doppler broadening and pressure broadening, [which are normally only observable with high resolution techniques (see Section 2)], are concealed by the extremely large saturation broadening of about 3 cm^{-1}. Therefore we conclude that, in the pressure range studied, collision-induced relaxation processes have only a minor influence on the observed line shapes.

4 Theoretical Considerations

Line shapes for saturated coherent Raman interactions with molecules on the basis of a two level model have been calculated in the work of Giordmaine and Kaiser [7], and Penzkofer, Laubereau and Kaiser [8]. Recent papers by Lucht and Farrow [4] and by Péalat, Lefèbvre, Taran and Kelley [2] discuss the numerical treatment of the equations which describe the measured spectra. The subsequent theoretical considerations follow closely the work of the above mentioned authors.

The polarization in the molecules induced by the electric field $\vec{E}$ is given by

$$\vec{P} = N\, Tr\left[\rho\, \overleftrightarrow{\alpha}(R)\right]\, \vec{E}\,. \tag{1}$$

Here N is the number density of the molecules, $\overleftrightarrow{\alpha}(R)$ is the electronic polarizability of a molecule which depends only upon the internuclear separation R in the Born-Oppenheimer approximation and ρ is the density matrix. The conversion of the incoming fields into the CARS field is small in a BOXCARS configuration. Therefore $\vec{E}$ is solely given by the external fields

$$\vec{E} = \frac{\vec{\varepsilon}}{2}\left[A_{P_1}(t)\, e^{-i\omega_{P_1}t} + A_{P_2}(t)\, e^{-i\omega_{P_2}t} + A_S(t)\, e^{-i\omega_S t} + cc\right]. \tag{2}$$

Parallel linear polarization is assumed for all waves. $A_{P_1}(t)$ and $A_{P_2}(t)$ are the complex slowly varying amplitudes of the pump beams, $A_S(t)$ is the corresponding amplitude of the Stokes beam. The part of $\vec{P}$ oscillating with the frequency $2\omega_P - \omega_S$ represents the CARS polarization. The density matrix ρ, written with respect to the eigenstates of the Hamiltonian H^0 which describes the unperturbed molecular dynamics, divides into pieces ($\tilde{\rho}$) of two states $|\Psi_1\rangle = |vJM\rangle$ and $|\Psi_2\rangle = |v'J'M'\rangle$, respectively, connected by the interaction with the external field $\vec{E}$. Each molecular state is described by the quantum numbers of vibration (v) and rotation (J,M). In the case of the electronic ground state of hydrogen there is no electronic contribution to the total angular momentum.

The density operator ρ obeys the evolution equation

$$i\hbar\dot{\rho} = \left[H^0 + H^{int}, \rho\right]. \tag{3}$$

Here H^{int} contains the interaction with the external field. The equation for the matrix elements of $\tilde{\rho}$ with respect to $|\Psi_1\rangle$ and $|\Psi_2\rangle$ can be written as

$$\begin{aligned}
\dot{\tilde{\rho}}_{11} &= \frac{1}{i\hbar}\left(H_{12}^{int}\tilde{\rho}_{21} - H_{21}^{int}\tilde{\rho}_{12}\right) - \left(\tilde{\rho}_{11} - \tilde{\rho}_{11}^{eq}\right)\gamma_1 \\
\dot{\tilde{\rho}}_{12} &= -\frac{1}{i\hbar}\left[\left(\hbar\omega_0 + H_{22}^{int} - H_{11}^{int}\right)\tilde{\rho}_{12} + H_{12}^{int}\left(\tilde{\rho}_{11} - \tilde{\rho}_{22}\right)\right] - \tilde{\rho}_{12}\gamma_2 \\
\dot{\tilde{\rho}}_{21} &= \frac{1}{i\hbar}\left[\left(\hbar\omega_0 + H_{22}^{int} - H_{11}^{int}\right)\tilde{\rho}_{21} + H_{21}^{int}\left(\tilde{\rho}_{11} - \tilde{\rho}_{22}\right)\right] - \tilde{\rho}_{21}\gamma_2 \\
\dot{\tilde{\rho}}_{22} &= -\frac{1}{i\hbar}\left(H_{12}^{int}\tilde{\rho}_{21} - H_{21}^{int}\tilde{\rho}_{12}\right) - \left(\tilde{\rho}_{22} - \tilde{\rho}_{22}^{eq}\right)\gamma_1\,.
\end{aligned} \tag{4}$$

As the states of the various transitions are connected by relaxation processes, rates for the population (γ_1) and amplitude relaxation (γ_2) have been added empirically. The thermal equilibrium populations for the two levels involved are given by $\tilde{\rho}_{11}^{eq}$ and $\tilde{\rho}_{22}^{eq}$, respectively.

The component of the polarization $\vec{P}$ in direction $\vec{\varepsilon}$ is given by

$$P = N E\, Tr\left[\rho\left(\alpha_\perp(R) + \Delta\alpha(R)\cos^2\Theta\right)\right]. \tag{5}$$

The anisotropy of the polarizability $\Delta\alpha(R) := \alpha_{\|}(R) - \alpha_{\perp}(R)$ contains the components of the polarizability tensor $\overleftrightarrow{\alpha}(R)$ perpendicular $[\alpha_{\perp}(R)]$ and parallel $[\alpha_{\|}(R)]$ to the molecular axis. The angle between the incoming electric field $\vec{E}$ and the internuclear axis is given by Θ. Expressed in components of the density matrix $\tilde{\rho}$ Eq. (5) takes on the form

$$\begin{aligned} P = NE \sum_{vv'JJ'MM'} \Big(& \tilde{\rho}_{12}[\langle v' \mid \alpha_{\perp}(R) \mid v\rangle\, \delta_{JJ'}\delta_{MM'} \\ & + \langle v' \mid \Delta\alpha(R) \mid v\rangle \langle J'M' \mid \cos^2\Theta \mid JM\rangle] \\ & + \tilde{\rho}_{21}[\langle v \mid \alpha_{\perp}(R) \mid v'\rangle\, \delta_{JJ'}\delta_{MM'} \\ & + \langle v \mid \Delta\alpha(R) \mid v'\rangle \langle JM \mid \cos^2\Theta \mid J'M'\rangle] \Big) . \end{aligned} \tag{6}$$

The polarizability tensor components are expanded around the equilibrium value R_0:

$$\alpha(R) = \alpha(R_0) + q\frac{d\alpha}{dq}(R_0) + \cdots . \tag{7}$$

The displacement from the equilibrium value is defined by

$$q \stackrel{\text{def}}{=} R - R_0 . \tag{8}$$

Keeping only terms up to linear order one gets for the polarization

$$P = NE \sum_{vv'JJ'MM'} (\tilde{\rho}_{12} + \tilde{\rho}_{21})\, \tilde{\alpha}(v, v', J, J', M, M') . \tag{9}$$

Here $\tilde{\alpha}(v, v', J, J', M, M')$ is defined as follows:

$$\begin{aligned} \tilde{\alpha}(v, v', J, J', M, M') := & \frac{d\alpha_{\perp}}{dq}(R_0)\langle v \mid q \mid v'\rangle\, \delta_{JJ'}\delta_{MM'} \\ & + \left(\Delta\alpha(R_0)\delta_{vv'} + \frac{d\Delta\alpha}{dq}(R_0)\langle v \mid q \mid v'\rangle \right) \\ & \times \langle JM \mid \cos^2\Theta \mid J'M'\rangle . \end{aligned} \tag{10}$$

For the variable $\xi := \tilde{\rho}_{12} + \tilde{\rho}_{21}$ one derives from Eq. (4)

$$\ddot{\xi} + 2\gamma_2\dot{\xi} + \omega^2\xi = \frac{\omega}{\hbar}\, \tilde{\alpha}(v, v', J, J', M, M')\, E^2\, (1 - 2\tilde{\rho}_{22}) . \tag{11}$$

Here the assumption was made that the sum $\rho_{11} + \rho_{22}$ is constant during the presence of the external fields. Furthermore, the Hamiltonian H^{int} of the electric dipole interaction has been taken to first order in q:

$$H^{int} = -\frac{1}{2}\vec{E}\, \overleftrightarrow{\alpha}(R)\vec{E} . \tag{12}$$

The frequency of the unperturbed Raman transition ω_0 is modified according to the Stark shift of the two levels involved.

$$
\begin{aligned}
\omega &\stackrel{\text{def}}{=} \omega_0 + \frac{1}{\hbar}(H_{22}^{int} - H_{11}^{int}) \\
&= \omega_0 - \frac{1}{2\hbar}E^2\Bigg[(\langle v' \,|\, q \,|v'\rangle - \langle v \,|\, q \,|v\rangle) \frac{d\alpha_\perp}{dq}(R_0) \\
&\quad + \left(\Delta\alpha(R_0) + \frac{d\Delta\alpha}{dq}(R_0)\langle v' \,|\, q \,|v'\rangle \right) \langle J'M' \,|\cos^2\Theta\,|\, J'M'\rangle \\
&\quad - \left(\Delta\alpha(R_0) + \frac{d\Delta\alpha}{dq}(R_0)\langle v \,|\, q \,|v\rangle \right) \langle JM \,|\cos^2\Theta\,|\, JM\rangle \Bigg] .
\end{aligned}
\tag{13}
$$

An additional equation for $\tilde{\rho}_{22}$ can be derived from Eq. (4):

$$
\ddot{\tilde{\rho}}_{22} + \gamma_1(\tilde{\rho}_{22} - \tilde{\rho}_{22}^{eq.}) = \frac{1}{2\hbar\omega}\tilde{\alpha}(v, v', J, J', M, M')\, E^2\, (\dot{\xi} + \gamma_2\xi) . \tag{14}
$$

The ansatz

$$
\xi = \frac{1}{2}\, [Q(t)\, exp(-i\omega_v t) + cc] \tag{15}
$$

defines the slowly varying amplitude $Q(t)$. Equations (11) and (14), expressed in the new variable $Q(t)$, become simplified if one neglects all terms that do not oscillate around the difference frequency $\omega_v := \omega_P - \omega_S$:

$$
\begin{aligned}
\dot{Q} + \left(\gamma_2 + i\frac{\omega^2 - \omega_v^2}{\omega_v}\right) Q &= \frac{i\omega}{2\hbar\omega_v}\tilde{\alpha}(v, v', J, J', M, M') A_P A_S^* (1 - 2\tilde{\rho}_{22}) \\
\dot{\tilde{\rho}}_{22} + \gamma_1 (\tilde{\rho}_{22} - \tilde{\rho}_{22}^{eq}) &= \frac{i\omega_v}{8\hbar\omega}\tilde{\alpha}(v, v', J, J', M, M')\, (A_P A_S^* Q^* - A_P^* A_S Q) .
\end{aligned}
\tag{16}
$$

This system of coupled differential equations is well known as the optical Bloch equations.

5 Numerical Simulation of the CARS Spectrum

The temporal behaviour of the laser pulses was recorded by a streak camera with a resolution of 200 psec. Figure 7(a) shows the pump pulse at 532 nm. Actually, this picture is a composite of five time windows with a width of 7 ns each. The shape of the pulse changes significantly from shot to shot. Nevertheless, some of the structures, for example the two double peaks, appear quite often. The sharpest peak has a width of about 500 psec. A similar pulse shape was measured for the dye laser [see Fig. 7(b)]. Both pulse forms were used to describe the temporal behaviour of the laser fields in the simulation. For the determination of the pulse amplitudes one also needs to know the shape and diameter of the laser beams within the interaction region. The diameter of the laser beams was measured with a pinhole with a diameter of 5 μm which was moved through the focus. The observed diameter of 30 μm coincides well with that calculated from diffraction limitations and the beam divergence. At a first glance, the intensity distribution across the laser beam is Gaussian.

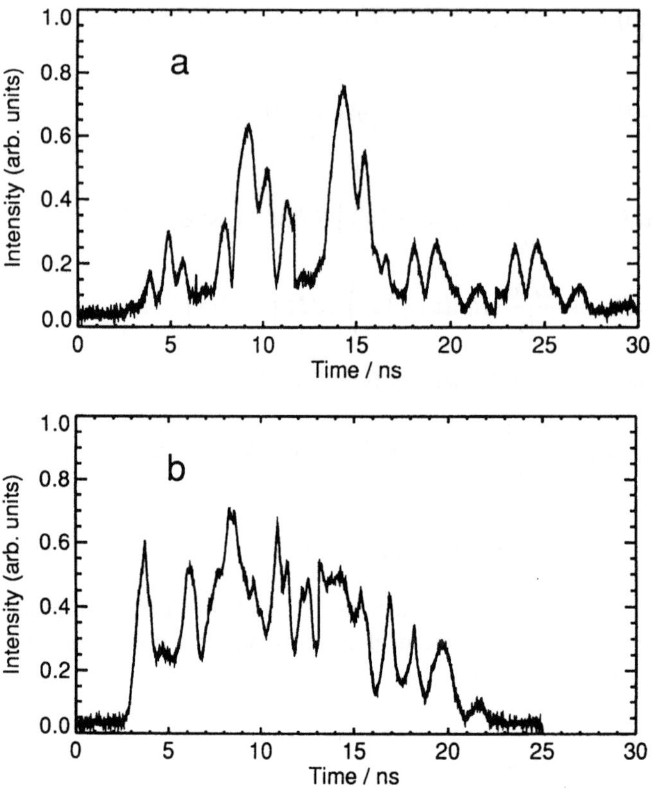

Fig. 7 Temporal shape of a) the pump and b) Stokes laser pulse, respectively.

Tab. 1 Calculated Stark shift and Rabi frequency for the Q(1) and the pure rotational S(1) transition in hydrogen ($A_P = 2{*}10^4 kg^{1/2}/s\,m^{1/2}, A_S = 8{*}10^3 kg^{1/2}/s\,m^{1/2}$, polarizability see [11].

$Q(1)\ v = 0 \rightarrow v' = 1$	$S(1)\ v = 0 \rightarrow v' = 0$
Stark shift:	
$\Delta\omega(M = 0) = 200\,\mathrm{MHz}$	$\Delta\omega(M = 0) = -40\,\mathrm{MHz}$
$\Delta\omega(M = \pm1) = 140\,\mathrm{MHz}$	$\Delta\omega(M = \pm1) = 130\,\mathrm{MHz}$
Rabi frequency:	
$\Delta\omega(M = 0) = 50\,\mathrm{MHz}$	$\Delta\omega(M = 0) = 45\,\mathrm{MHz}$
$\Delta\omega(M = \pm1) = 40\,\mathrm{MHz}$	$\Delta\omega(M = \pm1) = 35\,\mathrm{MHz}$

To get an impression of the importance of the different effects, the Stark shift and the Raman-Rabi frequency are calculated for the Q(1) and S(1) transition in hydrogen employing stengths of the pump and Stokes fields which are typical for those used in the experiment (see Table 1). Values of the polarizability were obtained from theoretical data of Kolos and Wolniewicz [11]. The matrix elements $\langle v \mid q \mid v \rangle$ were evaluated with a Morse potential between the H nuclei [2,9]. The matrix element $\langle v' \mid q \mid v \rangle$ for the transition $v = 0 \rightarrow v' = 1$ was calculated in the harmonic oscillator approximation for the vibrating H_2-molecule. Terms up to second order in the expansion of the polarizability were taken into account. With all field polarizations parallel, only transitions with no change in the magnetic quantum number are allowed. The Q(1) as well as the S(1) line split into two lines since the degeneracy of magnetic sublevels with opposite sign is not removed. For Q-branch transitions, the transition Stark shift is due only to the variation of the polarizability with vibrational quantum number, while for pure rotational transitions the Stark shift is due to the absolute value of the polarizability. The Stark shift scales with the sum of the intensities and results for Q-branch transitions in a general shift towards lower Raman frequency. The splitting of the Q(1) line is on the order of 60 MHz. For the pure rotational S(1) transition the line belonging to $M = \pm 1$ is shifted by 130 GHz towards lower frequency, while the transition belonging to $M = 0$ is shifted by 40 GHz in the opposite direction. The Raman-Rabi frequency scales with the product of the pump and Stokes laser amplitude and is for all transitions very similar.

With the mesasured temporal behaviour of the laser fields the system of coupled differential equations (17) has to be solved numerically to get the vibrational amplitude $Q(t)$ and the population of the excited state $\tilde{\rho}_{22}$. The pulses were divided into a number of relatively small, discrete time intervals (≈500) with constant field strength. A Runge-Kutta-Verner fifth-order method has been used for the numerical quadrature. The states of the various transitions are connected by relaxation processes. However, under the given experimental conditions the rotational transfer rates are small compared with the inverse laser pulse time (cf. Table 2). Therefore, it is justified to assume that the sum of the populations of the lower and upper levels $\tilde{\rho}_{11} + \tilde{\rho}_{22}$ is constant during the pulse time. The relaxation constants γ_1 and γ_2 from [10] were adapted for the experimental molecule density of 0.276 amagat. With no Stark effect and the laser tuned to the Raman resonance the real part of the vibrational amplitude is zero. The imaginary part oscillates with a frequency which is proportional to the product of pump and Stokes field amplitude. Changes in the laser amplitudes are reflected by changes in the oscillation period [see Fig. 8(a)]. The population in the excited state oscillates in a similar way between zero and one [see Fig. 8(b)].

The polarization [see Eq. (9)] enters the Maxwell equation as a source term. Assuming that phase matching is perfect and that the interaction length is much smaller than the coherence length, the CARS amplitude is calculated by time integrating the product of vibrational and pump laser amplitude. The nonresonant background is small compared to the interference of neighbouring lines and is therefore neglected. Furthermore, contributions from different velocity groups were not considered as the resulting effects (Doppler broadening) could not be resolved with our spectral resolution (cf. section 2). With the assumptions mentioned above, the CARS spectrum was calculated for a number of discrete wavelengths of the Stokes laser.

Figure 9(a) shows the result of the calculation for the Q(1) line neglecting the Stark shift term and interference with neighbouring lines. Obviously no asymmetry occurs.

Tab. 2 Rotational relaxation rates of H_2 calculated from [10] for a molecular density of 0.276 amagat.

$\gamma_1^{J=0} = (125\text{ns})^{-1}$	$\gamma_1^{J=1} = (410\text{ns})^{-1}$	$\gamma_1^{J=2} = (135\text{ns})^{-1}$
$\gamma_2^{J=0} = (93\text{ns})^{-1}$	$\gamma_2^{J=1} = (135\text{ns})^{-1}$	$\gamma_2^{J=2} = (82\text{ns})^{-1}$

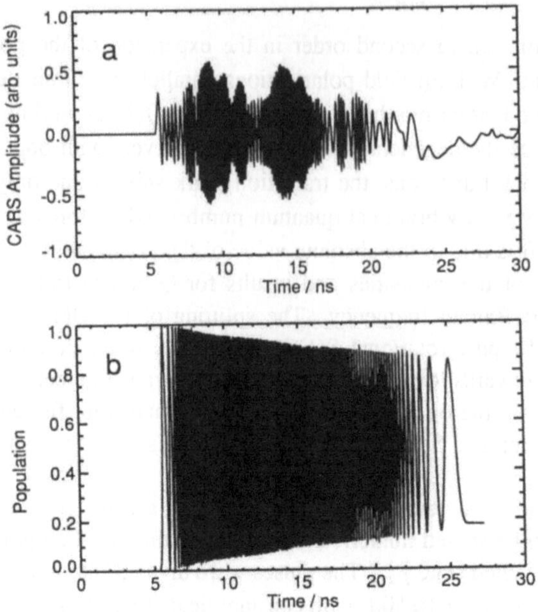

Fig. 8 a) Imaginary part of the vibrational amplitude Q and b) population of the upper level without Stark effect and the Stokes laser tuned to resonance.

Figure 9(b) illustrates how the interference with the lines Q(0) and Q(2) leads to a small asymmetry of Q(1) and a strong distortion of Q(0) and Q(2). Finally when the Stark shift term is included the asymmetry of the lineshape increases as shown in Fig. 9(c). A comparison with the measured spectrum shows qualitatively good agreement (see Fig. 10). Especially the strong distortion of Q(0) and Q(2) is described very well by the calculation. Strong discrepancies remain for the depth and width of the dip. A similar good agreement between simulation and measurement is obtained for the pure rotational transition S(1) (see Fig. 11).

As mentioned before, the energy of the laser pulses was fairly constant while the measured temporal shape and therefore the intensity changed drastically from shot to shot. To take this into account we recorded the temporal shape of the pump and Stokes beams simultaneously for ten laser shots. For each spectral increment of the Stokes laser one pair of pulses was selected randomly as input for the calculation of the CARS spectrum. The calculation indicates (see Fig. 12) that shot to shot laser intensity variations lead to strong fluctuations of the overall width of the CARS-signal. Because the Stark effect results in a general reduction of the Raman frequency for a pure vibrational transition, this additional source of signal fluctuation is more pronounced at the low frequency side of the Q(1) Raman transition. This behaviour is clearly discernible in the observed spectrum (see Fig. 2). The dip of the Q(1)-line is nearly washed out by the varying intensities.

To check whether there arises an additional effect from the spatial intensity distribution of the laser beams, the CARS signal was computed by summing all spatial contributions, assuming a Gaussian intensity distribution for Stokes and pump beams.

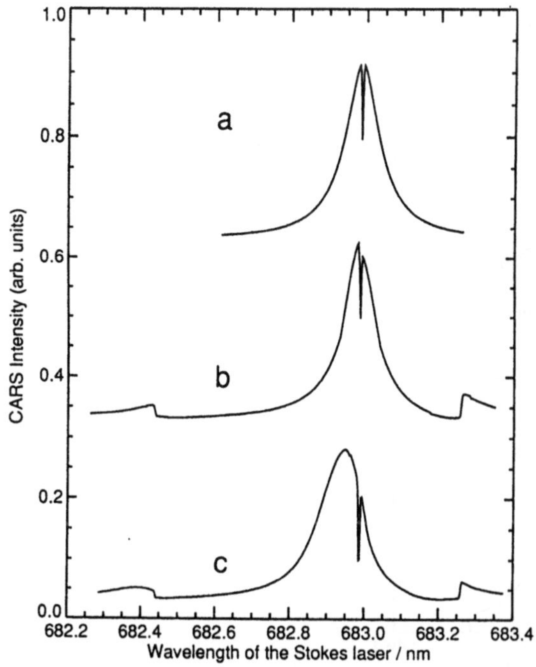

Fig. 9 Calculated H_2 CARS spectra: a) Q(1) transition saturated by stimulated Raman pumping, b) including interference between the first three Q lines, and c) additionally including the Stark effect.

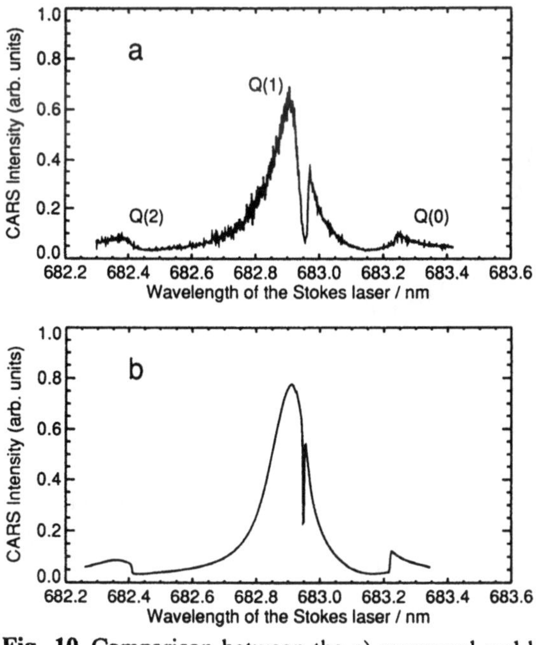

Fig. 10 Comparison between the a) measured and b) calculated line shape of the Q(1) line.

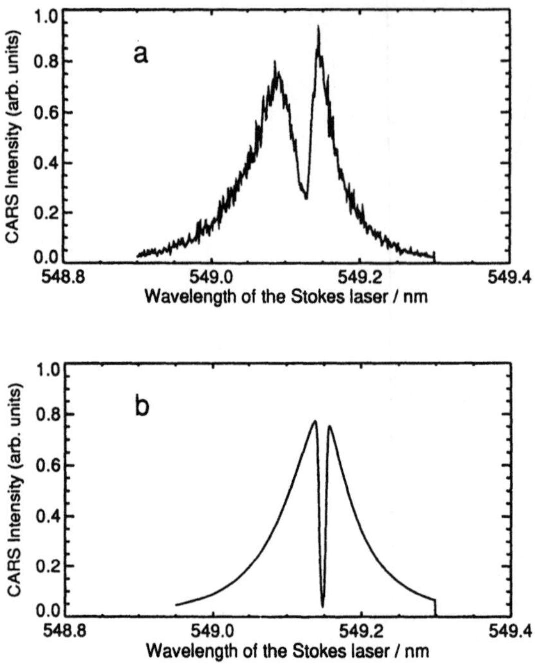

Fig. 11 Comparison between the a) measured and b) calculated line shape of the S(1) line.

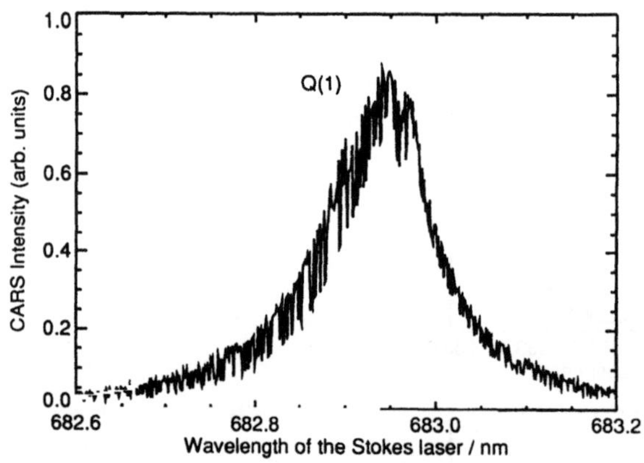

Fig. 12 Simulation of the Q(1) line shape taking into account the effect of shot to shot fluctuations in the field strength.

The result points out the predominant role of CARS light generated in zones of moderate laser intensity. However, no change in the width of the gap between the two components of the calculated Q(1) line could be observed.

6 Conclusion

The Stark effect and saturation through stimulated Raman pumping represent serious limitations to H_2 CARS spectroscopy. The line shifts and line shape distortions may not be seen at first glance with the nowadays more popular broadband techniques due to i) their inherently lower resolution and ii) the much lower spectral power density of the broadband Stokes laser. Nevertheless, our tests with a broadband setup revealed severe consequences for H_2 thermometry, since most of the models do not account for line shifts. Another problem connected to broadband H_2 CARS thermometry is the possibility of simultaneous stimulated Raman pumping of various transitions. This leads to a severely perturbed vibrational population distribution resulting in incorrect vibrational temperatures [12].

Acknowledgements

The authors would like to thank R. Knochenmuss and A. Stampanoni for their assistance in taking the streak camera data and P. Stalder for continued technical support. The sponsorship of the Swiss Federal Office of Energy (BEW) is gratefully acknowledged.

References

[1] A.C. Eckbreth, in: Laser Diagnostics for Combustion Temperature and Species, (Abacus Press, Tunbridge Wells, Kent and Cambridge, Mass, 1988).

[2] M. Péalat, M. Lefèbvre, J.-P.E. Taran and P.L. Kelley, Phys. Rev. A 38 (1988) 1948.

[3] R. Bombach, B. Hemmerling and W. Hubschmid, Chem. Phys. 144 (1990) 265.

[4] R.P. Lucht and R.L. Farrow, J. Opt. Soc. Am. B5 (1988) 1243.

[5] A.M. Toich, D.W. Melton and W.B. Roh, Optics Comm. 55 (1985) 406.

[6] R.L. Farrow and R.P. Lucht, Opt. Lett. 11 (1986) 374.

[7] J.A. Giordmaine and W. Kaiser, Phys. Rev. 144 (1966) 676.

[8] A. Penzkofer, A. Laubereau and W. Kaiser, Prog. Quantum Electron. 6 (1979) 55.

[9] L.A. Rahn, R.L. Farrow, M.L. Koszykowski and P.L. Mattern, Phys. Rev. Lett. 45 (1980) 620.

[10] R.L. Farrow and D.W. Chandler, J. Chem. Phys. 89 (1988) 1994.

[11] W. Kolos and L. Wolniewicz, J. Chem. Phys. 46 (1967) 1426.

[12] R. Bombach, B. Hemmerling and W. Hubschmid, Appl. Phys. B 51 (1990) 59.

Linear and Nonlinear Continuum Resonance Raman Scattering in Diatomic Molecules: Experiment and Theory

M. Ganz, W. Kiefer, E. Kolba, J. Manz, and J. Strempel

Institut für Physikalische Chemie, Universität Würzburg, Marcusstr. 9–11, W-8700 Würzburg, Fed. Rep. of Germany

1. INTRODUCTION

Resonance Raman spectra of the diatomic halogen molecules iodine and bromine with excitation above the dissociation limit of excited electronic states have been topics of experimental and theoretical interest for some time [1,2]. Excellent agreement between experiment and time-independent numerical calculations based on the dispersion relations found by Kramers and Heisenberg and derived by Dirac using second-order perturbation theory has been obtained particularly for the bromine molecule. This system is also of special interest since it shows scattering via two interfering excited states [3]. We became reinterested in this type of continuum resonance Raman scattering for the following reasons. First, the introduction of a time-dependent approach [4] allows the numerical calculation of continuum resonance Raman spectra without the summation over continuous states and therefore offers an alternative method which in addition nicely illustrates the scattering process in an instructive wavepacket picture. Second, the time-dependent approach also gives quantitative information on the scattering time of this type of resonance Raman scattering. Third, continuum resonance Raman spectra are extremely sensitive to changes in the potential functions [1-3,5] and represent therefore valuable experimental data for the precise determination of diatomic excited states. Fourth, transition to the repulsive potentials of electronic excited states induces competitive resonance Raman scattering and unimolecular dissociative processes. The dissociation itself can be monitored via electronic Raman scattering by the produced atoms [6]. In this paper we give illustrative examples for each of the points mentioned.

Among the nonlinear Raman spectroscopic techniques, CARS has already been shown to be a superior tool for Raman spectroscopy when performed under off-resonance as well as under resonance conditions. Resonance enhancement can be obtained when one or both of the intermediate levels of the CARS process are close to or even coincide with real levels of excited electronic states [7]. Here we restrict ourself to discussions on continuum resonance CARS work performed experimentally in iodine vapour. For the interpretation of the observed spectra, calculations applying the time-independent approach are reported. This work has already been published earlier [8]. Here, we review some of it in connection with the discussion on linear continuum resonance Raman scattering.

2. EXPERIMENTAL

The linear resonance Raman spectra were excited with an argon ion laser (Spectra Physics model 2035) and a krypton ion laser (Spectra Physics model 2025). The power used in a single laser line was generally approximately 1 W. The spectra were obtained with a Spex model 1404 double monochromator, a cooled Burle model C31034-02A photomultiplier and a photon counting/AT-computer-system or a Photometrics model RDS 200 CCD Raman detection system. Scattering experiments were performed on the isotopically pure (≈99%)

Springer Proceedings in Physics, Volume 63 **Coherent Raman Spectroscopy**
Eds.: G. Marowsky and V.V. Smirnov

molecules $^{79}Br_2$, $^{35}Cl_2$, and $^{127}I^{35}Cl$. The vapors were prepared from $Na^{79}Br$, $Na^{35}Cl$, and $^{127}I_2$, respectively. Information on the experimental setup for continuum resonance CARS excitation can be obtained from Ref. [8].

3. THEORETICAL APPROACH TO LINEAR CONTINUUM RESONANCE RAMAN SCATTERING IN DIATOMIC MOLECULES

In order to simulate and analyse the spectra, we evaluate the Raman intensities I_{fi} for transitions from initial to final vibrational-rotational states on the ground electronic surface, denoted $|i>$ and $|f>$, via excited electronic surfaces labelled e, depending on the incident photon frequency ω_I. In addition we determine theoretically the corresponding average delay times t_{fi} for transitions from $|i>$ to $|f>$ via electronic surfaces e. The fundamental expressions for I_{fi} and t_{fi} are adapted from detailed derivations in Ref [9] (see also Refs [4,5,10-14]), in a slightly different, i.e. more comprehensive and general way.

In the time-dependent approach [4,5,9-14], both quantities I_{fi} and t_{fi} are expressed in terms of the scattering amplitude $\alpha_{f,ei}(t)$ for transitions from $|i>$ to $|f>$ via surface e, at delay time t, and mediated by transition dipole operators μ_{Fe} and μ_{eI} for scattered and incident electromagnetic fields, respectively. Specifically,

$$\alpha_{f,ei}(t) = \langle f | \mu_{Fe} \exp(-i H_{eiI} t/\hbar) \mu_{eI} | i \rangle, \qquad (1)$$

where H_{eiI} is the molecular Hamiltonian operator H_e for the electronic state e, scaled such that molecular energies equal to the sum of the energies of the initial state $\hbar\omega_i$ plus incident photon $\hbar\omega_I$ are set to zero, thus

$$H_{eiI} = H_e - \hbar(\omega_i + \omega_I). \qquad (2)$$

Using the Condon approximation, μ_{Fe} and μ_{eI} are constants, and expression (1) is simplified to

$$\alpha_{f,ei}(t) = \mu_{Fe}\, \mu_{eI} \langle f | e,i(t) \rangle, \qquad (3)$$

where the virtual state

$$| e, i(t) \rangle = \exp(-i H_{eiI} t/\hbar) | i \rangle \qquad (4)$$

represents the initial state $|i>$ propagated till time t on the excited surface e. In practice, $| e, i(t) >$ is evaluated by Fast-Fourier-transform (FFT) propagations of $| i>$ on surface e, as in Refs [5,9,14]. From expression (1), we derive [9]

$$t_{fi} = N_{fi} \int_0^{\infty} dt \cdot t \cdot \frac{d}{dt} \left| \int_0^{t} d\tau \sum_e \alpha_{f,ei}(\tau) \right|^2 \qquad (5)$$

with proper normalization,

$$N_{fi}^{-1} = \left| \int_0^{\infty} \sum_e \alpha_{f,ei}(t)\, dt \right|^2. \qquad (6)$$

Expressions (3) - (6) imply that long time evolutions of overlaps $< f | e, i(t) >$ yield long Raman scattering delay times t_{fi}, and vice versa, see the results below.

Likewise, $\alpha_{f,ei}(t)$ yields the resonance Raman amplitudes [4,5,9-14]

$$\alpha_{fi} = -\frac{i}{\hbar} \sum_{e} \int_{0}^{\infty} \alpha_{f,ei}(t)\, dt \qquad (7)$$

and the resulting Raman intensities,

$$I_{fi} \sim | \alpha_{fi} |^2, \qquad (8)$$

with proper weighting factors for different temperatures, nuclear spins, and rotational states, as in Refs [1-3]. Below we show simulated spectra of the observed $I_{\parallel}$ intensity component [15].

In the more traditional Kramers-Heisenberg-Dirac approach [1-3], expression (7) is rewritten as

$$\alpha_{fi} = \sum_{e,n} \frac{\langle f \mid \mu_{Fe} \mid e,n\rangle\langle e,n \mid \mu_{eI} \mid i\rangle}{\hbar(\omega_{e,ni} - \omega_I)}, \qquad (9)$$

where $| e,n \rangle$ denote states of the excited electronic surface with energies $\hbar\omega_{e,n}$. The derivation of (9) from (1), (7) is straightforward - one simply inserts the closure relation $1 = \sum_n | e,n \rangle \langle n,e |$ together with the Schrödinger equation $\langle n,e | H_{eiI} = \langle n,e | \hbar (\omega_{e,ni} - \omega_I)$ into Eqs. (1) and (7), see Refs [4,5,9-14].

4. RESULTS AND DISCUSSION FOR LINEAR CONTINUUM RESONANCE RAMAN SCATTERING

4.1 Comparison of the Time-Dependent and the Time-Independent Approach: $^{79}Br_2$

In most cases it is very difficult or even impossible to calculate Raman spectra from the KHD expression numerically. Baierl and Kiefer [1,2] have shown by experimental and theoretical studies that for the case of the $^{79}Br_2$ and $^{81}Br_2$ molecules good agreement of observed and calculated spectra can be achieved with help of the traditional time-independent KHD calculations. The bromine system therefore offers the rare opportunity for comparison with the time-dependent approach. The time-dependent calculations on the same scattering system also show excellent agreement between experiment and theory as well as with the traditional time-independent calculations. As an example we display in Fig. 1 the observed spectrum together with the KHD calculations on the left side and the time-dependent calculations for the fundamental vibration of $^{79}Br_2$ for the excitation wavelength $\lambda_0 = 488.0$ nm on the right side. For both cases, time-independent and time-dependent theory, we show the simulated spectra ($I_{\parallel}$-component) which have been evaluated for contributions from the $B(^3\Pi_{0+u})$ state alone (spectrum D in Fig. 1), the $^1\Pi_{1u}$ state (abbreviated with Π) alone (spectrum C), and for contributions from both states (B and Π, spectrum B). The agreement is illustrated exemplarily for the rotationally unresolved Q-branch transitions of the fundamental vibrational region in Fig. 1. For further details we refer to Refs. 1,5 and 14. The very broad and weak band on the high frequency side of the Q-branches is due to unresolved $\Delta J = +2$ (S-branch) transitions. 0-branches ($\Delta J = -2$) contribute only very weakly on the low frequency side.

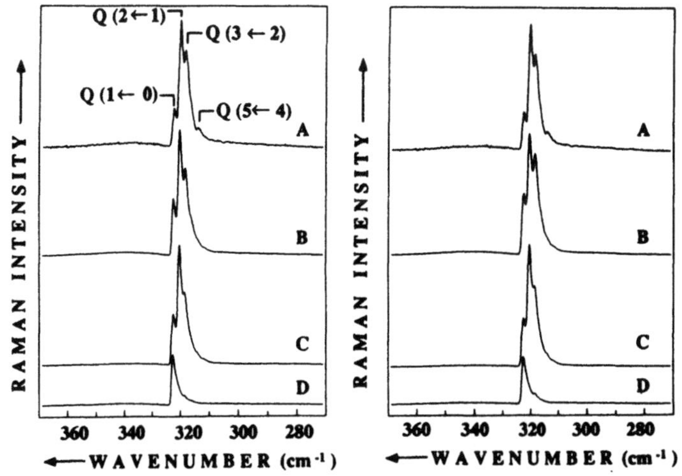

Fig. 1: *Left side:* experimental and KHD-calculated continuum resonance Raman spectra of $^{79}Br_2$ for λ_0 = 488.0 nm excitation. *Right side:* time-dependent calculations. For further details see text.

4.2 Scattering Time in Continuum Resonance Raman Scattering: $^{79}Br_2$

The time-dependent approach to resonance Raman scattering [4,5,10-13], Eqs. (1)- (7), is illustrated in Fig. 2. Exemplarily, we consider the vibrational $|i\rangle = |1\rangle$ to $|f\rangle = |6\rangle$ transition (first hot band of the fourth overtone) of $^{79}Br_2$ at 488.0 nm excitation. This process is described by initial (t = 0) electronic excitation from $|i\rangle$ to (essentially) the virtual state $|e, i(t{=}0)\rangle = |\Pi, 1(t{=}0)\rangle$, (panel A), followed by dissociative time evolution of $|e, i(t)\rangle$ till delay times t (panels B, C) where $|e, i(t)\rangle$ may be de-excited to the final state $|f\rangle$. Marginal interfering transitions via electronic state $B(^3\Pi_{0+u})$ are not shown in Fig. 2 for clarity of presentation, cf. Refs. [1-3,5,14].

From the sequence of snap-shots, panels A, B, C of Fig. 2, it is obvious that the most efficient resonance Raman transitions occur within ultra-short times, corresponding to the average delay time t_{61} = 19.2 fs (see Eqs. (5) and (6)). Thus, resonance Raman is faster than many competing processes. Extrapolating to larger systems, e.g. metal-organic complexes, with similar properties, e.g. repulsive surfaces of electronically excited states, we may anticipate that the corresponding Raman t_{fi}'s are even shorter than the corresponding times of intramolecular vibrational energy redistribution (IVR), t_{IVR} < ps. This suggests accurate time-dependent representations of resonance Raman scattering of such systems restricted to reactive degrees of freedom ("promoting modes"), see Ref. [16].

In Table 1 we have compiled average Raman delay times for transition 1←1 (Rayleigh) up to transition 6 ←1 (first hot band of the fourth overtone of $^{79}Br_2$) when excited with 488.0 nm. We notice an increase from 11.1 to 19.2 fs, which is clear because the final, stationary wavefunction increases in width for higher vibrational number causing a longer time-overlap with the propagating wavefunction $|i(t)\rangle$.

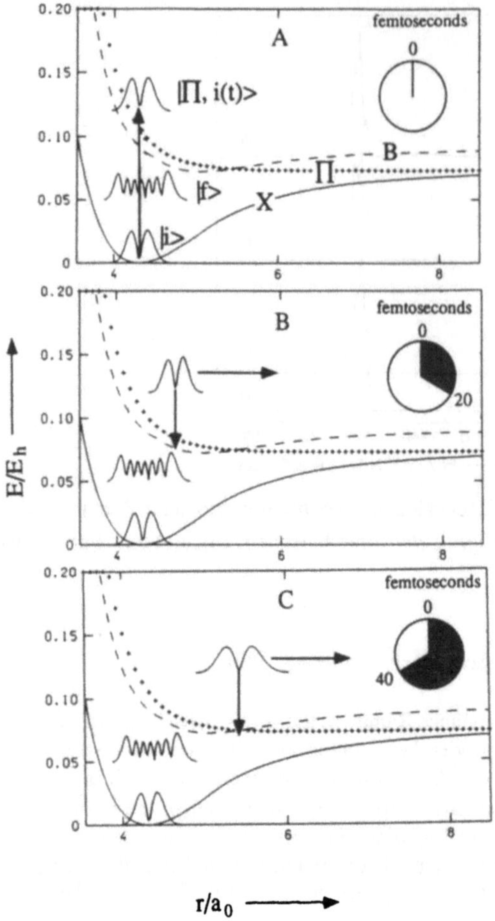

Fig. 2: Time-dependent approach to resonance Raman scattering of $^{79}Br_2$ at $\lambda_0 = 488.0$ nm. Exemplarily, panels A, B, C show the potential curves of the relevant ground ($X = {}^1\Sigma_g{}^+$, continuous line), and the excited ($B = {}^3\Pi_{0+u}$, dashed line, as well as $\Pi = {}^1\Pi_{1u}$, dotted line) electronic states, together with the absolute values of the coordinate representations of the initial state $|i\rangle = |1\rangle$, final state $|f\rangle = |6\rangle$, and the dominant [1-3,5] virtual state $|e,i(t)\rangle = |\Pi,1(t)\rangle$ at times t = 0, 20, 40 fs, respectively. Excitation and de-excitation processes as well as the related unimolecular dissociations are indicated schematically by vertical and horizontal arrows. For clarity of presentation, the energy gap between states $|i\rangle$ and $|f\rangle$ is blown up by some factor.

Table 1: Raman delay times for $^{79}Br_2$ and 488 nm excitation

Transition	Type of transition	τ [fs]
1 ← 1	Rayleigh	11.1
2 ← 1	First hot band of fundamental	13.5
3 ← 1	First hot band of first overtone	15.7
4 ← 1	First hot band of second overtone	17.9
5 ← 1	First hot band of third overtone	18.5
6 ← 1	First hot band of fourth overtone	19.2

4.3 Sensitivity of Resonance Raman Spectra to Changes in the Excited State Potential: $^{35}Cl_2$

Hartke has shown, applying the time-dependent theory to the $^{79}Br_2$- system, that the simulation of continuum resonance Raman scattering spectra is very sensitive with respect to the position and form of the excited state potential function [5]. In principle, the same high sensitivity can also be achieved using the time-independent KHD approach. Here we give, as an example, results from a Krypton laser-excited continuum resonance Raman scattering experiment in the $^{35}Cl_2$-system. In Fig. 3 (spectrum A) we show experimentally observed spectra of the first overtone transition region ($\Delta v = 2$) of $^{35}Cl_2$ excited with 413.1 nm. For this laser energy only the $^1\Pi_{1u}$-state gives rise to appreciable Raman intensity. Child and Bernstein [17] have determined an exponentially repulsive potential function:

$$V(R) = V_e + C \exp\{-\gamma(R-r_e)\} \qquad (10)$$

with $V_e = 20\,276$ cm^{-1}, $C = 10\,450$ cm^{-1}, $\gamma = 5.03$ Å^{-1}, and $r_e = 1.988$ Å. We have calculated a synthetic spectrum with the same potential but using $r_e = 2.013$ Å instead of 1.988 Å by a numerical evaluation of the KHD expression. The result is displayed as spectrum B in Fig. 3. Obviously, it matches with the observed spectrum very nicely. Notice that in both spectra (A and B) the ratio between the Q(2 – 0)- and the Q(3 – 1)-transition (first vibrational hot band) is approximately 1:4. In order to demonstrate the sensitivity of the spectra to the potential, we apply e.g. tiny shifts of – 0.05 Å and of +0.05 Å to the potential curve ($r_e = 1.963$ Å and $r_e = 2.063$ Å, respectively), thus changing the appearance of the Raman spectrum appreciably (see spectra C and D in Fig. 3, respectively).

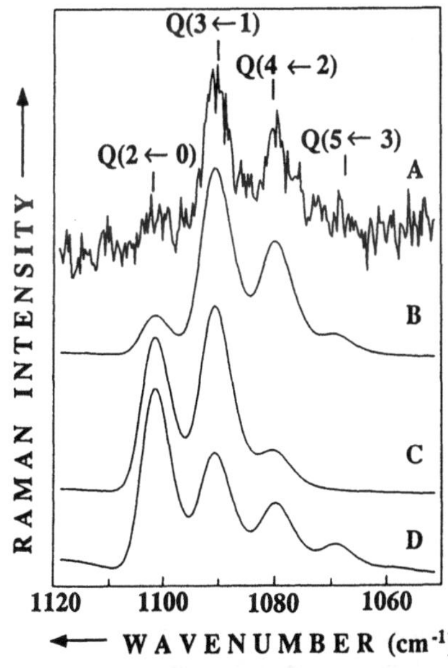

Fig. 3: *A:* Experimentally observed continuum resonance Raman spectrum of $^{35}Cl_2$ excited with 413.1 nm; *B:* Time-independently calculated spectrum using the excited potential curve given in Ref. 17 but using $r_e = 2.013$ Å instead of 1.988 Å; *C:* Same as in B, however, the potential curve is shifted by – 0.05 Å.; *D:* Same as in B, however, the potential curve is shifted by + 0.05 Å. Shown is the vibrational region of the first overtone including hot band transitions.

4.4 Electronic Raman Scattering on Iodine Atoms Produced During Photodissociation in $^{127}I^{35}Cl$

In principle, resonance Raman scattering should compete with unimolecular dissociation, see Section 4.2. The resulting products may also be observed via Raman scattering. As an illustrative example for simultaneous observation of corresponding continuum resonance Raman scattering and electronic Raman scattering we show in the left half of Fig. 4 the scattering processes in the $^{127}I^{35}Cl$ molecule and in the right half the electronic Raman scattering process between the atomic $^2P_{3/2}$ and $^2P_{1/2}$ state of the ^{127}I atom. The iodine molecule is formed by dissociation of ICl into I and Cl atoms and recombination of two iodine atoms to I_2.

As an example for the simultaneous observation of linear continuum resonance Raman scattering in the iodine and in the iodine chloride molecule we show in Fig. 5 the $\Delta v = 4$ transition region (third overtone) of ICl, where there is slight overlap with the $\Delta v = 7$ transitions in the I_2 molecule. The spectra displayed were excited with four different lines of the argon ion laser with wavelengths as indicated. The changes of the relative intensities between I_2 and ICl resonance Raman lines are due to different dissociation rates as well as due to different scattering cross sections when the energy of the dissociating and exciting laser line is varied. All observed rotational-vibrational band heads could be assigned. These results will be published elsewhere together with theoretical calculations [18].

Finally, Fig. 6 shows the observed $^2P_{1/2} \leftarrow {}^2P_{3/2}$ iodine atom electronic Raman transition at about 7603 cm^{-1} for excitation with $\lambda_0 = 457.9$ nm. This band is extremely weak and could only be observed with broad slits and long integration times when a scanning spectrometer system was employed. By means of a CCD camera we were able to use small slit widths and partially resolve the hyperfine splitting of the $^2P_{1/2} \leftarrow {}^2P_{3/2}$ transition [19].

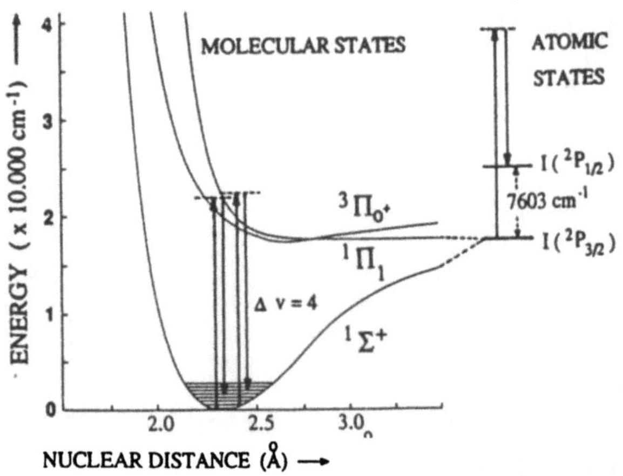

Fig. 4: Molecular states of ICl with $\Delta v = 4$ Raman transitions for the third overtone (4 - 0) and the corresponding first hot band (5-1) transition. On the right side the electronic Raman transition in the iodine atom is indicated.

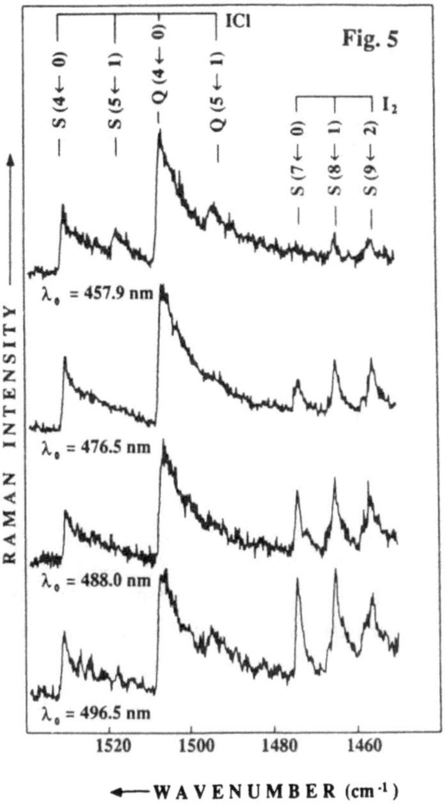

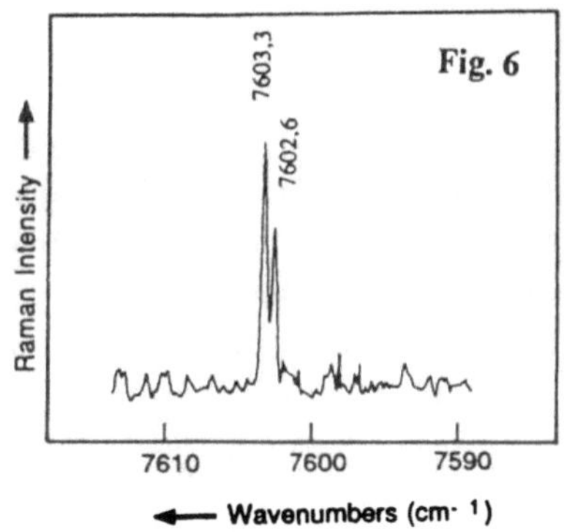

Fig. 5: Third overtone region in the resonance Raman spectrum of $^{127}I^{35}Cl$ excited with four different argon ion laser lines as indicated. Q- and S-band transitions are indicated. In addition, $\Delta v = 7$ S-band heads of I_2 are observed at the lower energy side of the spectrum.

Fig. 6: Electronic Raman scattering in iodine atoms. Shown is the $^2P_{1/2} \leftarrow {}^2P_{3/2}$ - Raman transition. The splitting is due to hyperfine structure.

5. CONTINUUM RESONANCE CARS

This section briefly summarizes theoretical and experimental results obtained earlier by Beckmann, Baierl and Kiefer [8].

5.1 Time-Independent Theory for Continuum Resonance CARS in Diatomic Molecules

If one is only interested in the spectral dispersion of the CARS intensity it is sufficient to calculate the third order non-linear susceptibility for the CARS process, $\chi^{(3)}_{CARS}$; it is well known, that

$$I_{CARS} \sim |\chi^{(3)}_{CARS}|^2. \tag{11}$$

Taran's group [20-22] has derived comprehensively third-order susceptibilities using the density operator formalism. Applying their results we write the general expression for the CARS susceptibility for one particular transition f ← i:

$$[\chi^{(3)}_{CARS}(-\omega_{AS}, \omega_L, \omega_L, -\omega_S)]_{if} = \frac{N}{\hbar^3} \frac{1}{\omega_{if} - \omega_L + \omega_S - i\Gamma_{if}}$$

$$\times \sum_{r'} \left(\frac{\mu^{\tau}_{ir'} \mu^{\sigma}_{r'f}}{\omega_{r'i} - \omega_{AS} - i\Gamma_{r'i}} + \frac{\mu^{\sigma}_{ir'} \mu^{\tau}_{r'f}}{\omega_{r'f} + \omega_{AS} + i\Gamma_{r'f}} \right)$$

$$\times \sum_{r} \left[\left(\frac{\mu^{\rho}_{fr} \mu^{\sigma}_{ri}}{\omega_{ri} - \omega_L - i\Gamma_{ri}} + \frac{\mu^{\sigma}_{fr} \mu^{\rho}_{ri}}{\omega_{ri} + \omega_S - i\Gamma_{ri}} \right) \rho_i \right.$$

$$\left. - \left(\frac{\mu^{\rho}_{fr} \mu^{\sigma}_{ri}}{\omega_{rf} - \omega_S + i\Gamma_{rf}} + \frac{\mu^{\sigma}_{fr} \mu^{\rho}_{ri}}{\omega_{rf} + \omega_L + i\Gamma_{rf}} \right) \rho_f \right]$$

$$+ \chi_{NR} \tag{12}$$

where $N\rho_i$ and $N\rho_f$ are initial number densities of the Raman active molecule in states |i> and |f>, respectively. The absorption frequencies from states |i> and |f> to states |r> (or |r'>; |r> and |r'> are the intermediate CARS states) are ω_{ri} and ω_{rf} (or $\omega_{r'i}$ and $\omega_{r'f}$) respectively, and the Γs are the corresponding damping factors; μ^{σ}_{ri} (for example) is the matrix component of the dipole moment operator $\mu^{\sigma}_{ri} = \langle r | \underline{p}\hat{e}_\sigma | i \rangle$, where $\hat{e}_\sigma$ is the unit vector in the direction of polarization of the ω_L field; μ^{ρ}_{fr} and μ^{τ}_{ir} involve interactions with ω_S and ω_{AS} fields, respectively ($\hat{e}_\rho$ and $\hat{e}_\tau$ are the unit vectors in the directions of polarisation of the ω_S and ω_{AS} fields, respectively). χ_{NR} is the non-resonant susceptibility.

For continuum resonance CARS excitation in diatomic molecules in the gas phase Eq. 12 can be modified to [8]

$$\chi^{(3)} = \frac{N}{\hbar^3} \sum_{\omega_{if}} \left| \frac{1}{(\omega_{if} + \omega_S - \omega_L - i\Gamma_{if})} \left\{ \left[\frac{\rho_i g_i}{(2J'' + 1)} \right. \right. \right.$$

$$\times \sum_{m', m''} \left(\sum_{r'} \frac{\langle i | \mu_\tau | r' \rangle \langle r' | \mu_\sigma | f \rangle}{\omega_{r'i} - \omega_{AS} - i\Gamma_{r'i}} \right) \times \left(\sum_{r} \frac{\langle f | \mu_\rho | r \rangle \langle r | \mu_\sigma | i \rangle}{\omega_{ri} - \omega_L - i\Gamma_{ri}} \right) \Bigg]$$

$$- \left[\frac{\rho_f g_f}{(2J' + 1)} \sum_{m', m''} \left(\sum_{r'} \frac{\langle i | \mu_\tau | r' \rangle \langle r' | \mu_\sigma | f \rangle}{\omega_{r'i} - \omega_{AS} - i\Gamma_{r'i}} \right) \right.$$

$$\left. \left. \left. \times \left(\sum_{r} \frac{\langle f | \mu_\rho | r \rangle \langle r | \mu_\sigma | i \rangle}{\omega_{rf} - \omega_S - i\Gamma_{rf}} \right) \right] \right\} \right| + \chi_{NR}. \tag{13}$$

Here, the abbreviation $\mu_\tau = \underline{p}\hat{e}_\tau$ and similar signs for the other components of the electric dipole moment are used. The sign Σ serves to indicate the inclusion of weak resonances from nearby discrete leves of bounded states. Note, that we have averaged over all degenerate initial states (m") and carried out the summation over all degenerate final states (m').

Inspection of Eq. 13 shows that the CARS susceptibility contains two products of two terms (the terms inside the round brackets) which are very similar to the terms in Eq. 9 when damping is neglected. Since it has been demonstrated above how to calculate these Raman polarizabilities, Eq. 13 can directly be employed for a first-principles calculation of the dispersion of the third-order CARS susceptibility and hence, by applying Eq. 11, of a complete CARS spectrum.

5.2 Results and Discussion for Continuum Resonance CARS in Iodine

For the iodine molecule CARS experiments with resonance with the continua of the $B(^3\Pi_{0+u})$ and $^1\Pi_{1u}$ states have been carried out with excitation frequencies for the pump laser

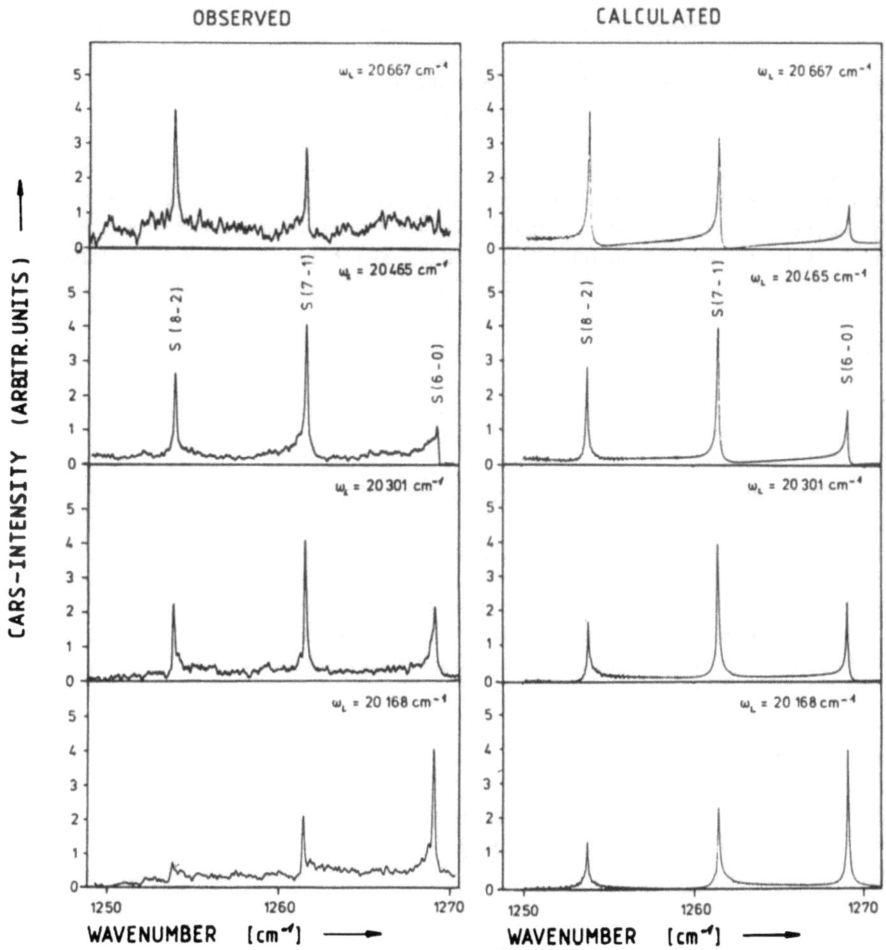

Fig. 7: Experimentally observed (left field) and theoretically calculated (right field) continuum resonance CARS spectra of the fifth overtone ($\Delta v = 6$) in iodine vapor for the pump laser frequencies ω_L = 20 667 cm^{-1}, 20 465 cm^{-1}, 20 301 cm^{-1}, and 20 168 cm^{-1} as indicated. S numbers refer to the initial and final vibrational-state assignments of the S-branches. The spectra for different ω_L are not scaled to each other.

in the range between the dissociation limit of the B-state and about 1000 cm^{-1} above this limit [8].

Typical experimentally obtained high resolution continuum resonance CARS spectra of iodine for $\Delta v = 6$ vibrational transitions are displayed in the left panel of Figure 7 for four excitation frequencies (ω_L = 20 667, 20 465, 20 301, and 20 168 cm^{-1}). Strong and sharp peaks at about 1269, 1261 and 1254 cm^{-1} are observed which have been assigned to S-branch transitions originating from the v" = 0,1,2 vibrational levels of the ground electronic state, respectively [8]. The right field of Figure 7 displays numerically calculated continuum resonance CARS spectra of the same spectral region. These spectra have been created applying the first-principles calculation described above. The final expression for $\chi^{(3)}$ as given in Eq. 13 was applied for this purpose. The intensities of the observed S-branch transitions are reproduced fairly well in the calculated spectra, although there are some slight discrepancies. However, the relative changes between different S-band heads for one particular excitation frequency are fully described by the time independent perturbation theory. For further details, particularly for the analysis of rotational (S, Q, O) transitions, discussions on the contributions from different electronic states, influence of the non-resonant susceptibility (χ_{NR}) and the Raman bandwidth (Γ), we refer to Ref. [8].

For the study described, iodine has been chosen because this is one of the few molecules which have continuous absorption in a spectral region easily accessible to pulsed dye-laser excitation and of which, simultaneously, the spectroscopic constants and potential functions are very well known. It is hoped that this model study in such a simple molecule can serve as a basis for more complex systems. Presently, Materny and Kiefer are carrying out similar (linear and nonlinear) experimental and theoretical studies on polydiacetylene single crystals. Results on these model systems will be published elsewhere [23].

ACKNOWLEDGEMENTS

The authors are very grateful to Dr. P. Baierl and Dr. B. Hartke for providing us with the computer programs for the simulation of the spectra. We also acknowledge the enthusiastic support by Mr. S. Görtler, T. Michelis, and Dr. H.-J. Schreier in producing a video-movie on the time-evolution of resonance Raman scattering, from which we adapted the snapshots presented in Fig. 2. We also thank the Fonds der Chemischen Industrie as well as the Deutsche Forschungsgemeinschaft (DFG) for financial support (projects C2/C3-SFB 347). E. Kolba should also like to thank the Studienstiftung des deutschen Volkes for a scholarship.

REFERENCES

[1] P. Baierl and W. Kiefer, J.Raman Spectrosc. 10 (1980) 197 and 11 (1981) 393.

[2] P. Baierl and W. Kiefer, J.Chem.Phys. 77 (1982) 1693.

[3] P. Baierl, W. Kiefer, P.F. Williams and D.L. Rousseau, Chem.Phys.Letters 50 (1977) 57.

[4] S.-Y. Lee and E.J. Heller, J.Chem.Phys. 71 (1979) 4777.

[5] B. Hartke, Chem.Phys.Letters 160 (1989) 538, J.Raman Spectrosc., in press, and Ph.D. thesis, University of Würzburg (1990).

[6] H. Chang, H.M. Lin and M.H. Hwang, J.Raman Spectrosc. 15 (1984) 205.

[7] W. Kiefer and D.A. Long, eds., Nonlinear Raman Spectroscopy and its Chemical Applications; Reidel, Dordrecht, 1982.

[8] A. Beckmann, P. Baierl and W. Kiefer, in Ref. 7, page 393.

[9] B. Hartke, W. Kiefer, E. Kolba, J. Manz, and H.-J. Schreier, to be published, see also B. Hartke, E. Kolba, J. Manz and H.H.R. Schor, Ber. Bunsenges. (1990) in press.
[10] V. Hizhnyakov and I. Tehver, Phys.Stat.Sol. 21 (1967) 755.
[11] J.B. Page and D.L. Tonks, Chem.Phys.Letters 66 (1979) 449.
[12] S.-Y. Lee and E.J. Heller, J.Chem.Phys. 71 (1979) 4777.
[13] D.J. Tannor and E.J. Heller, J.Chem.Phys. 77 (1982) 202.
[14] M. Ganz, B. Hartke, W. Kiefer, E. Kolba, J. Manz, and J. Strempel, Vibrational Spectroscopy 1 (1990) 119.
[15] P. Baierl and W. Kiefer, J.Raman Spectrosc. 15 (1984) 360.
[16] D.G. Imre and J. Zhang, Chem.Phys. 139 (1989) 89.
[17] M.S. Child and R.B. Bernstein, J. Chem. Phys. 59 (1973) 5916.
[18] M. Ganz and W. Kiefer, to be published.
[19] W.C. Hwang and J.V.V. Kasper, Chem.Phys.Lett. 13 (1972) 511.
[20] S.A.J. Druet, B. Attal, T.K. Gustafson, and J.P.E. Taran, Phys. Rev. A18, (1978), 1529.
[21] S.A.J. Druet, and J.P.E. Taran, in Chemical and Biochemical Applications of Lasers, Vol. 4, C.B. Moore, ed. Academic Press, New York, (1979), p. 187.
[22] S.A.J. Druet, and J.P.E. Taran, Progr. Quant. Electron. 7, 1 (1981).
[23] A. Materny and W. Kiefer, to be published.

Nonlinear Interferometry

G. Lüpke and G. Marowsky

Max-Planck-Institut für biophysikalische Chemie, Abt. Laserphysik,
Am Fassberg, W-3400 Göttingen, Fed. Rep. of Germany

Recent developments in the application of phase-difference measurements in coherent nonlinear optics are reviewed. Based upon a precise determination of the relevant coherence length, novel experimental procedures such as phase-controlled (nonlinear) interferometry or suppression of unwanted background contribution from different phases are possible. This concept is demonstrated for both sum- and difference-frequency generation for second- and third-order optical processes.

1. Introduction

Experiments performed in **Nonlinear Interferometry** are based upon a phase-controlled superposition of various, usually tensorial, field amplitudes $E(\omega^*)$ that can be derived from the respective nonlinear polarizations $P(\omega^*)$. The frequency ω^* may be a harmonic such as 2ω for second-harmonic generation (SHG) or 3ω for third-harmonic generation (THG) as well as a shorthand notation for second- or third-order difference-frequency generation (DFG) such as $\omega^* = 2\omega_1 - \omega_2$. In the latter case the frequency ω^* is the result of a nonlinear interaction between two incoming frequencies ω_1 and ω_2.

In all these processes the term *phase* will denote a relative phase difference between the incoming fundamental field amplitudes $E(\omega_i)$ ($\omega_i = \omega_1$, ω_2, ...) and the generated nonlinear signal amplitudes $E(\omega^*)$. Relative phases of nonlinear optical signals were first introduced by Chang et al. as early as 1965 [1]. Phases and phase-changes are of relevance in quite a number of effects in physics [2-7]: Structural anisotropies, dispersion-related phenomena (linear dispersion, absorption, resonances), experimental geometries for the observation of nonlinear signals from surface and bulk contributions. All these effects may in turn be used to introduce phase-changes.

This paper discusses the various aspects of phases in nonlinear optics, in both theory and experiment, in three independent sections:
Section 2 is devoted to a brief description of the concept of phases, their occurrence and sign change due to resonances, and to coherence lengths for the various nonlinear optical processes.
Section 3 deals with phase measurements in harmonic generation, both SHG and THG, and with the concept of optical heterodyning.

Springer Proceedings in Physics, Volume 63 **Coherent Raman Spectroscopy**
Eds.: G. Marowsky and V.V. Smirnov

Section 4 offers examples for DFG processes with applications for CARS-type experiments. Suppression of the nonresonant background and applications for evaluation of saturation measurements will be discussed in some detail .

2. The Concept of Phases

For the sake of simplicity we would like to discuss the occurrence of optical phases and the concomitant sign change upon passing through a resonance by considering a complex Lorentzian function [8]:

$$g_L = 1/(1 + ia\Delta\omega) \quad . \tag{1}$$

The Lorentzian g_L is non-normalized, $\Delta\omega$ denotes the normalized frequency detuning from resonance ($\Delta\omega = 0$), and a is the detuning parameter corresponding to the reciprocal linewidth. Equation (1) can be split into

$$g_L = \mathrm{Re}(g_L) - i\,\mathrm{Im}(g_L) \tag{2}$$

$$= \frac{1}{1 + a^2(\Delta\omega)^2} - i\,\frac{a\Delta\omega}{1 + a^2(\Delta\omega)^2} \quad . \tag{2a}$$

With $\tan\Phi = \mathrm{Im}(g_L)/\mathrm{Re}(g_L)$ the phase Φ is given by

$$\Phi = \arctan(a\Delta\omega) \quad . \tag{3}$$

Figure 1 shows the change of the phase Φ upon variation of the frequency detuning $\Delta\omega$ for a selection of values $a = 0.01 \ldots 1.00$. With respect to an externally set phase reference, a change in Φ by +180° should be observable by passing from $\Delta\omega = -\infty$ to $\Delta\omega = +\infty$. Next, let us consider the relation between *phases* and *coherence length*. A change in phase by 180° corresponds to passage of one coherence length (see Fig. 2), which will be derived from the oscillation pattern of any nonlinear optical experiment:

$$\sin^2(\Delta kL/2)/(\Delta kL/2)^2 \quad . \tag{4}$$

Hence, two adjacent peaks of an interference experiment are separated by

$$L_c = 2\pi/\Delta k \quad , \tag{5}$$

where the quantity Δk must be specialized for the various nonlinear processes taken into consideration. (Note that the definition according to eq. (5) has been used throughout Refs. [9,10], but is in disagreement with other definitions (e.g. [11]), using $L_c = \pi/\Delta k$ instead of $2\pi/\Delta k$!). The expressions for L_c for SHG, THG, and DFG processes are as follows (for details see [10]):

$$L_c\ (\mathrm{SHG}) = \lambda/2(n_{2\omega}-n_\omega) \tag{6a}$$

$$L_c\ (\mathrm{THG}) = \lambda/3(n_{3\omega}-n_\omega) \tag{6b}$$

$$L_c\ (\mathrm{DFG}) = 2\pi/\left|2k(\omega_p)-k(\omega_s)-k(\omega_{as})\right| \quad . \tag{6c}$$

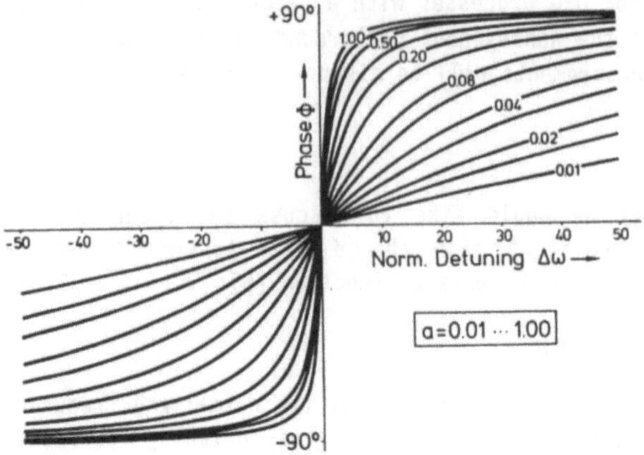

Figure 1: Variation of phase Φ versus normalized detuning Δω upon passing through a resonance (Δω = 0).

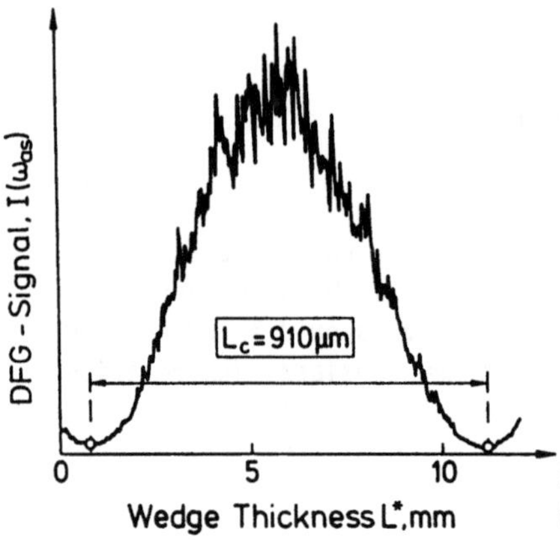

Figure 2: DFG signal $I(\omega_{as})$ versus wedge thickness $L^* \propto L$ from Ref. [12].

For harmonic generation, the quantity λ denotes the fundamental wavelength and n_ω, $n_{2\omega}$, $n_{3\omega}$ the respective indices of refraction. For $\chi^{(3)}$-type difference-frequency generation (DFG) ω_p, ω_s, and ω_{as} denote frequencies with $\omega_{as} = 2\omega_p - \omega_s$. In the latter case an observed interference pattern can be described by

$$I(\omega_{as}) \propto \sin^2(\pi L/L_c) \quad . \tag{7}$$

Figure 2 shows an example of an interferogram obtained by superposition of the radiation of a dye laser, tuned to $\lambda_s = 580$ nm, on the second-harmonic of a Nd:YAG laser ($\lambda_p = 532$ nm) in Schott BK-7 glass [12]. Computation of L_c with the available dispersion data for BK-7 resulted in $L_c = 907$ µm, whereas an experimental value of 910 µm was derived from the variation of the wedge thickness $L^* \propto L$. According to Ref. [13] the high

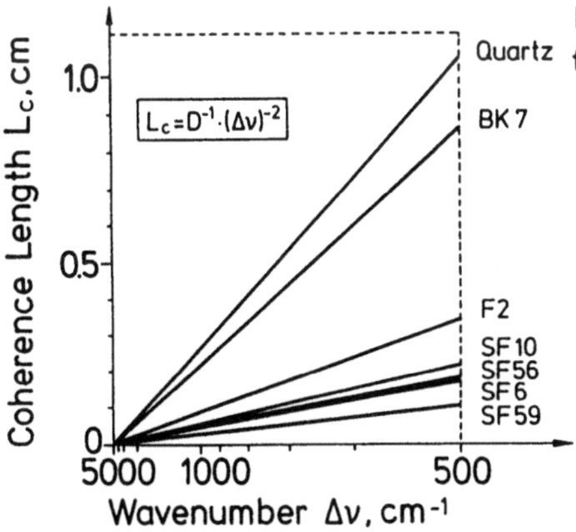

Figure 3: Coherence lengths L_c for a few selected materials.

degree of symmetry of ω_s and ω_{as} with respect to the center frequency ω_p permits a rather simple computation of L_c from available dispersion data for DFG processes. With

$$L_c = D^{-1}(\Delta\nu)^{-2} \quad , \tag{8}$$

where $\Delta\nu$ denotes the Stokes shift in cm^{-1}, the quantity D can be calculated as

$$D = \lambda_p^{\,3} \cdot \left. \frac{d^2 n(\lambda)}{d\lambda^2} \right|_{\lambda=\lambda_p} \quad . \tag{9}$$

Equation (9) is useful for a precise determination of the curvature of the dispersion n(λ) by third-order coherence-length measurements. Figure 3, taken from Ref. [13], shows a compilation of coherence lengths based upon dispersion data from the literature.

3. Harmonic Generation and Optical Heterodyning

In contrast to DFG experiments with rather large coherence lengths, L_c values for SH and TH processes are small, typically in the range of a few tens of micrometers [10], if the dispersion of solid-state materials is used. Relative phase differences between fundamental and harmonic signals can be measured if the weak dispersion of gaseous media, usually air under atmospheric conditions, is used. Figure 4 shows an example of an interference pattern from Ref. [3] using the technique of Ref. [14], which demonstrates that SH signals from a pure substrate and an adsorbate-substrate sample can be out of phase by 180°, depending on the spectral vicinity to a resonance and the particular experimental geometry. Different observation geometries such as total internal reflection [5,15] may introduce considerable changes in the phase of the nonlinear signal amplitude.

The principle of *optical heterodyning* is demonstrated with reference to Fig. 5. Two nonlinear experiments, E_I and E_{II}, are performed at locations (I) and (II), and are phase-controllably superimposed by means of the

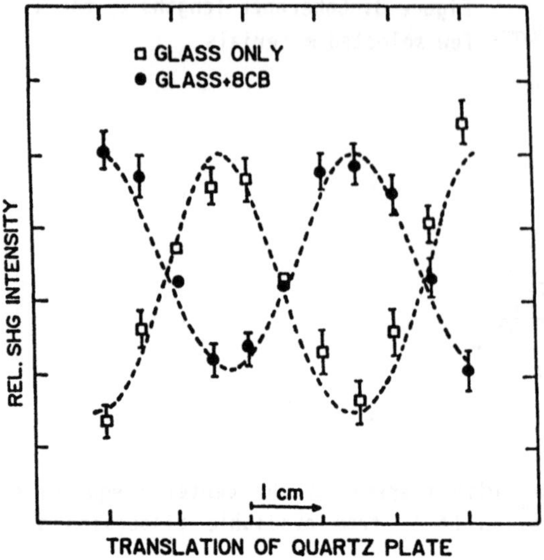

Figure 4: Comparison of interference patterns obtained when a SH-signal is generated from both sample and substrate or substrate only. A crystalline quartz plate mounted onto a translation stage served as a reference. The dashed curves are fits to the theoretical interference band shape according to eq. (7) with L_c = 3.6 ±0.1 cm.

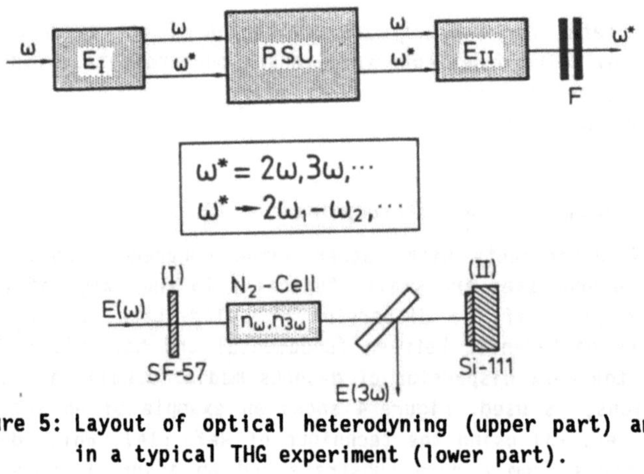

Figure 5: Layout of optical heterodyning (upper part) and its realization in a typical THG experiment (lower part).

phase-shifting unit P.S.U. The resulting frequency ω^* is observed after appropriate rejection of the incoming fundamental frequency. The lower part of Fig. 5 shows the details of a THG experiment. A fundamental wave of frequency ω and amplitude $E(\omega)$ produces a signal of frequency $\omega^* = 3\omega$ at positions (I) and (II). A thin plate of highly dispersive Schott SF-57 glass serves as a 3ω-source at position (I), a piece of a silicon wafer of (111)-orientation mounted on a stable substrate produces another 3ω signal in reflection at position (II). Both THG signals interfere at the silicon surface. Variation of the gas pressure of a N_2 cell between between the two signal locations produces an interference pattern in 3ω similar to that of Fig. 4. Coherent superposition of the signal contributions of different origin is typical for nonlinear optical experiments such as phase-controlled addition of isotropic and anisotropic contributions [2,15] or the addition of the various tensor elements, responsible for the respective nonlinear polarization $P_{NL}(\omega^*)$. Another example is the situation encountered in CARS-type configurations [16], where the resonant part of the nonlinear susceptibility $\chi^{(3)}(\Delta\omega)$ adds coherently to the nonresonant contribution $\chi^{(3)}{}_{NR}$:

$$\chi^{(3)} = \chi^{(3)}{}_{NR} + \chi^{(3)}(\Delta\omega) \quad . \tag{10}$$

4. DFG Processes and CARS Experiments

The determination of the coherence length and its relation to fundamental material properties has already been discussed in some detail in Section 2. In contrast to the other third-order process, third-harmonic generation, $\chi^{(3)}$-type difference-frequency generation offers the unique advantage of phase-matching by tuning of one of the fundamental wavelengths (usually λ_s) in combination with multi-dimensional setups. In addition, fringe-counting techniques can be applied to increase the experimental precision. Figure 6 shows as an example a coherence-length measurement, using a wedge pair prepared from Schott SF-56 glass. In this case the precision for a L_c-measurement is ultimately determined by the spectral bandwidth of the laser light sources used - in the examples of Figs. 2 and 6 typically 0.7 cm^{-1}.

Figures 7 and 8 summarize the experimental setup and experimental results of CARS-background suppression by phase-controlled subtraction of the unwanted nonresonant contribution [12]. As resonant signal, the rotational S(3)-line of molecular hydrogen with a Stokes shift of 1035 cm^{-1} was used. This signal was buried in the noise of 4 atm butane (left-side cell in Fig. 7 and upper spectrum in Fig. 8). By providing a phase-reversed nonresonant signal in the second cell filled with butane only, it was possible to completely suppress the nonresonant butane contribution. The lower spectrum of Fig. 8 indicates how the weak S(3)-CARS-signal due to 100 torr of hydrogen in the presence of 4 atm of butane could be regained. The successful recovery of the resonant signal strongly depends on the complete cancellation of the two nonresonant signals and perfect *linear* superposition of both the fundamental and the generated nonlinear signal. At least one order of magnitude can be gained in detection sensitivity of the resonant minority species.

The absence of phase distortion and the reproducibility of the nonlinear signal generation are prerequisites for the application of phase-sensitive measurements. Perfect superposition of two CARS-signals could only be achieved if the angular deviation of the two signal directions originating

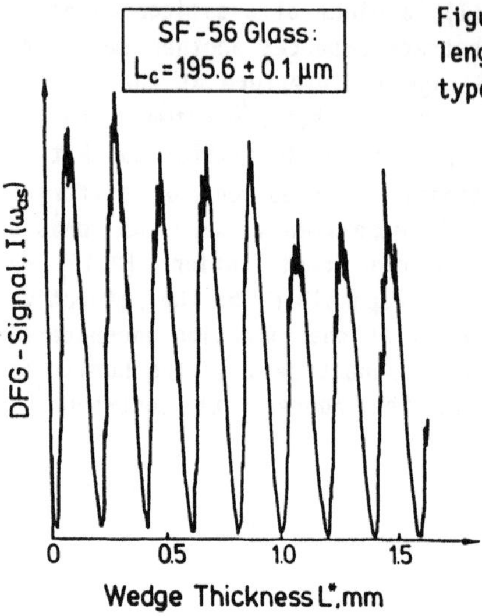

Figure 6: Determination of the coherence lenght L_c of SF-56 glass by $\chi^{(3)}$-type difference-frequency generation (DFG).

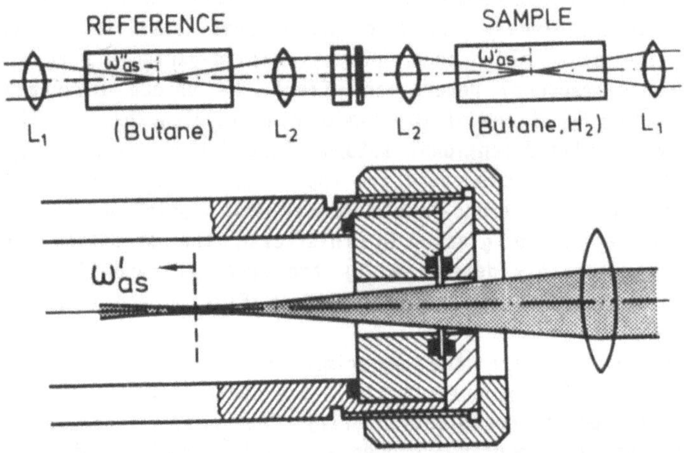

Figure 7: Optical-heterodyning experiment for CARS-background suppression. Lower diagram shows technical details of high-pressure gas cell with ultra-thin windows to avoid uncompensated nonresonant background.

from sources (I) and (II) did not exceed 50 seconds of arc and signal detection was restricted to paraxial rays. Best results in terms of complete cancellation of the two phase-reversed signals were obtained from identical gaseous samples. In fact, the Stokes shift of $\Delta\nu = 1555$ cm, which was used to record the interference pattern of Fig. 2, corresponded to the excitation of the center of the Q-branch ($\Delta J = 0$, $\Delta v = 1$) of atmospheric oxygen at both signal locations. Superposition of resonant and nonresonant contributions was much more difficult due to saturation in the line center

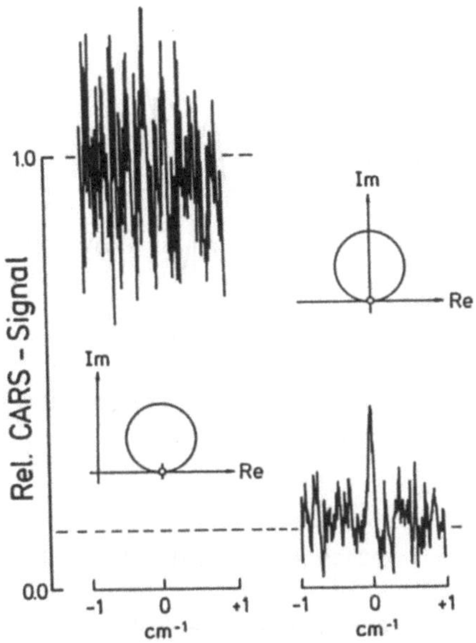

Figure 8: Nonresonant background reduction by phase-controlled nonlinear interferometry.

of the resonant Raman transition. Saturation and the concomitant reduction in contrast was inevitable for the narrowband resonances in molecular hydrogen for both rotational and vibrational Raman transitions. Recent results concerning the difficulties with H_2 thermometry due to saturation are reported in Refs. [17,18]. In the presence of saturation the $\Delta\omega$-dependent lineshape $I(\omega_{as})$ of a CARS-signal can be written as [19]

$$I(\omega_{as}) \propto \left| \frac{1 - ia\Delta\omega}{(1 + G) + (a\Delta\omega)^2} \right|^2 . \tag{11}$$

The quantity G is called the "saturation parameter" with $G = I_p I_s / I^2_{sat}$ according to Ref. [19]. The lineshape of eq. (11) reduces to the already discussed non-normalized Lorentzian (eq. (1)) in the limit $G \rightarrow 0$. For values $G > 0$ the shape of the $\Delta\omega$-dependent phase $\Phi(\Delta\omega)$, as has been discussed for various detuning parameters a in Fig. 1, remains unchanged. Using the relations

$$g = (1 + G)^{1/2} \quad , \tag{12a}$$

$$A = \tfrac{1}{2}(1/g - 1) \quad , \tag{12b}$$

$$B = \tfrac{1}{2}(1/g + 1) \quad , \tag{12c}$$

equation (11) can be rearranged into the format of two Lorentzian profiles

$$I(\omega_{as}) \propto \left| \frac{A}{g + ia\Delta\omega} + \frac{B}{g - ia\Delta\omega} \right|^2 . \tag{13}$$

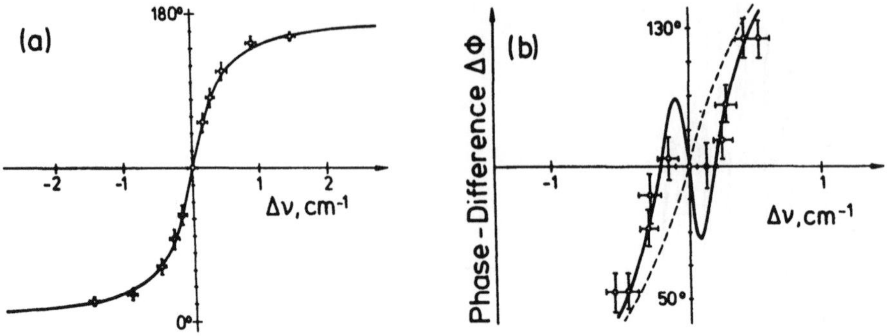

Figure 9: Phase-difference $\Delta\Phi$ versus spectral detuning for (a) low and (b) high input intensities I_p and I_s. (For details see text).

In any case, saturation means renormalization of the detuning parameter a or of the frequency $\Delta\omega$ by $g = 1/(1 + G)^{1/2}$.

Figure 9 shows in comparison phase measurements at low and high input-power levels:

$$(9a)\quad I_p = 6.0\ \mathrm{GW/cm^2},\ I_s = 0.9\ \mathrm{GW/cm^2}$$

$$(9b)\quad I_p = 20\ \mathrm{GW/cm^2},\quad I_s = 4.0\ \mathrm{GW/cm^2}\ .$$

From both figures the sign change in the phase upon passing a resonance is discernible. As the Raman transition, again the S(3) line of hydrogen was used. A butane cell of adjustable pressure delivered the nonresonant reference signal for the $\Delta\Phi$ phase-difference measurements. Figure 9a shows that a $\Delta\Phi$ value as high as ±80° with respect to resonance was achievable under these experimental conditions. The occurrence of the nonresonant contribution from 100 torr of hydrogen itself prevented us from recording $\Delta\Phi \to \pm 90°$. Figure 9b displays the result of a phase measurement under strong saturation. The dashed line shows a fit with $G = 1.5$ according to eq. (11), and the solid line an approach using eq. (13) with adjustable parameters A, B, and g. No approximation shows good agreement with the measured data points in the vicinity of the heavily saturated resonance at $\Delta\nu = 0$. We attribute this behavior to the onset of Rabi oscillations of the coherently driven two-level system by the high intensities I_p and I_s. In fact, interference visibility disappeared completely at resonance under these excitation conditions, indicating that a signal-phase is no well-defined physical quantity under transient excitation conditions.

5. Conclusion

The concept of phases in nonlinear optical experiments has been discussed for harmonic generation and third-order difference frequency generation. Phase-controlled nonlinear interferometry offers unique advantages concerning different items such as evaluation of dispersion properties of solid materials or background suppression in CARS-type experiments. Using the briefly described technique of optical heterodyning an experimental determination of selected elements of the respective susceptibility with respect to magnitude and sign is possible. Of fundamental interest seems to be the observation of phase changes in saturated or non-saturated optical transitions.

Acknowledgement

The experimental work has been supported by the Deutsche Forschungsgemeinschaft through the Leibniz-Prize program. We thank J. Jethwa for a critical reading of the English version of this manuscript and E. Heinemann for expert technical assistance.

References

[1] Chang, R.K., Ducuing, J., Bloembergen, N.: "Relative phase measurement between fundamental and second-harmonic light", Phys. Rev. Lett. 15 (1965) 6-8.

[2] Tom, H.W.K., Heinz, T.F., Shen, Y.R.: "Second-harmonic reflection from silicon surfaces and its relation to structural symmetry", Phys. Rev. Lett. 51 (1983) 1983-1986.

[3] Berkovic, G., Shen, Y.R., Marowsky, G., Steinhoff, R.: "Interference between second-harmonic generation from a substrate and from an adsorbate layer", J. Opt. Soc. Am. B 6 (1989) 205-208.

[4] Epperlein, D., Dick, B., Marowsky, G., Reider, G.A.: "Second-harmonic generation in centro-symmetric media", Appl. Phys. B 44 (1987) 5-10.

[5] Sieverdes, F., Lüpke, G., Marowsky, G., Bratz, G., Felderhof, U.: "Second-harmonic generation in total reflection", presented at the NATO Advanced Research Workshop 'Organic molecules for nonlinear optics and photonics', La Rochelle (France), Aug. 26 - Sept. 1, 1990.

[6] Sipe, J.E., Moss, D.J., van Driel, H.M.: "Phenomenological theory of optical second- and third-harmonic generation from cubic centrosymmetric crystals", Phys. Rev. B 35 (1987) 1129-1141.

[7] Guyot-Sionnest, P., Shen, Y.R.: "Local and nonlocal surface nonlinearities for surface optical second-harmonic generation", Phys. Rev. B 35 (1987) 4420-4426.

[8] Pantell, R.H., Puthoff, H.E.: "Fundamentals of quantum electronics", John Wiley, New York (1969).

[9] Yariv, A.: "Quantum electronics - Third edition", John Wiley, New York (1989).

[10] Lüpke, G., Marowsky, G., Steinhoff, R.: "Phase-controlled nonlinear interferometry", Appl. Phys. B 49 (1989) 283-289.

[11] Milonni, P.W., Eberly, J.H.: "Lasers", John Wiley, New York (1988).

[12] Marowsky, G., Lüpke, G.: "CARS-background suppression by phase-controlled nonlinear interferometry", Appl. Phys. B 51 (1990) 49-51.

[13] Lüpke, G.: "Phasen und Symmetrien in der Nichtlinearen Optik", Dissertation, University of Göttingen (1990).

[14] Hicks, J.M., Kemnitz, K., Eisenthal, K.B., Heinz, T.F.: "Studies of liquid surfaces by second harmonic generation", J. Phys. Chem. 90 (1986) 560-562.
[15] Lüpke, G., Marowsky, G., Sieverdes, F.: "Surfaces and adsorbate analysis by second-harmonic generation", Proceedings: NATO Advanced Research Workshop 'Organic molecules for nonlinear optics and photonics', La Rochelle (France), Aug. 26 - Sept. 1, 1990.
[16] Levenson, M.D., Kano, S.S.: "Introduction to nonlinear laser spectroscopy", Academic, Boston (1988).
[17] Bombach, R., Hemmerling, B., Hubschmid, W.: "Saturation effects and stark shift in hydrogen Q-branch CARS spectra", Chem. Phys. 144 (1990) 265-271.
[18] Bombach, R., Gerber, T., Hemmerling, B., Hubschmid, W.: "Aspects of hydrogen CARS thermometry", Appl. Phys. B 51 (1990) 59-60.
[19] Lucht, R.P., Farrow, R.L.: "Saturation effects in coherent anti-Stokes Raman scattering spectroscopy of hydrogen", J. Opt. Soc. Am. B 6 (1989) 2313-2325.

Evaluation of the CARS Spectra of Linear Molecules in the Keilson–Storer Model

S.I. Temkin and A.A. Suvernev

Institute of Chemical Kinetics and Combustion,
SU-630090 Novosibirsk, USSR

Abstract. The completion of the quasiclassical theory for CARS bandshape is presented, thus supplying a description of smoothed contour transformation with density. Simple analytical formulae are obtained, valid under arbitrary strength of collision, with Gordon and Langevin rotational relaxation models as limiting cases. The proposed experimental data processing allows to extract crosssections of stochastically independent vibrational dephasing and rotational energy relaxation.

The achievements of CARS probing of flows and flames are widely known. For applications in technical combustion systems, the pressure regime above 1 bar is of special interest [1]. On the other hand, direct CARS observation of Q-branch collapse is now possible [2]. For an isotropic spectrum with unresolved rotational structure the opportunity of simplified description arises : rotational quantum number j is treated as continuous variable. The aim of the following is to present the completion of the quasiclassical theory of CARS Q-branch transformation with density, the first version having been published a dozen years ago [3].

For molecules with comparatively low rotational temperature, one may essentially simplify the theoretical analysis of the density transformation of a smoothed contour, considering the rotation of a molecule in the framework of the classical theory. The process of intermolecular interaction in a gas may be considered as a succession of isolated collisions, therefore a good approximation is the impact description of the rotational relaxation, according to which angular momentum is a random Markovian variable J(t). The spectrum observed for the isotropic light scattering is determined by the correlation function of a scalar part of the molecular polarizability $\bar{\alpha}(t)$:

$$I_{IS}^{CARS}(\omega) \propto \left|\int_0^\infty K(t)e^{-i\omega t}dt\right|^2 \quad , \tag{1a}$$

$$K(t) = \frac{\langle\bar{\alpha}(0)\bar{\alpha}(t)\rangle}{\langle\bar{\alpha}^2\rangle} = \int_0^\infty K(t,J)dJ \ . \tag{1b}$$

In (1b) averaging over the time realizations of random process J(t) is perfomed strictly and leads to a standard kinetic equation [3] for a partial correlation function K(t,J) :

$$\frac{\partial}{\partial t} K(t,J) = (i\alpha_e J^2 - \frac{1}{\tau})K(t,J) + \frac{1}{\tau}\int_0^\infty f(J',J)K(t,J')dJ' \ . \tag{2}$$

Eds.: G. Marowsky and V.V. Smirnov

Here τ is the mean time between successive collisions, and $f(J',J)$ is the probability density of the change of angular momentum J' to J as a result of a single collision. In the present paper we restrict ourselves to the case of non-adiabatic change of the angular momentum at collisions. One of the most general phenomenological models of a kernel f, which satisfy this assumption, is the so-called Keilson-Storer model [3] :

$$f(J',J) = \frac{J}{d(1-\gamma^2)} I_0\left[\frac{\gamma JJ'}{d(1-\gamma^2)}\right] \exp\left[-\frac{J^2+\gamma^2 J'^2}{2d(1-\gamma^2)}\right] , \tag{3}$$

where $d = \frac{hkT}{8\pi^2 Bc}$; $|\gamma| \leq 1$ is the parameter determining the degree of the correlation of the angular momentum change. It is worth to point out, that (3) is the continuous analogue of "scaling laws" for the case of discrete rotational structure. Limiting cases of this model: non-correlated jumps ($\gamma=0$) and diffusion ($\gamma=1$) have been considered earlier in [3]. When the correlation of rotational relaxation is arbitrary, spectrum calculation reduces to the calculation of a continuous fraction. Analogously to our recent results for spherical molecules [4] , it can be easily shown that the Laguerre polinomials are eigenfunctions of an integral operator with (3) as a kernel :

$$\begin{aligned} &\int_0^\infty f(J',J)\varphi_B(J')L_n(J'^2/2d)dJ' = \gamma^{2n}\varphi_B(J)L_n(J^2/2d) , \\ &\varphi_B(J) = \frac{J}{d}\exp(-J^2/2d) . \end{aligned} \tag{4}$$

Then after substitution of the expansion

$$K(t,J) = \varphi_B(J) \sum_{n=0}^{\infty} a_n(t) L_n(J^2/2d) \tag{5}$$

into equation (2) and performing the Fourier transformation (1a), we obtain

$$a_0(\omega) = \frac{1}{i(\omega-\omega_Q)+i\omega_Q/b_0(\omega)} , \tag{6a}$$

$$b_n(\omega) = \frac{a_n(\omega)}{a_{n+1}(\omega)} = 2 + \frac{1}{n+1}\left[1 - \frac{\omega}{\omega_Q} + \frac{i}{\tau_E\omega_Q}\sum_{m=0}^{n}\gamma^{2m}\right] - \left(1+\frac{1}{n+1}\right)/b_{n+1}(\omega), \tag{6b}$$

where $\tau_E = \tau/(1-\gamma^2)$ is the time of rotational energy relaxation and $\omega_Q = \alpha_e d$ is the mean frequency of the Q-branch [5].

The aboveproposed algorithm can be realized also in the presence of vibrational dephasing. According to [6], taking into account the shift of vibrational phase $\Delta\varphi$ in each collision leads to the appearance of factor $\beta = \langle\exp(i\Delta\varphi)\rangle$ in front of the integral part of equation (2). Again repeating the derivation of (6b) from (5) and from (2), transformed in this way, we obtain a recurrent procedure, differing only by the substitutions in the right-hand part of (6b) $\omega \to \omega - i/\tau_{dp}$, $1/\tau_e \to 1/\tau_e - (1-\gamma^2)/\tau_{dp}$ $(1/\tau_{dp} = (1-\beta)/\tau)$:

$$a_0(\omega) = \left[i(\omega-\omega_Q) + 1/\tau_{dp} + i\omega_Q/b_0(\omega) \right]^{-1} , \qquad (7a)$$

$$b_n(\omega) = 2 + \frac{1}{n+1}\left(1 - \frac{\omega}{\omega_Q} + \frac{i\gamma^{2(n+1)}}{\omega_Q \tau_{dp}} + \frac{i}{\omega_Q \tau_E} \sum_{m=0}^{n} \gamma^{2m} \right) - \left(1 + \frac{1}{n+1}\right)/b_{n+1}(\omega). \qquad (7b)$$

The dephasing contribution that is linear with the density can be easily extracted, if the experiment provides pressure high enough for the rotational broadening contribution to be small ($\omega_Q^2 \tau_E^2 << 1$). In a general case parameters τ_E and τ_{dp} are determined by a self-consistent two-parameter fitting.

Due to the orthogonality property of the Laguerre polynomials, the fully averaged correlation function (1b) coincides with the first expansion coefficient in (4):

$$K(t) = a_0(t) \ , \ I_{IS}^{CARS}(\omega) \propto |a_0(\omega)|^2 . \qquad (8)$$

To improve the convergency of the recurrent procedures (6) and (7), it is convenient to use for the truncation at given N an asymptotic expansion of the solution of (6b) at N>>1:

$$b_N(\omega) \approx 1 - \left\{ \frac{1}{N}\left[\frac{i}{\omega_Q \tau_E} \sum_{m=0}^{N} \gamma^{2m} - \frac{\omega}{\omega_Q} \right] \right\}^{1/2} , \qquad (9)$$

$$N >> \mathrm{Sup}(L, 1/L),\ L = \left[\frac{\omega^2}{\omega_Q^2} + \frac{1}{\tau_E^2 \omega_Q^2} \right]^{1/2},\ \gamma^{2N} << 1 .$$

According to estimations (9), to realise calculation algorithm (6, 7), one has to perform tens of iterations for the calculation of the density transformation of the static contour into a fully collapsed one (Fig.1). For a quantitative comparison with the experiment it is often necessary to calculate the density dependence of the contour width at the half-height $\Delta\omega_{1/2}$. Fig.2 also presents a well-known result of the perturbation theory ($\omega_Q \tau_E >> 1$) treatment [7a], which is the same for Raman and CARS:

$$\Delta\omega_{1/2}^{CARS} = \Delta\omega_{1/2}^{RAM} = 2\omega_Q^2 \tau_E . \qquad (10)$$

It is interesting that the contour becomes insensitive to the degree of correlation of rotational relaxation at densities somewhat less than that required for formula (10) to hold. Using closeness of the "corridor" in Fig.2 with analogous $\Delta\omega_{1/2}^{RAM}(\Gamma)$ dependence for values $\Gamma > 3$, it is possible to estimate σ_E cross-section for N_2 with the help of stimulated Raman measurement for the density region from 13.5 up to 516 Amagat [8]: $\sigma_E = 32\ A^2$,

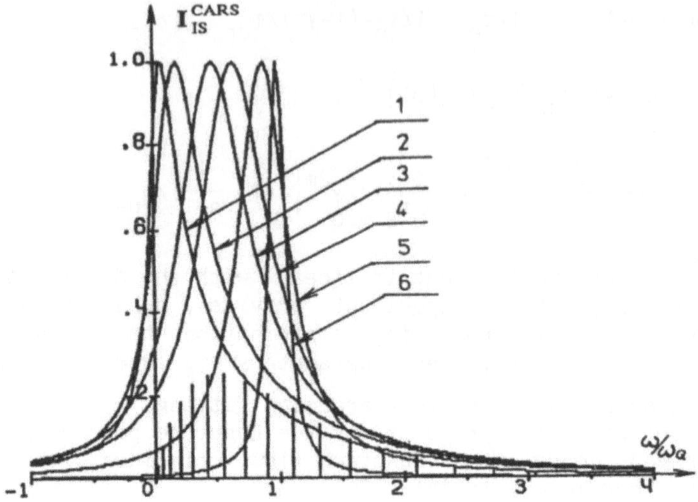

Fig.1. Density transformation of the CARS Q-branch (dephasing is absent). Contours are normalized by their maximum values, the bar-spectrum is given in arbitrary units. 1. Γ=0.05; 2. Γ=0.1; 3. Γ=0.5; 4. Γ=1.; 5. Γ=3.; 6. Γ=12 ($\Gamma = 1./\omega_Q \tau_E$, γ=0.7).

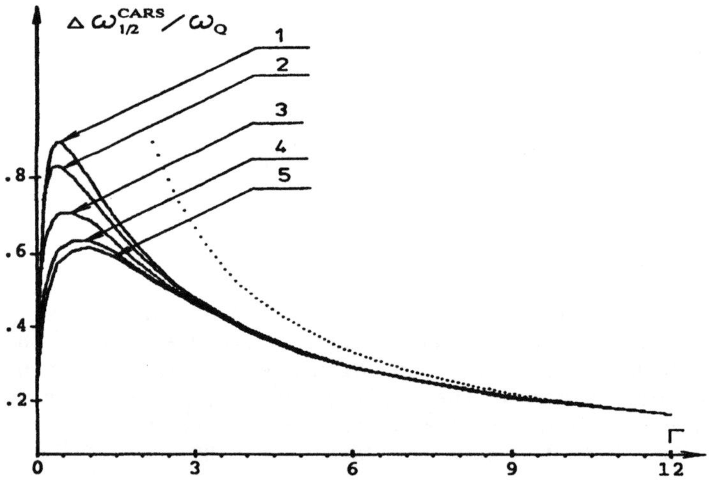

Fig.2. Density dependence of line widths for the various efficiency of collisions. 1. γ=0; 2. γ=0.4; 3. γ=0.7; 4. γ=0.9; 5. γ=1. Dots denote the result of the perturbation theory [7a].

$N\tau_E$ = 0.179 Amagat•ns, which agrees with our previous estimation [7b] σ_E = 39 A^2, $N\tau_E$ = 0.147 Amagat•ns, and with [8] σ_E = 35 A^2, $N\tau_E$ = 0.164 Amagat•ns.

Comparing quasiclassical theory and description with scaling in calculating smoothed CARS line shapes we conclude, that the evident advantage of the former is simple final analytical expressions and advantage of the latter - availability of non-phenomenological temperature dependence for the theoretical parameters.

REFERENCES

1. W.Stricker,M.Woyde. In Joint meeting of the German and Italian Sections of the Combustion Institute. 1989, Naples, Italy, 1.1
2. Th.Bouche,Th.Dreier,B.Lange,J.Wolfrum,E.U.Franck,W.Schilling: Appl.Phys. **B50**, 527 (1990)
3. S.I.Temkin,A.I.Burshtein: JETP Lett. **28**, 538 (1978)
4. S.I.Temkin,A.A.Suvernev,A.I.Burshtein: Opt.Spectr. **66**,69 (1989)
5. S.I.Temkin,A.I.Burshtein: JETP Lett. **24**, 86 (1976)
6. V.A.Alekseev,A.V.Malyugin: Zh.Eksp.Teor.Fiz. **80**, 897 (1981)
7. S.I.Temkin,A.I.Burshtein: (a)Chem.Phys.Lett. **66**, 52 (1979); (b) ibid. **66**, 57 (1979)
8. B.Lavorel, G.Millot, R.Saint-Loup, H.Berger: Poster report at Ninth European CARS Workshop, Dijon, France, March 19-20, 1990

Resonance-CARS Spectroscopy of Bio-molecules and of Molecules Sensitive to Light

W. Werncke, M. Pfeiffer, A. Lau, and Kim Man Bok

Central Institute of Optics and Spectroscopy, Rudower Chaussee 6, O-1199 Berlin, Fed. Rep. of Germany

It is demonstrated that resonance CARS is suitable for investigations of the light sensitive bio-molecules 20-chloro-chlorophyll and phytochrome both bearing a strong fluorescence. Short lived photoisomers of cyanine dyes exhibiting the same molecular structure but different lengths of the methine chains are investigated by time resolved resonance CARS spectroscopy. Using a normal coordinate analysis the photoisomers are identified as 1,2-mono-cis isomers which are generated independent of the chain length up to heptamethine.

1. Introduction

Time resolved resonance Raman spectroscopy is a very useful method to obtain vibrational spectra of short lived molecules bearing a strong fluorescence /1/. With an energy load less than 0.1 mJ, CARS spectra covering a frequency range of about 300cm^{-1} can be recorded with one single laser pulse /2/. Resonance enhancement in CARS is obtained if the wavelength of the weak CARS signal coincides with the electronic one photon transition while the pump-laser frequency may lie outside of resonance, so that saturation effects or sample modification by the latter may be avoided. Therefore CARS should be very useful for the investigation of unstable molecules and nonreversible photochemical reactions, especially if a continuous flow is excluded because of lack of substance for an expensive sample.

Generally an assignment of the observed changes in the vibrational spectra to structural changes of the molecules is difficult, if a strong coupling of vibrational modes in complex molecules occurs. This generally is the case in -

Springer Proceedings in Physics, Volume 63 Coherent Raman Spectroscopy
Eds.: G. Marowsky and V.V. Smirnov

conjugated systems with atoms having nearly equal masses (C, N, O). In this case the correct assignment requires a normal coordinate analysis.

2. Experimental technique

The scheme of the excite and probe CARS spectrometer used for our experimental investigations is shown in Fig. 1. It consists of a laser source (nitrogen laser-EL2, dye laser-DLE) used for initiation of the photophysical or photochemical processes. The nitrogen laser is synchronized with an XeCl-excimer laser delivering the pumping radiation for the two dye lasers of the CARS spectrometer. The parameters and some experimental details of the setup are summarized in Table 1. For further information about the experimental technique see Ref. /3/.

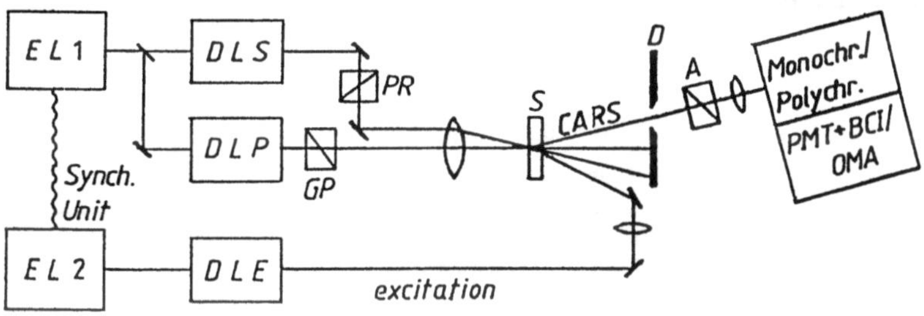

Fig. 1 Excite and probe CARS spectrometer for time resolved investigations
EL1 - excimer laser for dye laser pumping, DLS - Stokes dye laser, DLP - pump dye laser, PR - polarization rotator, GP - Glan prism, S - sample, A - analyser, D - diaphragm, monochr. - monochromator, polychr. - polychromator , PMT - photomultiplier tube, BCI - boxcar integrator, OMA - optical multichannel analyser, EL2 - nitrogen laser for excitation, DLE - dye laser for excitation

Table 1. Parameters of the ns-excite and probe CARS spectrometer

operation:	broad band / narrow band / one shot operation
repetition:	1Hz / 5 Hz
pulse length:	10ns
time resolution:	10ns / 2ns *)
time delay between excite and probe pulses:	0 (± 5ns) - 100 ms
laser power:	≤ 100kW
tuning range for excitation λ_{EXC}:	750nm -360nm, 337nm, 308nm
tuning range for probing λ_P:	700nm - 360nm
spectral width:	
narrow band dye laser:	< $1cm^{-1}$
broad band dye laser:	≈ $200cm^{-1}$-$300cm^{-1}$
CARS cell:	
windows:	d= 0.1mm
sample:	d= 0.2-1mm
volume:	≈ 1µl

*) using a ps-gate of the boxcar integrator

3. Experimental results

3.1 20-chloro-chlorophyll-a **)

Contrary to chlorophyll-a, 20-chloro-chlorophyll-a is photochemically degraded if it is exposed to daylight. In a first step after irradiation some of the molecules are converted into a photoproduct exhibiting a red shift of the lowest electronic (Q_Y) transition from 667nm to 686nm and a decrease of optical density of the Soret band at 432nm /4/. Therefore only resonance Raman scattering which is selectively enhanced by the Q_Y-transition gives the spectroscopic informa-

**) In cooperation with A. Struck and H. Scheer; Botanisches Institut der Universität München, Federal Republic of Germany

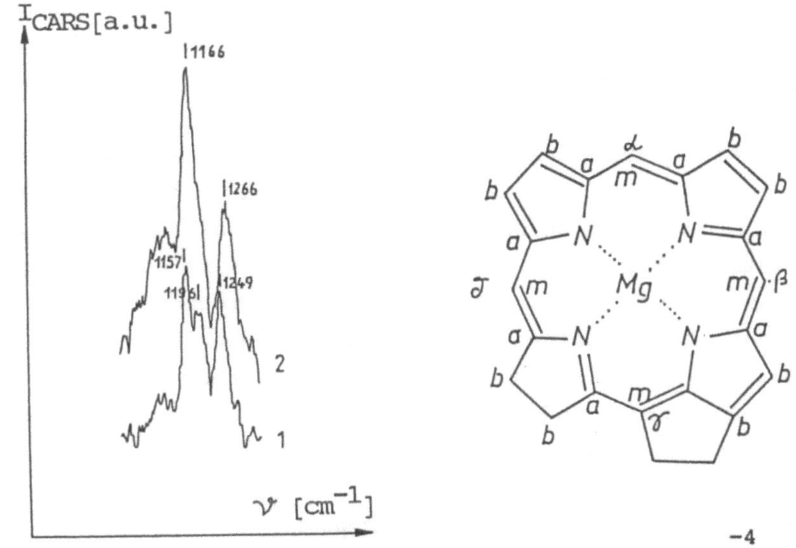

Fig. 2 Single pulse resonance CARS spectra of 10^{-4} mol/l 20-chloro-chlorophyll-a dissolved in acetone (1) and of the photoproduct (2) after irradiation (frequency range $1100cm^{-1}$ $-1300cm^{-1}$), /left side/.
molecular structure of chlorophyll; in the case of 20-chloro-chlorophyll-a the H-atom is replaced by a Cl-atom at the δ-position, /right side/.

tion with respect to the photoproduct. Spontaneous resonance Raman scattering is difficult to apply here because of the strong fluorescence and the photochemical instability of the molecules.

It was possible to obtain single shot CARS spectra of the parent molecules and, after irradiation with a white light source, of the photoproduct too, as shown in Fig 2. In this case we used a pumping wavelength λ_P =685nm, as this turned out to be the most favorable resonance condition with respect to the photoproduct.

The only significant difference we obtained in the frequency range from $900cm^{-1}$ to $1700cm^{-1}$ is the strong decrease of intensity at $1196cm^{-1}$, as can be seen in spectrum 2. With this exception other lines of chlorophyll-a /5/, 20-chloro-chlorophyll-a and of the new species do not show any significant differences of frequencies or relative intensities. As it is one of the most intense lines of the parent molecule the strong decrease of intensity of the CARS line at $1196cm^{-1}$ indicates that the spectrum 2 really originates mainly from the photoproduct.

The nearly unchanged vibrational pattern supports the assumption of a still cyclic porphyrin and the change of intensity (or strong shift) of the $1196cm^{-1}$ skeleton vibration, which arises to a high degree from C-N stretching and C_m-H bending modes /6/, seems to be in agreement with an assumption of a structure similar to a 20-oxonia-chlorophyll /4/. However, this assumption has to be further confirmed by extended experiments and detailed calculations of vibrational modes.

3.2 Phytochrome in its native environment***)

Phytochrome is one of the most important photoreceptors in higher plants. It acts as a sensor for the light environment and transduces the spectral information in such a way that plants can respond by modification of growth and development /7/. It has been shown that phytochrome can exhibit a P_R and a P_{FR} modification absorbing either in the red (λ_{max} ≈667nm) or in the infrared (λ_{max} ≈730nm) respectively. P_R can be converted into P_{FR} by irradiating into the red absorption band or vice versa from P_{FR} into P_R using infrared radiation. To explain the mechanism of phototransformation either a Z,E photoisomerization /8/ or an intramolecular proton transfer /9/ has been suggested. Preresonance spontaneous Raman spectra at λ= 752nm have been obtained at very high concentrations of phytochrome P_R /10/.

With λ_P = 694nm we recorded resonance CARS spectra under strict resonance conditions, as under these conditions at least the CARS signal coincides with the maximum of the absorption band of P_R /11/. Spectra of deuterated and nondeuterated phytochrome in the frequency range of $1500cm^{-1}$ - $1700cm^{-1}$ are shown in Fig. 3 together with a possible representation of the molecular conformation of this molecule.

***) In cooperation with G. Hermann and E. Müller; Sektion Biologie, Friedrich Schiller-Universität Jena, Federal Republic of Germany

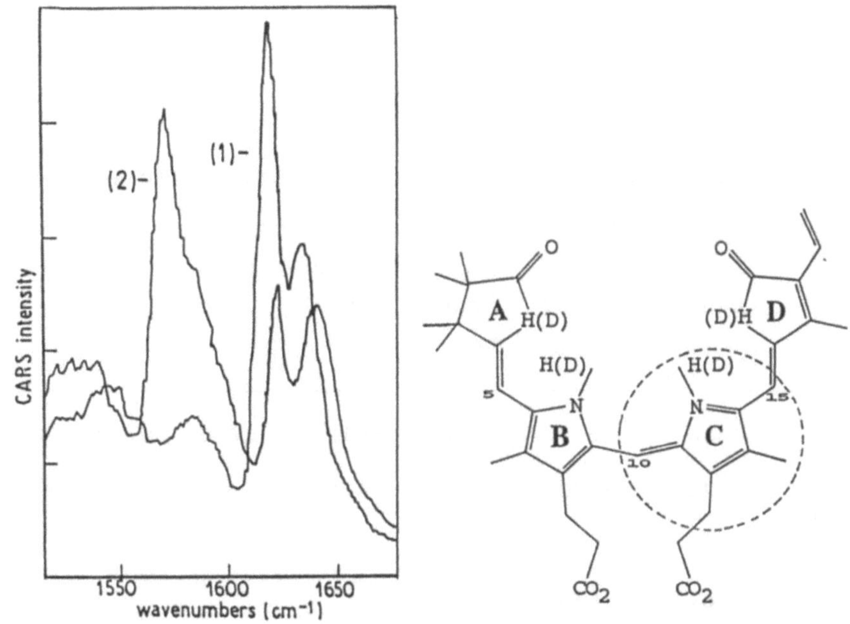

Fig. 3 Resonance CARS spectra of deuterated (1) and undeuterated (2) P_R dissolved in 50mM tris/HCl buffer with 1mM EDTA at pH 7.8 obtained at room temperature /left side/.
Representation of one possible structure of phytochrome.
Calculations include the ring structure indicated by a dashed line /right side/.

Comparing the spectra and performing respective normal coordinate calculations, we assigned the observed frequencies to specific local modes. The effect of deuteration on the spectra can be sufficiently well described by taking into account only one structure element of the whole molecule, consisting of a five-atom ring with one hydrogen (or deuterium) atom at the N- site and four exocyclic C-atoms at the C-sites. Experimental and calculated frequencies together with the respective assignments are given in Table 2.
In Fig. 4 it is demonstrated that the application of the CARS measurement does not lead to a conversion of P_R. But irradiation into the 660nm band completely removes the formerly observed CARS spectrum, as can be seen in spectrum (1). After

Table 2. Experimental and calculated frequencies (cm^{-1}) and assignments of phytochrome P_R

Undeuterated P_R			<======>	Deuterated P_R		
Assign-ment	Experi-mental	Calcu-lated	Local symme-try	Experi-mental	Calcu-lated	Assign-ment
v_{CC}/wr	1636 s	1636	a1	1631 s	1635	v_{CC}/wr
v_{CC}/wr	1622 s	1622	a1	1616 s	1615	v_{CC}/wr
$v_{CN}+r_{NH}+$						
v_{CC}/br	1564 s	1564	b1			
v_{CC}/wr	1513 vw	1513	b1	1525 m	1517	v_{CC}/wr
			b1	1500 m	1475	$v_{CN}+v_{CC}$/br
				1430 m		
				1329 w		
$r_{NH}+v_{CC}$/br	1321 m	1321	b1	1319 w		
$v_{CN}+v_{CC}$/br	1262 vw	1262	a1	1260 vw	1253	$v_{CN}+v_{CC}$/br
			b1		1081	r_{ND}

v-stretch, r-in plane rock, wr-within the ring, br-bridging bonds, vw-very weak, w-weak, m-medium, s-strong

infrared irradiation it reappears (2) due to the conversion of P_{FR} into P_R.

Experiments toward probing of P_{FR} with a near infrared CARS are now under way in our laboratory.

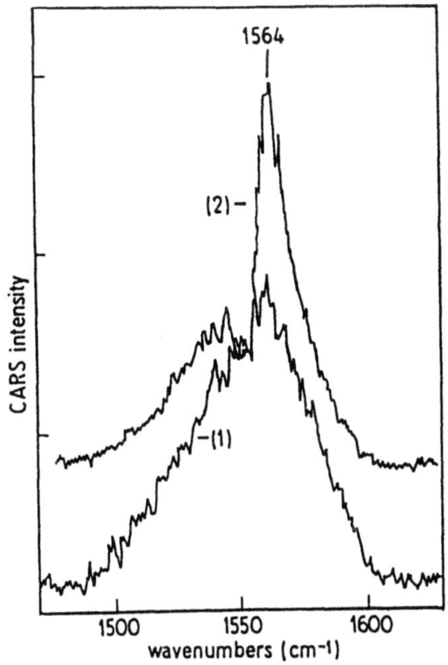

Fig. 4. Photoreversible disappearance- and reappearance of the resonance CARS band at 1564cm^{-1}

3.3 Photoisomerization of bisdimethylamino-trimethineperchlorate, -pentamethineperchlorate and -heptamethineperchorate

Photoisomerization at low temperatures of bisdimethylamino-trimethineperchlorate, -pentamethineperchlorate and -heptamethineperchorate was reported in an early paper of Scheibe et al. /12/. At room temperature these photoisomers are generated after some picoseconds, exhibiting a lifetime of some milliseconds /13/.

A schematic representation of the all-trans configuration of the parent molecules is shown in Fig. 5.

However, although this type of molecule has already been intensively studied, time resolved resonance CARS investigations are interesting because they provide vibrational information about the short lived intermediates /15,16/ and photoisomerized forms.

The problem with photoisomerization is that in the case of dyes with a long methine chain numerous mixed E,Z photoisomers in principle may be generated, which have to be distinguished.

Bisdimethylamino-

-trimethine

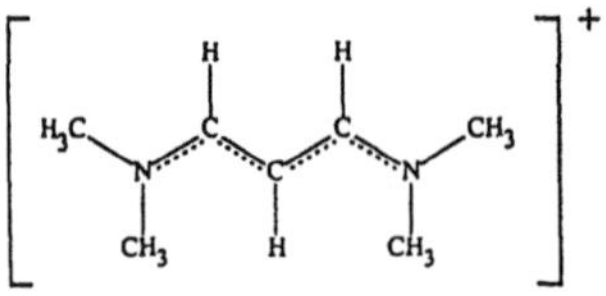

-pentamethine

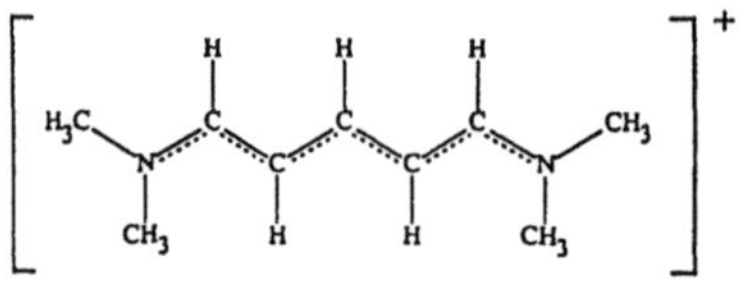

-heptamethine

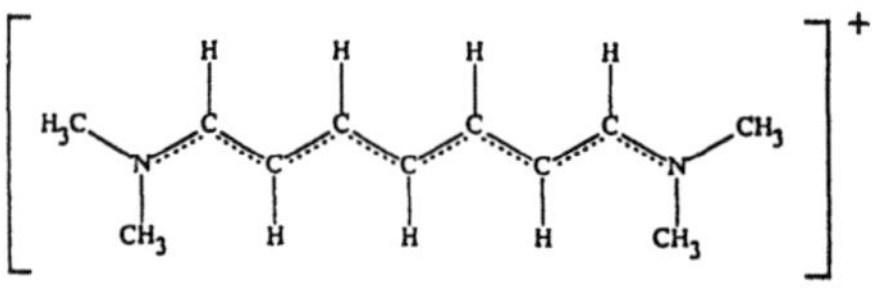

Fig. 5. All-trans configuration of bisdimethylaminotrimethine, -pentamethine and heptamethine

Here we report on photoisomerization studies of the three bisdimethylamino-(tri, penta, hepta)-methineperchlorate dyes by time resolved resonance CARS after preceding photo-excitation.

The absorption maxima of the three dyes, of their photoisomers and the wavelengths used for excitation λ_{EXC} and for CARS probing λ_P are summarized in Table 3.

Table 3. Absorption maxima of bisdimethylaminomethine dyes /12/ ; excitation- and CARS- probing wavelengths (λ_{EXC},λ_P) [nm]

	TRI	PENTA	HEPTA
all-trans	315	414	514
photoisomer	340	442	543
λ_{EXC}	308	400	510
λ_P	370	464	580

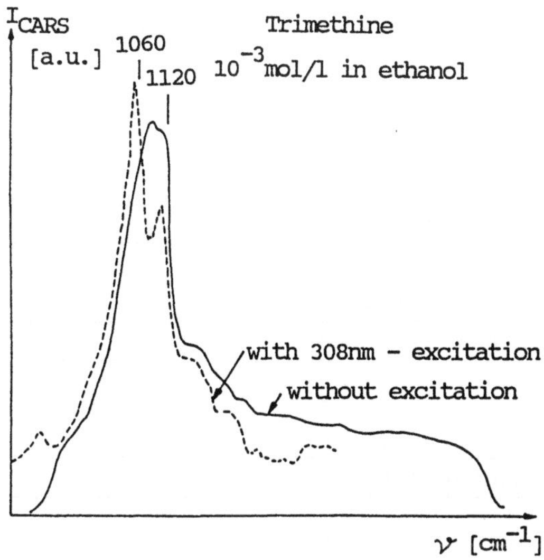

Fig. 6 Resonance CARS spectrum of bisdimethylaminotrimethine-perchlorate (10^{-3} mol/l dissolved in ethanol) and of its photoisomer after 308nm irradiation

A resonance CARS spectrum of the trimethine before and immediately after 308nm irradiation is shown in Fig. 6.

The main characteristic feature is the appearance of the strong $1060cm^{-1}$ CARS line together with a line at 1120 cm^{-1} which remains nearly unshifted with respect to the frequency of the parent molecule. As shown in Table 4, the appearance of one intense Raman band near $1070cm^{-1}$ is also observed for the other dyes. Other alterations within the investigated range of $900cm^{-1}$-$1700cm^{-1}$ are rather small and do not show any significant effect.

Our results on the heptamethine have already been published partially in /15/. A normal coordinate calculation was carried out including a force constant adjustment to reproduce the observed vibrational frequencies of the heptamethine and the ^{15}N-substituted form of it . A fit for the increments of the bond polarizabilities of the conjugated chain was done to reproduce the observed Raman intensities. Transferring both force constants and bond polarizability increments to the photoisomers and taking only their altered geometries into consideration, the corresponding Raman spectra for all possible plane mono-cis configurations were calculated. From

Table 4. Parent molecule and photoisomer CARS frequencies of bisdimethylaminomethine dyes in the range $1000cm^{-1}$ - $1200cm^{-1}$

parent molecule	photoisomer
bisdimethylaminotrimethine	
1120 s	1120 s
	1060 s
bisdimethylaminopentamethine	
1125 s	1125 s
1082 w	1072 s
bisdimethylaminoheptamethine	
1022 s	1022 s
1085 w	1070 s

these only the 1,2 mono-cis configuration results in spectra corresponding to the experimental finding in the $1000cm^{-1}$ - $1200cm^{-1}$ range. In contrast to this result, quantum chemical calculations of the potential energies in the excited singlet state suggest a 2,3 mono-cis configuration for the hepta-methine /14/ and for the pentamethine /12/. Extension of the measurements to the series of the methines provides further information about the photoisomerization process.

In the trimethine only the 1,2 mono-cis isomer can occur. The bi-cis isomer can be excluded and twisted isomers are very unlikely, as an interruption of the -system due to a twisting should result in a shift to shorter wavelengths of the absorption band for the photoisomers. This is not observed /12/. The CARS spectrum of the trimethine photoisomer therefore should show the essential features of a 1,2 mono-cis isomer. An interpretation of the similar spectral picture for penta- and hepta-methines calls for a comparative analysis of the vibrational modes in the three molecules based on a normal coordinate analysis.

The force constants obtained from the spectral fit for the molecules in the all-trans state are given in Table 5 for the π-chain bonds.

Table 5 shows a slight reduction of the $N=C_1$ force constant accompanied by an enhancement of the $C_1=C_2$ constant

Table 5. Force field of the chain bonds for the bis-dimethyl-amino-methine series obtained by fit of the observed IR- and Raman-spectra (f_i-bond stretching constants, f_{ij}-interaction constants). Force constants f in mdyne/Å.

	$f_1 = K_{N=C_1}$	$f_2 = K_{C_1=C_2}$	$f_3 = K_{C_2=C_3}$	f_{12}	f_{23}	f_{13}
TRI	7.95	6.34	6.34	0.37	0.30	-0.25
PENTA	7.89	6.36	4.93	0.41	0.41	-0.17
HEPTA	7.86	6.38	4.93	0.42	0.12	-0.14

in going from tri- to heptamethines and a tendency to dimerization in the C=C-C unit (bond alternation) in lengthening the chain. This behavior and the signs of the interaction constants ($f_{i,i+1}>0$, and $f_{i,i+2}<0$) confirm the π-bond character of the chain and the parameters seem to be a realistic representation of the molecule.

The Raman spectrum calculated for the trimethine on the basis of the parameters of Table 5 coincides rather well with the experimental results in the range 1000cm^{-1}-1200cm^{-1}. The appearance of the band 1060cm^{-1} is a consequence of the lowering of the C_{2v}-symmetry in the isomer.

Comparing the calculated normal coordinates for the larger molecules shows the appearance of a normal mode analogous to that of the 1060cm^{-1} vibration of trimethine in all the 1,2-mono-cis forms for which an analogous spectroscopic appearance

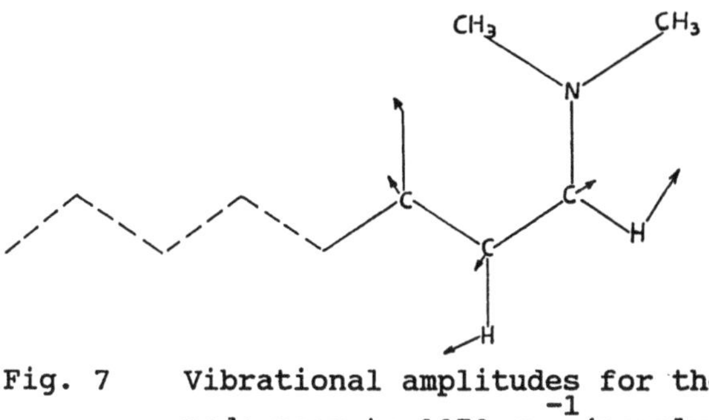

Fig. 7 Vibrational amplitudes for the characteristic isomer mode near to 1070 cm^{-1} (nearly equal appearances for tri-, penta- and hepta-methines)

is to be expected. This sustains the assumption of a 1,2-mono-cis character of isomerization in all three investigated isomers. Furthermore, the calculated force field gives information about the bond strengths within the π-system including interaction effects between neighboring bonds.
The corresponding vibrational mode is characterized in Fig.7.

4. Conclusions

We have shown several examples, where resonance CARS is successfully applied for investigations of molecules easily destroyed or converted by light and of their photoproducts. The pertubation due to the irradiation needed for CARS generation can be minimized so far that this effect could be neglected. For the assignment of the vibrations and of the observed spectral changes a normal coordinate analysis is necessary. As a result the photoisomers of bisdimethylamino-methine dyes are assigned to their respective 1,2 mono-cis configurations up to the heptamethine, for which a characteristic vibrational mode could be identified.

References:

/1/ W. Werncke, H.-J. Weigmann, J. Pätzold, A. Lau, K. Lenz, M. Pfeiffer, Chem. Phys. Lett. **61**, 105 (1979).

/2/ A. Lau, K. Kneipp, W. Werncke, K. Lenz, H.-J. Weigmann, D. Fassler and G. Hinzemann, Adv. Mol. Rel. Interaction Proc. **24**, 27 (1982).

/3/ A. Lau, W. Werncke, M. Pfeiffer; Spectrochimica Acta Rev. **13**, 191 (1990)

/4/ A. Struck, E. Cmiel, S. Schneider, H. Scheer; Photochem. Photobiol. **51**, 217 (1990)

/5/ E. Höxtermann, W. Werncke, I.N. Stadnichuk, A. Lau, P. Hoffmann; studia biophys. **92**, 147 (1982); studia biophys. **115**, 85 (1986)

/6/ M. Lutz, B. Robert; in Biological Applications of Raman Spectroscopy, Vol. 3 (Th.G. Spiro ed.), Wiley & Sons, New York, Chichester, Brisbane, Toronto, Singapore, 347 (1988)

/7/ W. Rüdiger; Structure and Bonding **40**, 101 (1980)

/8/ F. Thümmler, W. Rüdiger, E. Cmiel, S. Schneider; Z. Naturforschg. 38c, 359 (1983)

/9/ P.S. Song, Q. Chae, J.G. Gardner, Biochim. Biophys. Acta 576, 479 (1979)

/10/ St.P.A. Fodor, C.G. Lagarias, R.A. Mathies; Photochem. Photobiol. 48, 129 (1988)

/11/ W. Werncke, Kim Man Bok, A. Lau, G. Hermann; Abstracts "Fourth Symposium Optical Spectroscopy" p. 126, October 1986, Reinhardsbrunn, GDR
G. Herman, E. Müller, W. Werncke, M. Pfeiffer, Kim Man Bok, A. Lau; Biochem. Physiol. Pflanzen 186, 135 (1990)

/12/ G. Scheibe, J. Heiss, K. Feldmann; Ber. Bunsenges. Phys. Chem. 70, 52 (1966)

/13/ S. Rentsch, R.V. Danelius, R.A. Gadonas, Chem. Phys. 59, 119 (1981).

/14/ H. Hartmann, P. Wähner; Abstracts "Fourth Symposium Optical Spectroscopy" October 1986, Reinhardsbrunn, GDR

/15/ W. Werncke, A. Lau, M. Pfeiffer, H.-J. Weigmann, W. Freyer, Tschö Jong Tscholl, Kim Man Bok, Chem. Phys. 118, 133 (1987)

/16/ A. Lau, W. Werncke, M.Pfeiffer, H.-J. Weigmann and Kim Man Bok, J. Raman Spectrosc. 19, 517 (1988)

Part II

High-Resolution Spectroscopy

High-Resolution CARS-IR Spectroscopy of Spherical Top Molecules

D.N. Kozlov, V.V. Smirnov, and S.Yu. Volkov

General Physics Institute, Academy of Sciences,
Vavilov St. 38, SU-117942 Moscow, USSR

Abstract. The report provides a detailed description of a high-resolution CARS-IR spectrometer for recording of Raman and infrared spectra. The obtained rotationally resolved structures of CARS spectra of the ν_1 band Q-branch and IR spectra of the ν_3 band Q^0-branch of $^{28}SiH_4$ and $^{74}GeH_4$, as well as the results of theoretical treatment of these spectra, are presented and discussed.

1. Introduction

Over the last decade the progress in lasers has resulted in the appearence of new methods of molecular spectroscopy [1-3], based on tunable lasers and providing high spectral resolution and sensitivity. As a result, investigations of the fine rotational structure of excited vibrational states of polyatomic molecules became possible, being of interest not only in itself, but also providing information necessary for study of processes of resonant interactions of radiation with molecules, as well as for applied research in laser physics, non-intrusive gas diagnostics, etc.

Polyatomic molecules are characterized by a high density of excited vibration-rotational (VR) energy states with various interactions between them, making the simple model "harmonic oscillator - rigid rotator" invalid for theoretical analysis of high-resolution experimental spectra. Taking intermolecular interactions into account is in general a complicated problem, which can be simplified if symmetry properties of the molecule under study are considered [4].That is why study of the structure of interacting energy states in highly symmetric spherical top molecules is the first step in the construction of theoretical models describing polyatomic molecules.

The dyad of vibrational states ν_1(A1) and ν_3(F2), quasi degenerate in such tetrahedral molecules as GeH_4 and SiH_4, is a remarkable example of a group of interacting vibrational states in a spherical top molecule. Theoretical study of interaction between the ν_1 and ν_3 states of tetrahedral molecules was carried out for the first time in [5, 6]. Further analysis has shown (see [7]) that experimental data on the rotational structure of the ν_1 band, with only Raman transitions of the Q-branch being allowed, as well as of the ν_3 band in IR-spectra, should be processed simultaneously in an adequate theoretical treatment. It should be noted that while the ν_3 state was investigated with moderate or high resoluti-

Eds.: G. Marowsky and V.V. Smirnov

on by traditional means of absorption spectroscopy in [8-10] (SiH_4) and [11-13] (GeH_4), the resolution 0.05-0.005 cm^{-1} in Raman spectra, needed for study of the rotational structure of the ν_1 band, was achieved only by means of modern methods of coherent spectroscopy: CARS [14] and SRS/IRS [15,16].

In the present report, a detailed description of a CARS-IR spectrometer used for simultaneous recording with high resolution of Raman and infrared spectra is given. The obtained rotationally resolved structures of CARS spectra of the ν_1 band Q-branch and IR spectra of the ν_3 band Q^0-branch of $^{28}SiH_4$ and $^{74}GeH_4$ are presented and discussed. The results of theoretical treatment of these spectra, carried out in collaboration with D.Sadovskii and B.Zhilinskii on the basis of the approach developed in [7,17] and taking into account intramolecular interaction between the ν_1 and ν_3 states, are also presented.

2. Experimental

The scheme of high-resolution CARS-IR spectrometer is given in Fig.1. The spectrometer may operate either in the cw mode, with an apparatus spectral resolution of better than 6 MHz (0.0002 cm^{-1}), or in a pulsed mode, with a repetition rate of 10 Hz and a pulse duration of 6 nsec, with an apparatus spectral resolution of 80 MHz (0.003 cm^{-1}). The continuous frequency scanning range is

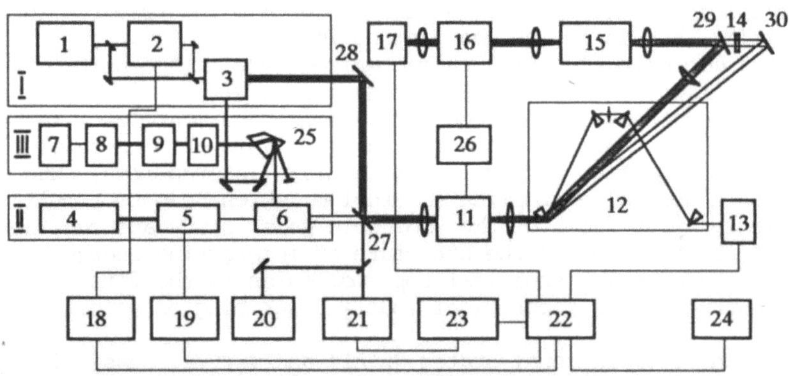

Fig.1. High-resolution CARS-IR spectrometer.
1, 4 - argon-ion lasers; 2, 5 - dye lasers; 3,6 - dye amplifiers; 7,8 - Nd:YAG laser and amplifier; 9 - frequency doubling crystal; 10 - frequency tripling crystal; 11, 16 - gas cells; 12 - filter; 13 - photomultiplier; 14 - $\lambda/2$ plate; 15 - $LiIO_3$ crystal; 17 - InSb photodetector; 18,19 - dye laser controllers; 20 - FP interferometer; 21 - wavemeter; 22 - CAMAC crate; 23 - LSI-11/23 microcomputer; 24 - recorder; 25 - Pellin-Broca prism; 26 - gas filling system; 27-30 - pump beams adjustment mirrors.

1 cm^{-1}. The actual experimental spectral resolution is limited by the Doppler effect and typically amounts to 0.005 cm^{-1}.

2.1. General Description

The main components of the spectrometer are: pump lasers with amplifiers; a spectral analysis and frequency calibration system; the difference-frequency source based on a non-linear crystal; a computer-based monitoring, data acquisition and processing system. Pump sources I and II beams at ω_1 and ω_2, polarized in the same plane, are directed collinearly by means of mirrors 27 and 28 to cell 11 containing the gas under study where coherent anti-Stokes radiation arises at $\omega_3 = 2\omega_1 - \omega_2$. The anti-Stokes emission is isolated by means of the four 60-degree dispersing prisms of filter 12 and detected by photomultiplier unit 13. After the dispersing element separates the pump beams in the filter 12, one beam passes through a $\lambda/2$ plate 14. Then the pump beams with perpendicular polarization are combined by mirrors 29 and 30 and are focused into non-linear optical crystal 15, where IR emission at the difference frequency $\omega_{IR} = \omega_1 - \omega_2$ is generated. After passing through cell 16 with the investigated gas, this emission is registered by photodetector unit 17. An important feature of this scheme is the absence of relative frequency detuning between IR and CARS spectra.

2.2. The Pump Lasers System

The biharmonic laser pump system consists of three laser units (see Fig.1): unit I, the source of frequency-stable radiation ω_1, unit II, used to obtain tunable radiation ω_2, and unit III, designed for pulsed amplification of radiation from cw lasers 1 or 2 and 5.

Laser unit I includes argon laser 1 (Spectra-Physics, M 171) pumping dye laser 2 (Spectra Physics, M 580). Single-frequency TEM_{00} radiation from either argon laser 1 (1-2 W) or dye laser 2 (20-40 mW), is used. The argon laser has a linewidth of 6 MHz and a long-term stability of better than 40 MHz over a 20 minute period. The dye laser has an effective lasing linewidth of 10 MHz. A frequency stabilization system, locked to absorption lines of I_2 vapor [18,19], is employed to improve its long-term stability. The high density of the I_2 lines (1-5 lines/cm^{-1}) makes it possible to use this configuration across the entire dye tuning range.

Laser unit II includes single-frequency dye laser 5 (Spectra Physics, M 380D) pumped by an argon laser 4 (Spectra Physics, M 171). The dye laser has a power of up to 1 W and linewidth less than 2 MHz.

Laser unit III consists of a Q-switched Nd:YAG laser with a repetition rate of 10 Hz, an amplifier and two successively arranged nonlinear optical crystals: for frequency doubling (KD*P), to pump the dye laser amplifier, and tripling (KDP), to pump the argon laser amplifier. Pulse energies obtained are 180 mJ (λ=1064 nm), 40 mJ (λ=532 nm) and 10 mJ (λ=355 nm). Pulsed energies of amplified narrowband radiation reach 5 mJ (λ=580 nm) and 0.5 mJ (λ=515 nm).

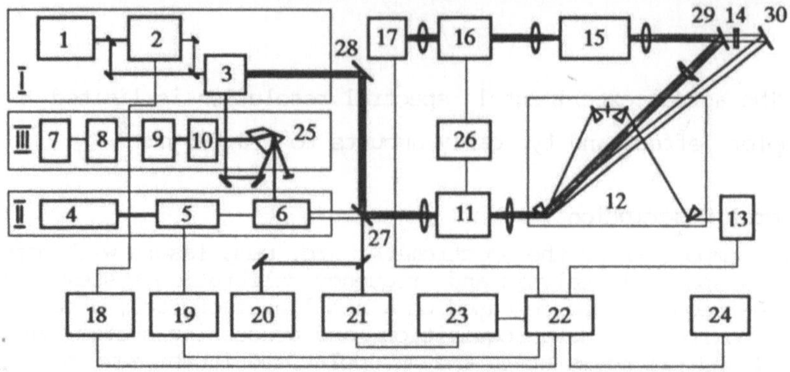

Fig.2. Scheme of a three-stage pulsed amplifier of a cw dye laser.

The design features of one dye laser pulsed amplifier channel are shown in Fig.2. The amplifier uses R6G dye and is pumped by 2nd harmonic radiation of a Nd:YAG laser. The first stage (dye concentration 120 mg/l) is pumped by 4% of the total pump power which is focused by a cylindrical lens CL_1 (f=22 cm) onto cell C_1. The transverse dimensions of the beam in the gain region are 100 μm and the amplification length is 5 mm. The input signal polarized in the vertical plane is focused into a cell by lens L_1 (f=22 cm). A spatial frequency filter, consisting of lens L_2, diffraction grating G, mirror M and diaphragm D, is installed in order to eliminate amplified spontaneous emission. The filter transmission width is 0.3 nm with maximum transmission coefficient equal to 60%. With a continuous tuning range of the cw dye laser of 1 cm^{-1} the dispersion of the filter does not cause significant spatial detuning of the amplified signal path in subsequent channels of the amplifier.

The second amplifier stage (dye concentration 120 mg/l) is pumped by 40% of the total pump power delayed by 1.5 nsec with respect to the amplified signal pulse. Pump emission is focused onto cell C_2 with amplification length of 5 mm by means of cylindrical lens CL_2 (f=22 cm). The amplified emission is focused by lens L_2 (f=20 cm) so that the beam size in the cell is 300 μm.

The third stage utilizes longitudinal pumping, the amplification region being 1 mm in diameter and 10 mm in length. Dye concentration is 50 mg/l, providing pump saturation conditions. The pump pulse is also delayed by 1.5 nsec with respect to the amplified signal pulse and is focused onto cell C_3 by lens L_4 (f=50 cm). The amplified emission is focused by lens L_3 (f=33 cm) and collimated by lens L_5. The amplitude instability of the output emission pulses is 10%. The power ratio between the amplified signal and the spontaneous emission is 10^4. The amplified signal linewidth is 0.003 cm^{-1}, being limited by the duration of the pulse.

2.3. Spectral Analysis and Frequency Calibration System

This system consists (see Fig.1) of a scanning confocal Fabry-Perot interferometer 20 with free spectral range (FSR) of 0.06 cm^{-1} and a wavemeter 21, based on four Fizeau interferometers (FI) and linear multichannel detectors.

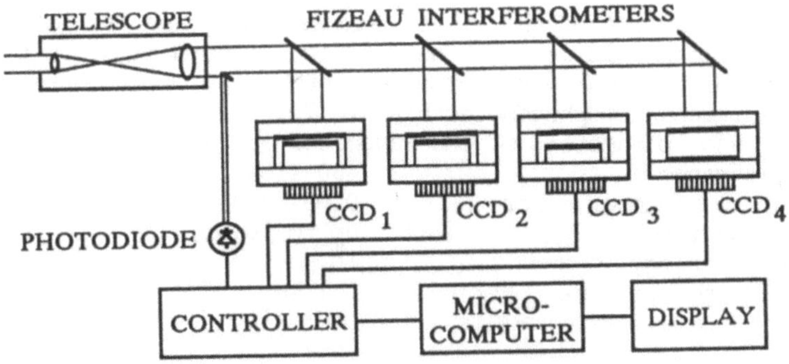

Fig.3. Scheme of laser radiation wavemeter.

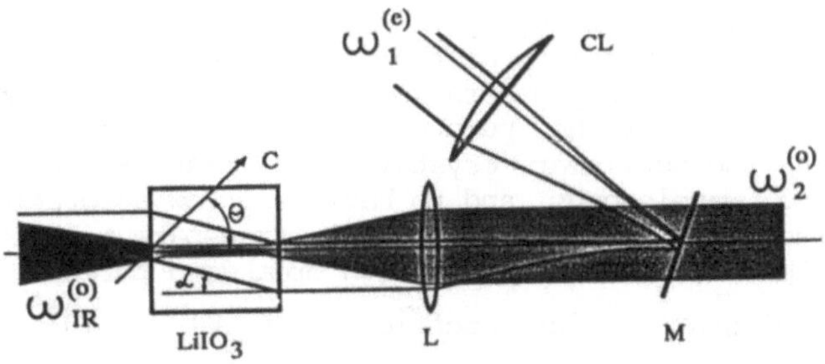

Fig.4. Optical scheme of pump-beam mixing for difference frequency generation;
θ - angle of synchronism (collinear phase matching angle),
α - drift angle.

The computer-controlled wavemeter is used for precise frequency calibration of both CARS and IR spectra, as well as for real-time analysis of emission spectra of employed cw or pulsed lasers. The wavelength measurement principle consists in step-by-step refinement of the unknown value, obtained tentatively from spatial fringe period measurement with an error less than half of the FSR of the FI with the smallest base.

The design of the wavemeter is shown in Fig.3. The input radiation passes through the telescope and beamsplitters and is incident on four Fizeau interferometers. A portion of the beam is split to the photodiode of the high-speed exposure correction circuit. The bases of FIs 2, 3 and 4 are equal to 0.1, 2.0 and 40 mm, corresponding to a wavelength refinement factor of 20. The first FI acts as an auxiliary to the second one, with a base difference of 5 μm. The wedge angle of FIs 2, 3 and 4 is 8" and is selected to match the spatial period of the fringe pattern and the length of the photodetector. The wedge angle of the FI 1, used for the preliminary wavelength measurement, is equal to 16". The interferometers have silver coated mirrors with reflectivity of 90% at 600 nm and flatness better than $\lambda/50$. The interferometer parts are fabricated from fused silica and placed in optical contact.To reduce the dispersion of the gas medium inside the interferometers and increase the stability of the silver coating, they are filled with helium at atmospheric pressure and sealed. The optical unit temperature stabilization system provides stability of better than 0.1^{o} C. Multichannel photodetectors are 1024 pixel (14 mm) CCD arrays. The major parameters of the wavemeter are as follows:

spectral range	400-1100 nm,
absolute wavenumber error	0.002 cm^{-1}
repetition rate	up to 10 Hz,
threshold energy	10 μJ,
maximal exposure time	0.1 sec.

2.4. Narrowband Difference-Frequency Source

Mixing of laser radiation at fixed frequency ω_1 in a nonlinear crystal with radiation at frequency ω_2 from a tunable laser produces tunable infrared emission at the difference frequency $\omega_{IR} = \omega_1 - \omega_2$. Difference frequency generation in cw mode, characterized by the low power level of the pump sources and the need for hard focusing, requires the nonlinear crystal to be transparent at all three interacting wavelengths and to have very low absorption of the pump laser radiation. The $LiIO_3$ crystal, transparent up to 5.5 μm, has been selected, since it has a high breakdown threshold and a rather high nonlinear optical coefficient $d_{15}=1.8 \cdot 10^{-8}$ CGSE units [20]. The weak temperature dependence of the refractive index in $LiIO_3$ eliminates the need for crystal temperature stabilization.

The e-o=o-type interaction and collinear phase matching of the wave vectors with angle tuning was employed in our experiment (see Fig.4). Both mixed beams, combined by dichroic mirror M, are focused by lens L (f=14 cm) in the $LiIO_3$ crystal with a confocal parameter equal to the crystal length (1.7 cm). In order to achieve maximum overlap of the beams, the extraordinary beam is expanded in the drift plane by cylindrical lens CL (f=25 cm), forming the telescopic system together with the spherical lens L. It's important that in this scheme the difference frequency emission has near-axial symmetry with a near-Gaussian profile, being easily transformed by the optical elements of the spectrometer. The IR-emission power is 1 μW in cw mode and 1 kW in pulsed mode.

2.5. Monitoring, Data Acquisition and Processing System

The spectrometer monitoring, data acquisition and processing system includes (see Fig.1) the dye laser controllers 18 and 19, the wavemeter 21, the CARS and IR photodetector units 13 and 17, CAMAC spectrometer/computer interface 22 and the LSI-11/23 microcomputer 23.

In cw mode the anti-Stokes scattering signal is recorded by a cooled photomultiplier for photon counting and a counter with an analog output to the recorder. The infrared emission is recorded by Ge:Au or InSb photodetectors, cooled to T=77 K, and a lock-in amplifier. In pulsed mode the same detectors are used with the one difference that the derived voltage pulses are input to the gated BOXCAR integrators.

3. Results and Discussion

The CARS spectra of the Q_{01}-branch and the IR spectra of the Q^{o}_{03}-branch with a resolved rotational structure were obtained for $^{28}SiH_4$ and $^{74}GeH_4$ molecules at room temperature. The experiments employed chemically pure SiH_4 with a natural silicon isotope concentration (92% ^{28}Si, 5% ^{29}Si, 3% ^{30}Si), so that strongly expressed spectral lines refer to VR transitions of the $^{28}SiH_4$ molecule. The monoisotopic $^{74}GeH_4$ gas was prepared in laboratory from enriched germanium with 91% of ^{74}Ge. The recorded CARS spectrum of the Q_{01}-branch (p = 5 torr, T = 295 K) and IR spectrum of the Q^{o}_{03}-branch (p=0.4 torr, T=295 K) of $^{74}GeH_4$ molecules are shown in Fig.5.

The tetrahedral splittings in the Q_{01}-branches of $^{74}GeH_4$ and $^{28}SiH_4$ are such that the spectral components corresponding to transitions from the various rotational sublevels with the same J form a separate line or, as J increases, a group of lines. There is no overlapping of components with different values of J in this case. The value of J for each group of lines can be easily determined from their intensities and average frequency positions. The IR spectra of the Q^{o}_{03}-branches of these molecules have a more complicated structure, since mixing of components with different values of J begins to occur at J>4 in $^{74}GeH_4$ and at J>5 in $^{28}SiH_4$.

A preliminary analysis of the obtained spectra yields the following results. First, tetrahedral splittings of the rotational le-

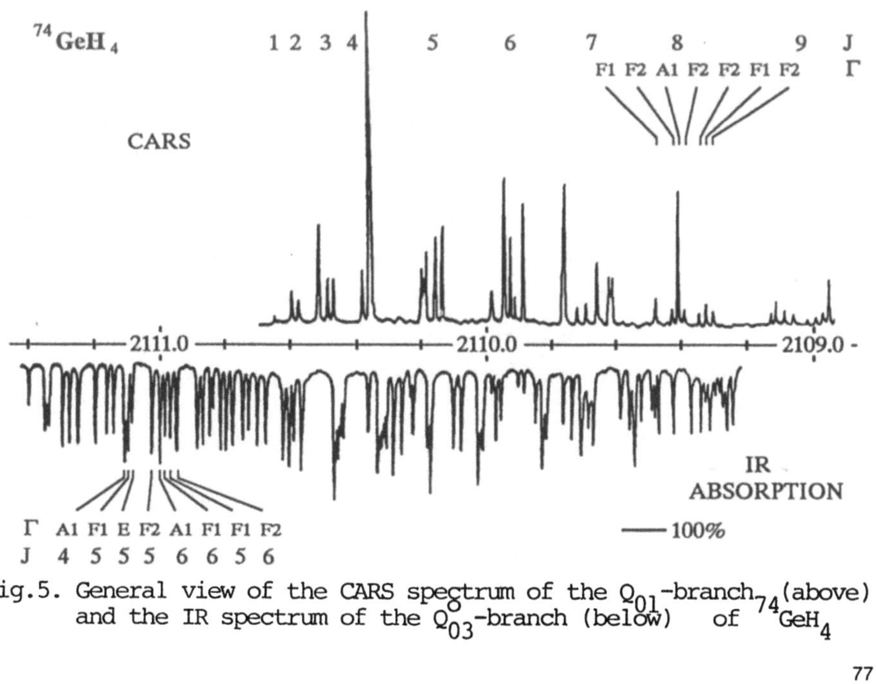

Fig.5. General view of the CARS spectrum of the Q_{01}-branch (above) and the IR spectrum of the Q^{o}_{03}-branch (below) of $^{74}GeH_4$

vels of the excited vibrational states of $^{28}SiH_4$ and $^{74}GeH_4$ molecules (which are most clearly expressed in $^{74}GeH_4$) exceed tetrahedral splittings in the ground state by a factor of 2-10. Second, while the average frequency positions of J-line tetrahedral components in both CARS and IR spectra are described with sufficient accuracy (particulary in $^{28}SiH_4$, where tetrahedral splittings are smaller) by the relation $\nu(J) - \nu_i = \alpha_i J(J+1)$, $i = 1,3$, typical for rigid spherical rotors, the rotational energies in the excited vibrational states are not described by simple relations providing tetrahedral splittings within the scope of the model of isolated ν_1 and ν_3 states. Moreover, the forbidden VR transitions to the ν_3 state are quite intensive in IR-spectra. These facts indicate that perturbations exist in the rotational structure of the ν_1 and ν_3 states, which are strongly dependent on J and the symmetry type of VR levels. Accounting for the fact that these states in $^{28}SiH_4$ and $^{74}GeH_4$ molecules are more then 200 cm^{-1} from the remaining vibrational states in the $\{\nu_1, \nu_3, 2\nu_2, 2\nu_4, \nu_2+\nu_4\}$ group and are close together ($\nu_{31} = \nu_3^o - 2(B\xi)_{eff} - \nu_1 \approx 2.3$ cm^{-1} in the $^{28}SiH_4$ molecule and $\nu_{31} \approx 0.41$ cm^{-1} in the $^{74}GeH_4$ molecule), we may naturally assume that the Coriolis resonance between the ν_1 and ν_3 states is of major importance. Although this interaction becomes allowed only in the second order of perturbation theory, the existing quasi degeneracy of these states in $^{28}SiH_4$ and $^{74}GeH_4$ makes this interaction rather strong.

The above considerations formed the basis of theoretical models used earlier for treatment of the experimental data for $^{74}GeH_4$ and other isotopes of this molecule [12,16,25] and $^{28}SiH_4$ [21-24]. Detailed information about the results of these treatments is given in Tables 1 and 2. Tables 3 and 4 present tensor VR operators, included in the effective Hamiltonian for the $\{\nu_1, \nu_3\}$ dyad and corresponding parameters, obtained by different authors for $^{74}GeH_4$ and $^{28}SiH_4$. The values of spectroscopic parameters are reduced to standard coefficients denoted $u_{k,m}^{\Omega(k,\Gamma)}$, which have the same meaning as $t_{k,m}^{\Omega(k,\Gamma)}$ from [26], but are determined as coefficients by the corresponding tensor operators, without any additional factors [4]. The relations between $u_{k,m}^{\Omega(k,\Gamma)}$ and other parameters are given in [17,25,27].

The determination of effective Hamiltonian parameters as a result of experimental data treatment is carried out by varying the initial set of parameters to minimize a functional of differences between calculated and experimental values of frequencies of observed VR transitions. The functional used is the mean squared deviation of theoretically calculated energies of the $\{\nu_1, \nu_3\}$ dyad VR levels, determined using experimentally observed frequencies of as-

Table 1. Theoretical analysis of high-resolution spectra of the $\{\nu_1, \nu_3\}$ dyad of the $^{74}GeH_4$ molecule.

Theor. studies	Exp.	Number of variable parameters ν_1	interaction	ν_3	total	J_{max}	Assigned: transitions	levels	rms deviation in 10^{-3} cm^{-1}
[12]	[12]	2	1	7	10	10	135		5.9
[16]	[16], ν_1	2	1	6	9	11	26	26	3
	[12], ν_3					10	135		6
[25]	[14], ν_1	2	1	5	8	9		40	9
	ν_3					9		30	9
Pres.	[14]*, ν_1	2	1	7	10	12	75	70	2
study	[13,14], ν_3					14	125	97	3

* In the present study refined and completed data of [14] are used.

Table 2. Theoretical analysis of high-resolution spectra of the $\{\nu_1, \nu_3\}$ dyad of the $^{28}SiH_4$ molecule.

Theor. studies	Exp.	Number of variable parameters ν_1	interaction	ν_3	total	J_{max}	Assigned: transitions	levels	rms deviation in 10^{-3} cm^{-1}
[21,22]	[9,10]	4	1	10	15	16	500		7
[15]	[15], ν_1	4	1	11	16	14	97	84	2.4
	[9,10], ν_3					16	500		7.2
[23]	[9,10]	2	2	7	11	16	500		7
[24]	[8]					19	2046	559	0.68
Pres.	[14]*, ν_1	3	1	11	15	13	62	62	3.2
study	ν_3					15	163	157	6.5

* In the present study refined and completed data of [14] are used.

signed transitions and energies of ground state rotational levels. The parameters of the ground state were taken from [28, 29] (GeH_4) and [30] (SiH_4). The same resolution and precision of frequency calibration of CARS and IR spectra meant that it was not necessary to insert any weights for different energy levels. The search for the minimum of the functional was carried out in the course of an iterative procedure, where at each step the effective Hamiltonian matrix was built using a given set of parameters. After diagonalization the energy values obtained, as well as the corresponding wavefunctions, were used to calculate the functional and its gradient.

Table 3. Parameters of the third-order effective VR Hamiltonian obtained by different authors for the $\{\nu_1, \nu_3\}$ dyad of the $^{74}GeH_4$ molecule.

n	$U^{\Omega(K,\Gamma)}_{k,m}$ k,m Ω(K,Γ)	Values of spectroscopic parameters [16]	[12]	Pres.study	cm^{-1}
0	ν_1 - 2110	0.7052(18)	0.702(3)	0.7020*	
	ν_3 - 2110	1.1423(16)	1.143(1)	1.1445(44)	x 1
1	3,3 1(1,F1)	0.3249(4)	0.3250(2)	0.3250(55)	
2	1,1 2(0,A1)	0.7781(4)	0.7726(26)	0.77690(79)	
	3,3 2(0,A1)	1.1032(15)	1.0995(15)	1.104(12)	
	3,3 2(2,E)	0.1556(21)[a]	0.1487(25)[a]	0.1586(83)	$x\ 10^{-2}$
	3,3 2(2,F2)	-0.2476(17)[a]	-0.2586(21)[a]	-0.261(10)	
	1,3 2(2,F2)	0.4638(9)	0.4571(55)	0.4672(12)	
3	3,3 3(1,F1)	0,69(21)	0.58(16)	0.6(15)	$x\ 10^{-5}$
	3,3 3(3,F1)	0*	0.329(14)	0.22(29)	

* Parameters not included or fixed in the analysis.

a For each pair of constants parameter α_{224} is determined in [12, 16], while parameter α_{220} is fixed.

Ground state parameters:

$u_o^{2(0,A1)} = -1.1673$, $u_o^{4(0,A1)} = -6.2630\ 10^{-6}$ [28]

$u_o^{4(4,A1)} = -1.5478\ 10^{-6}$ [29].

For the analysis of $^{74}GeH_4$ data the parameters of [12] were chosen as initial values.Only diagonal 3d order terms were included in the Hamiltonian for the ν_3 state. The use of the 4th order terms was found to be of no importance because of good correspondence between experimental and calculated data. The obtained parameters and estimates of their root mean square deviations are presented in Table 3. They are in good agreement with the values given by other authors. Thus, only 10 parameters allow us to describe about 200 experimentally measured frequencies of transitions between rotational states with momentum number J up to 12 (in CARS spectra) and 14 (in IR spectra) with the average accuracy 0.003 cm^{-1}. Figure 6, with a portion of the observed and calculated IR-spectrum of the ν_3 band Q-branch, shows satisfactory correspondence between experiment and theory.

Figure 7a shows calculated positions of energy levels for the $\{\nu_1, \nu_3\}$dyad in $^{74}GeH_4$. In the ν_3 state the pattern of three sub-

Table 4. Parameters of fourth-order effective VR Hamiltonian obtained by different authors for the $\{\nu_1,\nu_3\}$ dyad of the $^{28}SiH_4$ molecule.

n	k,m	Ω(K,Γ)	[23]	[15]	[21,22]	Pres.study	cm^{-1}
	$U_{k,m}^{\Omega(K,\Gamma)}$		Values of spectroscopic parameters				
0	ν_1	− 2186	0.8679(35)	0.873(2)	0.867(10)	0.8751(12)	x 1
	ν_3	− 2189	0.1943(78)	0.1901(12)	0.1865(14)	0.1846(08)	
1	3,3	1(1,F1)	−7.747(18)	−7.728(10)	−7.7281(73)	−7.731(14)	$x10^{-2}$
2	1,1	2(0,A1)	7.667(14)	7.759(10)	7.701(82)	7.807(15)	$x10^{-3}$
	3,3	2(0,A1)	11.005(63)	10.908(22)	10.781(37)	11.060(16)	
	3,3	2(2,E)	1.7553(63)	1.849(23)	1.799(17)	1.823(17)	
	3,3	2(2,F2)	−3.0553(87)	−3.047(20)	−3.006(18)	−3.039(20)	
	1,3	2(2,F2)	5.330(11)	5.165(27)	5.349(12)	5.349(17)	
3	3,3	3(1,F1)	−0.87(58)	0*	0*	0*	$x10^{-6}$
	3,3	3(3,F1)	3.16(40)	3.29(58)	4.06(39)	5.73(57)	
	1,3	3(3,F2)	−1.91(81)	0*	0*	0*	
4	1,1	4(0,A1)	0*	−0.318(52)	2.2(16)	3.44(36)	$x10^{-7}$
	3,3	4(0,A1)	0*	−0.584(78)	−8.1(20)	1.99(42)	
	1,1	4(4,A1)	0*	1.46(27)	−1.20(41)	0*	
	3,3	4(4,A1)	0*	−0.87(30)	0.93(20)	0*	
	3,3	4(2,E)	0*	1.18(34)[a]	1.01(29)[a]	2.39(54)	
	3,3	4(2,F2)	0*	1.44(41)[a]	1.23(36)[a]	−0.19(83)	
	3,3	4(4,E)	0*	1.69(24)[b]	0*	0.35(64)	
	3,3	4(4,F2)	0*	−2.76(39)[b]	0*	−0.99(66)	

* Parameters not included or fixed in the analysis.
a-b For each of two pairs of constants one parameter G_{220} and G_{224} accordingly is determined in [15, 22].

branches E_J^-, E_J^o and E_J^+ of VR energy levels, provided by Coriolis interaction of vibrational and rotational angular momenta in a triply degenerate vibrational state, is clearly observed. A relatively large Coriolis constant

$$B\,\xi_3 = -(6)^{-0.5}\; u_{3,\,3}^{1(1,\,F1)} = -\,0.1327\ cm^{-1}$$

results in a large separation of these syb-systems as J grows. The

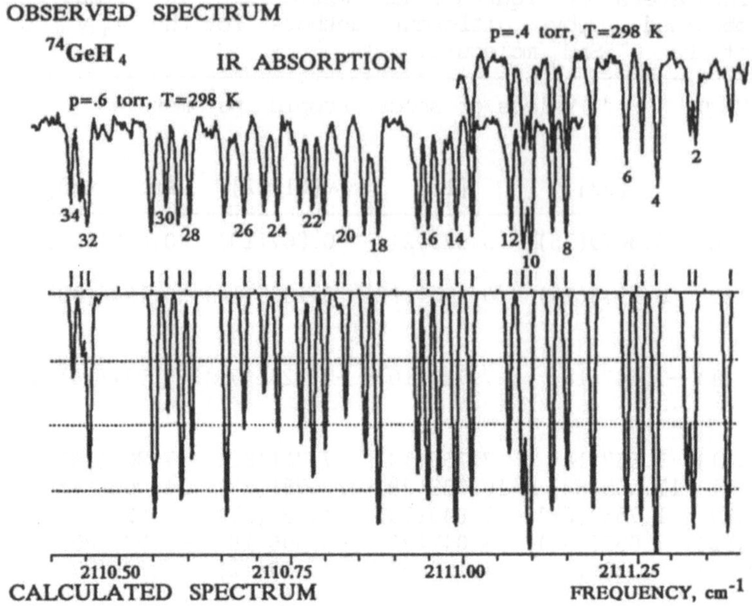

Fig.6. Portion of the experimental (above) and calculated (below) IR spectra of the Q^{o}_{03}-branch of the $^{74}GeH_4$ molecule.

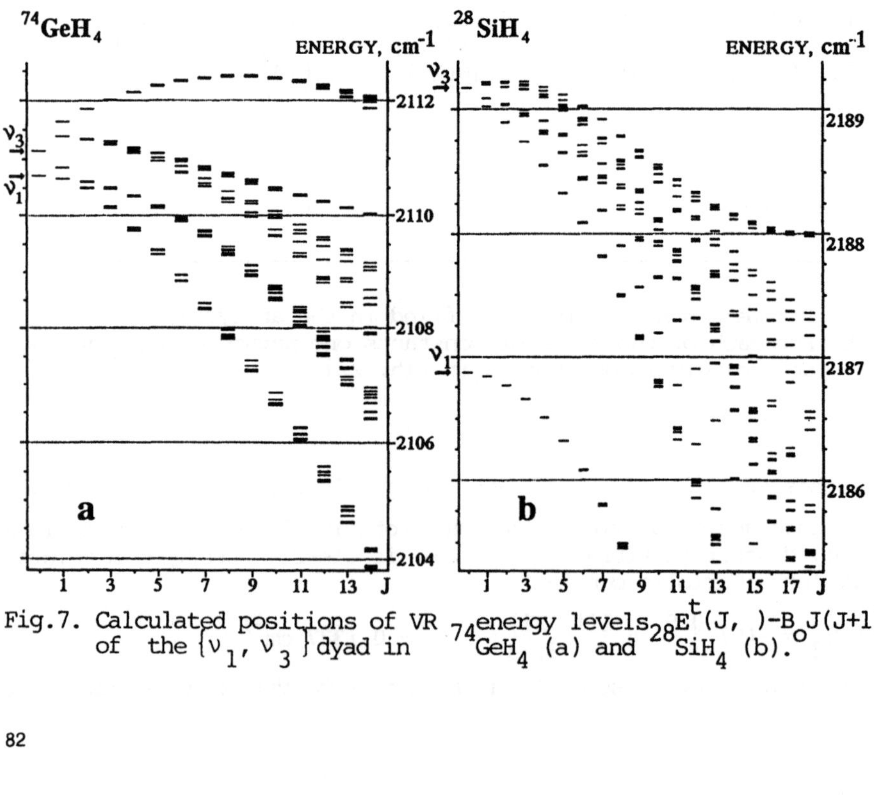

Fig.7. Calculated positions of VR energy levels $E^{t}(J, \ )-B_{o}J(J+1)$ of the $\{\nu_1, \nu_3\}$ dyad in $^{74}GeH_4$ (a) and $^{28}SiH_4$ (b).

strongest tetrahedral splittings are experienced by the central E_J^o sub-system, providing upper levels for Q_{03}^o-branch VR transitions. The lower sub-system E_J^+ of the ν_3 state intersects ν_1 levels at J=2. The calculations confirm that without doubt the tetrahedral rotational structure of the ν_1 state is completely determined by VR interaction with the ν_3 state.

Despite the distance between the interacting ν_1 and ν_3 states in $^{28}SiH_4$ (see Fig.7b) being larger than in $^{74}GeH_4$, experimental spectra of the ν_3 band and the rotational structure of the corresponding VR states appear to be significantly more complicated in $^{28}SiH_4$ than in $^{74}GeH_4$. In this connection the diagonal 4th order terms have been included in the Hamiltonian in addition to terms of the orders 0-3. The reduced effective Hamiltonian with one 3rd order term has been used [17] for a more correct approach to the inverse spectroscopic problem. Table 4 presents effective VR operators for the $\{\nu_1, \nu_3\}$ dyad of $^{28}SiH_4$, the obtained values of spectroscopic parameters and their rms deviations.

As a result of our calculations with 16 variable parameters for the $\{\nu_1, \nu_3\}$ dyad of $^{28}SiH_4$, 230 VR transitions in the Q-branch spectra of ν_1 and ν_3 bands were assigned and reproduced with mean standard deviation of 0.006 cm^{-1}. The largest considered J number was 13 in CARS spectra and 15 in IR spectra. Parts of observed and calculated IR spectra are shown in Fig.8 for comparison of theory with experiment.

Comparing parameters obtained in the present study with those given by other authors, one can see that they are in good agreement for the orders 0-2. The variations in higher order parameters are connected with the differences in models used for calculations and ambiguity of the effective Hamiltonian. Besides, the difference in number and type of the experimental data taken for the analysis is also of importance. In [15, 22-24] data on R- and P-branches of the ν_3 band absorption spectra have been used, with addition of some data on forbidden dipole transitions of the ν_1 band, arising due to mixing of the ν_1 and ν_3 states. Spectra of the ν_1 band Q-branch, similar to those obtained in our experiments, were employed for the calculations only in [15].

The spectroscopic parameters obtained allow us to reproduce rotational fine structure of the $\{\nu_1, \nu_3\}$ dyad of the $^{28}SiH_4$ molecule and probabilities of spectroscopic processes in which these states may participate. The corresponding diagram is presented in Fig.7b. For the ν_1 state interaction with the ν_3 state results in relatively small (less than 0.1-0.25 cm^{-1}) tetrahedral splittings of the levels and the appearence of forbidden transitions of the ν_1 band in IR-absorption spectra. In the rotational structure of the ν_3 state the complete violation of characteristic regular pattern of the three Coriolis-split sub-systems E_J^+, E_J^o and E_J^- can be observed as J grows. While levels of the state and of the lo-

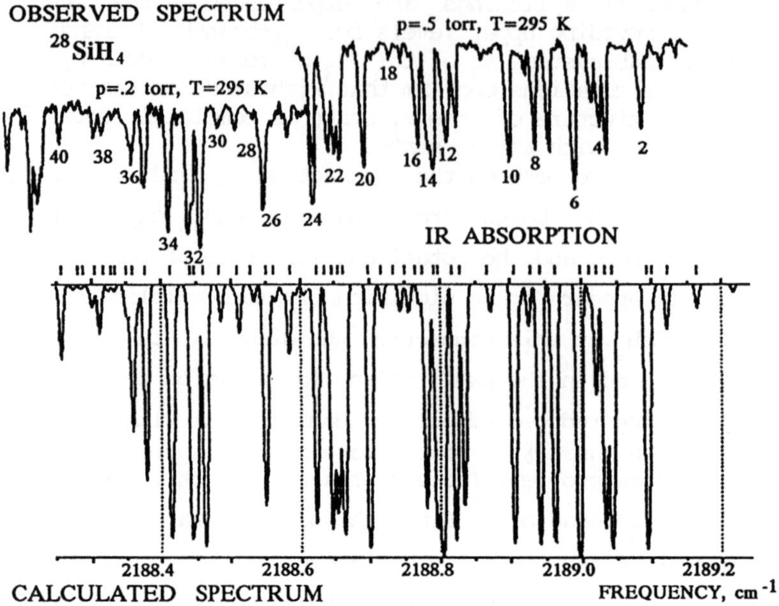

Fig.8. Portion of the experimental (above) and calculated (below) IR spectra of the Q^{O}_{03}-branch of the $^{28}SiH_4$ molecule.

wer sub-system E^{-1}_{J} of the ν_3 state are developing rather separately, the levels of E^{O}_{J} and E^{+}_{J} sub-systems become mixed with each other and then disappear, starting from J=6-7, the new regularities of rotational levels being formed with further growth of J.

The reorganization of levels of the E^{+} and E^{O} sub-systems takes place in the energy range 1-3 cm^{-1}, significantly exceeding that of rotational splittings in the ν_1 state. Hence, these changes should be attributed primarily to the rotational dynamics of the ν_3 state itself. The reorganization results in the appearence of a large number of transitions to the E^{+}-levels observed in the Q-branch spectrum of the ν_3 band. As was already noted in [22], this strong distortion of the ν_3 state and corresponding transition probabilities (selection rules) can formally be explained by the anomalously small value of the Coriolis interaction parameter $B\zeta_3$=0.03155 cm^{-1}. In this case the structure of the ν_3 state rotational energy levels is determined by second and higher order VR interactions at relatively small rotational excitation.

4. Conclusion

In conclusion, fully resolved rotational structure of Q-branches in CARS spectra of the ν_1 band and IR spectra of the ν_3 band of $^{74}GeH_4$ and $^{28}SiH_4$ molecules have been obtained, using a high-resolution CARS-IR spectrometer. The rotational structure of spectra and transition probabilities for the dyad $\{\nu_1, \nu_3\}$ of quasidegenerate states of GeH_4 and SiH_4 molecules were described using the effective Hamiltonian, which takes into account the higher-order VR interactions between these vibrational states. As a result of simultaneous treatment of CARS and IR spectra, precision of the theoretical description of observed line frequencies has been realized close and, in the case of $^{74}GeH_4$ equal to the experimental precision of absolute frequency calibration, estimated as $(2\text{-}3).10^{-3}$ cm^{-1}. The calculated intensities of observed transitions, including numerous forbidden lines, are in satisfactory agreement with our experimental data.

References

1. W.Demtroder: Laser Spectroscopy, Springer Ser.Chem. Phys., Vol.5 (Springer, Berlin, Heidelberg 1981).
2. S.A.Akhmanov, N.I.Koroteev: Methods of Nonlinear Optics in Light Scattering Spectroscopy (Nauka, Moscow 1981).
3. A.Weber (ed.): Raman Spectroscopy of Gases and Liquids, Topics Curr. Phys., Vol.11 (Springer, Berlin, Heidelberg 1979).
4. B.I.Zhilinskii, V.I.Perevalov, V.G.Tyuterev: Method of Irreducible Tensor Operators in the Theory of Molecular Spectra (Nauka, Novosibirsk 1987).
5. J.Susskind: J.Chem. Phys. 56, (1972) 5152.
6. F.W.Birss: Mol. Phys. 31, (1976) 491.
7. L.Ya.Baranov, B.I.Zhilinskii, D.N.Kozlov, A.M.Prokhorov, V.V.Smirnov: JETP 79, (1980), 46.
8. G.Guelachvili: New Rev. Opt. Appl. III, 6, (1972) 317.
9. A.Cabana, L.Lambert, C.Pepin: J.Mol.Spectr., 93, (1972) 429.
10. G.Pierre, R.Saint-Loup: C.R.Acad.Sci. Paris, Ser.B, 276, (1973), 937.
11. S.J.Daunt, G.V.Hasley, K.Fox, R.J.Lovell, N.M.Gailar: J.Chem. Phys., 68, (1978) 1319.
12. P.Lepage, J.P.Champion, A.G.Robiette: J.Mol.Spectrosc. 89, (1981) 440.
13. R.P.Schaeffer, R.W.Lovejoi: J.Mol.Spectrosc. 113, (1985) 310.
14. G.Ya.Zueva, D.N.Kozlov, P.V.Nickles, A.M.Prokhorov, V.V.Smirnov, S.M.Tchuksin: Opt.Commun. 35, (1980) 218.
15. A.Owyoung, P.Esherick, A.G.Robiette, R.S.M.McDowell: J. Mol. Spectrosc. 86, (1981) 209.
16. S.Q.Mao, R.Saint-Loup, A.Aboumajd, P.Lepage, H.Berger, A.G.Robiette: J.Raman Spectrosc. 13, (1982) 257.
17. D.A.Sadovskii, B.I.Zhilinskii: Opt. Spektrosk. 58, (1985) 565.
18. S.Gerstenkorn, P.L.Luc: Atlas du spectre d'absorption de la molecule d'iode. (CNRS, Paris 1978).
19. S.Gerstenkorn, P.L.Luc: Rev. Phys. Appl. 14, (1979) 791.
20. A.J.Campillo, C.L.Tang: Appl.Phys.Lett. 16, (1970) 242.
21. A.Cabana, D.L.Gray, I.M.Mills, A.G.Robiette: J.Mol.Spectrosc. 66, (1977) 174.
22. A.Cabana, D.L.Gray, A.G.Robiette, G.Pierre: Mol.Phys. 36, (1978) 1503.

23. G.Pierre, J.P.Champion, D.N.Kozlov, V.V.Smirnov: J. de Phys. 43, (1982) 1429.
24. K.Bouzouba: These, Dijon (1984).
25. L.Ya.Baranov, S.Yu.Volkov, B.I.Zhilinskii, D.N.Kozlov, S.I.Mednikov, D.A.Sadovskii, V.V.Smirnov: Preprint of Gen. Phys. Inst., No.169, Moscow (1985).
26. J.P.Champion, G.Pierre: J.Mol.Spectrosc. 79, (1980) 225.
27. D.A.Sadovskii, B.I.Zhilinskii: J.Mol.Spectrosc. 115, (1986) 235.
28. R.H.Kagann, I.Ozier, G.A.McRae, M.C.L.Gerry: Can.J.Phys. 57, (1979) 593.
29. W.A.Kreiner, R.Opferkuch, A.G.Robiette, P.H.Turner: J. Mol. Spectrosc. 85, (1981) 442.
30. I.Ozier, R.M.Lees, M.C.L.Jerry: Can. J.Phys. 54, (1976) 1094.

High Resolution Coherent Raman Spectroscopy: Studies of Molecular Structures

B. Lavorel, G. Millot, and H. Berger

Laboratoire de Spectronomie Moléculaire et Instrumentation Laser, U.R.A. CNRS n° 777, Université de Bourgogne, 6, Bd. Gabriel, F-21000 Dijon, France

1. Introduction

One of the main advantages of the non-linear coherent Raman techniques is the high resolution that can be achieved in rovibrational spectroscopy. Typically an instrumental function of the order of several thousandths of a wavenumber is routinely achieved in SRS or CARS experiments. Since the first recording of the stimulated Raman spectrum of $^{12}CH_4$ in 1978 [*1*], numerous studies of molecules have been performed [*2-8*]. We have built a stimulated Raman experiment in Dijon in which particular attention has been paid to the frequency measurement of the Raman lines [*9*]. We will describe the application of our experiment to a wide variety of molecules over the last few years : linear molecules, spherical top molecules, and also symmetric top molecules.

2. Experimental

The instrumental apparatus is shown in fig.1 in the Raman gain configuration : an amplification of the probe laser, a stabilized krypton ion laser, is induced by the pulsed pump laser in the gas cell. The latter is a dye laser amplified by a frequency doubled Nd-YAG laser in a four-stage dye amplifier. The probe and pump powers are respectively 0.3-0.5 W and 0.5-1.0 MW. The detection is ensured by a fast photodiode and a boxcar system. The signal to pump power ratio is formed and averaged over a given number of shots. In the inverse Raman mode, the Kr^+ laser is replaced by an Ar^+ laser stabilized on saturated absorption lines of I_2.
The resolution of the apparatus is mainly limited by the Fourier transform of the pump pulse : 0.0022 cm^{-1} (F.W.H.M.).

The frequency calibration of the Raman spectra is provided by an accurate wavemeter and a temperature stabilized étalon as follows : the frequency of the tunable dye laser is measured at each end of the scan. The standard deviation for a set of measurements is about 3 MHz. The Raman shift at each point of the scan is interpolated from transmission fringes of the étalon (300 MHz F.S.R.).

The reference source for the wavemeter is either the Ar^+ laser or an I_2 - stabilized HeNe laser. The wavemeter is a Michelson-type interferometer with two corner cubes, one

Springer Proceedings in Physics, Volume 63 **Coherent Raman Spectroscopy**
Eds.: G. Marowsky and V.V. Smirnov

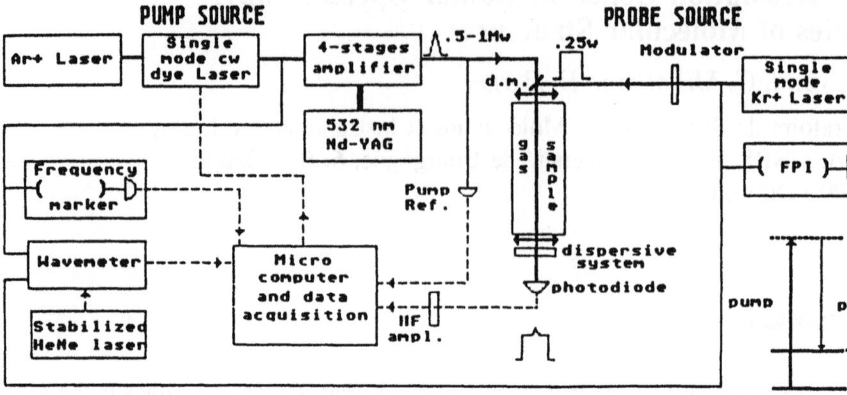

Fig.1 Stimulated Raman spectrometer at Dijon University (Raman gain configuration)

of them travelling in an evacuated vertical tube. The frequency of the probe laser is also measured. In the case of the Ar^+ laser, this allows a check of the wavemeter, the two lasers being actively stabilized with an absolute reference : the hyperfine absorption lines of I_2.

A test with the CO molecule, whose Raman frequencies can be accurately deduced from infrared data, has demonstrated a statistical standard deviation of 10 MHz on Raman frequency measurements [*9*]. This test included interpolation and fit of the peaks. This value can be considered as the limit of our measurements. The estimated accuracy generally lies in the range 10 - 30 MHz, depending on the signal to noise ratio and on the lineshape.

Two methods can be applied to measure line frequencies : the simplest way is the determination with a polynomial of the frequency of the maximum of the Raman peaks. This procedure works well on isolated lines, but for blended lines, one generally uses a profile fitting procedure taking into account the main causes of broadening in the lineshape. This necessitates *a priori* information such as number of lines in the blend and intensities. A typical example of the second method is displayed in fig.2 for the ν_1 Q-branch of silane where 4 lines have been fitted to the experimental spectrum.

An important feature of spectroscopy is the sensitivity of the experiment. In stimulated Raman spectroscopy and at the shot noise limit, the S/N ratio is proportional to the pump power and to the square root of the probe power. An increase of laser power is one means of improvement, but phenomena such as saturation and Stark effect arise which perturb the Raman spectrum. An alternative is the use of a multiple-pass cell (M.P.C.) for investigation of weak Raman cross-section bands.

The M.P.C. built in Dijon consists of two concentric spherical mirrors separated by 1 m [*10*]. The entrance and output of the beams are allowed by slits in the mirrors. Accurate adjustments of the two mirrors are neccessary; but the main problem which we have encountered is the damage of the coating of mirrors. Indeed, when the laser beams are

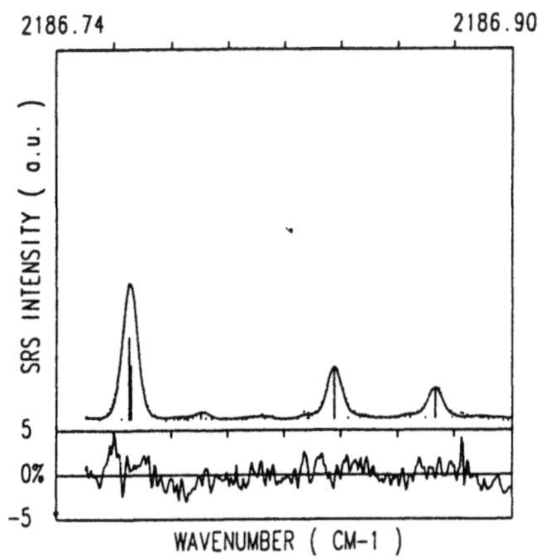

Fig.2 Fit of the calculated Raman profile (full line) to the experimental data (dots)

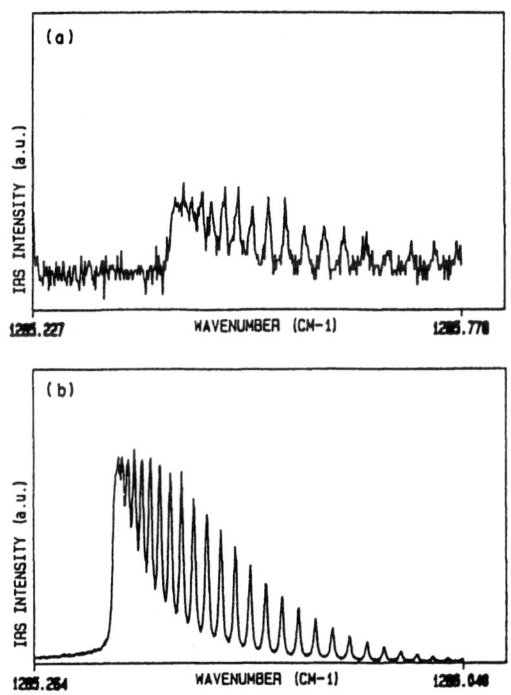

Fig.3 (a) Single-pass and (b) multiple-pass SRS spectrum of the $2\nu_2$ band of CO_2 at .03 atm

not exactly matched to the cell, an oscillation of the spot size on the mirrors appears [*10*]. Thus the intensity can reach the damage threshold of the coating for the minimum size.

A test on the N_2 Q-branch at 24 Torr pressure has shown an improvement of x22 for 44 passes, whereas the calculated value was x24. A striking example of the enhancement is the very sharp Q-branch of the $2\nu_2$ band of CO_2 shown in fig.3 at a 23 Torr pressure, for one (fig.3a) and 44 passes (fig.3b). Such a cell offers the possibility of lowering the pressure and resolving the band at 1.5 Torr. A Voigt profile of the spectral shape was then performed and Raman frequencies were employed to adjust molecular constants [*10*].

3. Linear Molecules

Two linear molecules which are present in combustion media have been investigated: N_2 and O_2. N_2 is the usual probe molecule of CARS thermometry which involves computation of the Raman spectrum of hot bands. The oxygen molecule could also become an important probe. We have studied the first hot bands of these molecules in order to improve molecular constants, especially of O_2, for which only a small amount of data was available.

Numerous spectra of the fundamental of O_2 have been obtained during the study of collisional linewidths and lineshifts [*11*]. Only one is shown at room temperature and 0.4 atm in fig.4a. The first and second hot bands were recorded at 1 atm and 1350 K in a furnace [*12*]. These spectra exhibit the overlap of the fundamental and first hot band (fig.4b), and the overlap of the three bands (fig.4c). All the Raman line frequencies have been corrected from the pressure lineshift accurately measured [*11*].

The data have been analysed using the well-known expression of the Q-branch line positions, with the assumption of a $^1\Sigma_g^+$ theory in spite of the fact that O_2 has $^3\Sigma_g^-$ electronic ground state :

$$\nu_{v,n} = \Delta G(v,v+1) + \Delta B(v,v+1) N(N+1) - \Delta D(v,v+1) N^2(N+1)^2 + \cdots \tag{1}$$

The ΔD coefficients have been found to be non-significant. The resulting constants fill in gaps in accurate data on the rovibrational structure of the ground state of O_2. The results can also be expressed in terms of Dunham coefficients.

N_2 has been studied with the same goal, but only the fundamental and the first hot bands were recorded. A set of molecular constants has also been refined [*13*]. In fig.5 appears a part of the recorded spectrum for which Raman lines intensities have been measured. The 17 data of this resolved spectrum have been used to fit the temperature : we found a value of 1211 $\pm$ 84 K which is close to the value indicated by the thermocouple : 1260 K.

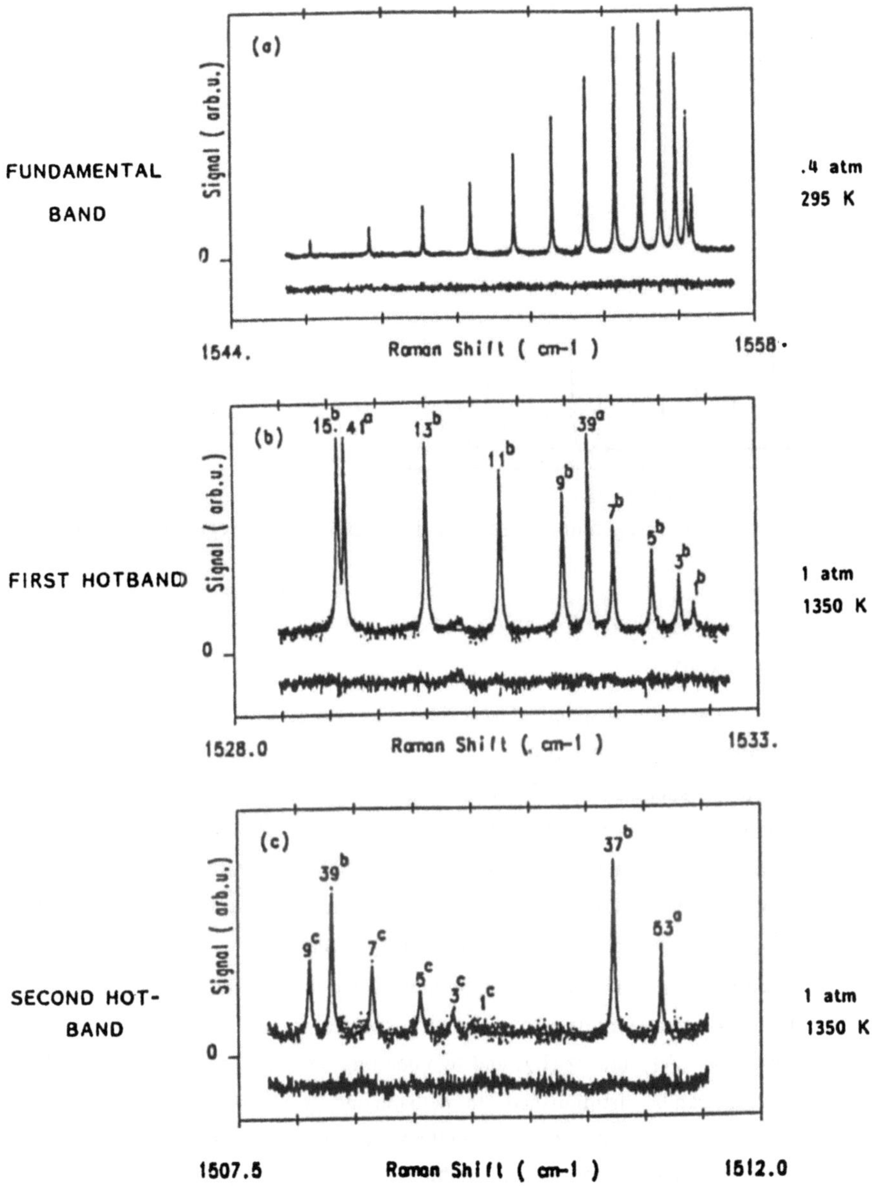

Fig.4 Experimental and calculated Raman spectra of the fundamental and first hot bands of O_2

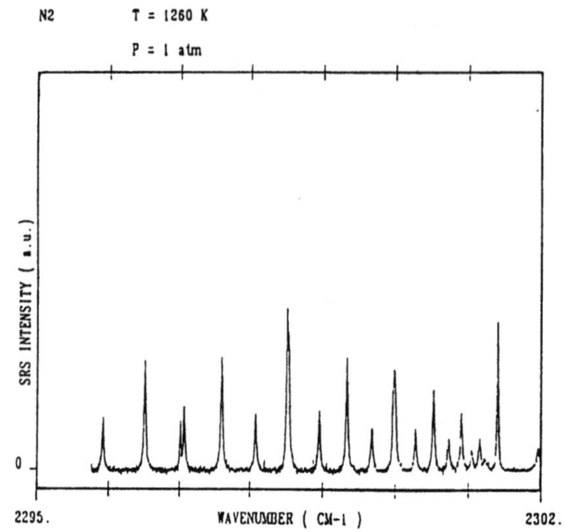

Fig.5 SRS spectrum of nitrogen in atmospheric air at 1260 K in the region of the first hot band

4. XY4 Spherical Tops

We now focus on the study of spherical tops XY_4. The interest of such experiments lies in the fact that some vibrational bands of XY_4 molecules are forbidden by selection rules in infrared absorption spectroscopy. This drawback is partially overcome through the interaction between levels which produces a mixing, so that non-allowed lines become observable. But numerous energy levels remain inaccessible by infrared spectroscopy (for example : low J values). Raman spectroscopy, thanks to the accuracy of the frequency measurements, can provide very useful complementary data.

An XY_4 molecule has a tetrahedral symmetry and is characterized by four normal vibrational modes corresponding to symmetry species in the T_d point group : two stretching modes ν_1 (A_1) and ν_3 (F_2) and two bending modes ν_2 (E) and ν_4 (F_2). Two types of behaviour can be distinguished : In methane and its isotopic species, the bending and stretching modes are close in energy, and the stretching modes' energy is twice that of the bending modes. So the harmonic and combination levels $2\nu_2$, $\nu_2+\nu_4$, $2\nu_4$ fall near the ν_1 and ν_3 levels. These vibrational levels form a polyad of interacting bands usually called a pentad. In silane and germane, the five levels are dissociated into a triad ($2\nu_2$, $\nu_2+\nu_4$, $2\nu_4$) and a dyad (ν_1 , ν_3) which can be considered separately.

$^{13}CD_4$

The first stimulated Raman spectrum recorded in the laboratory with absolute frequency calibration was that of $^{13}CD_4$ [*14*]. A part of this spectrum is shown in fig.6, including the

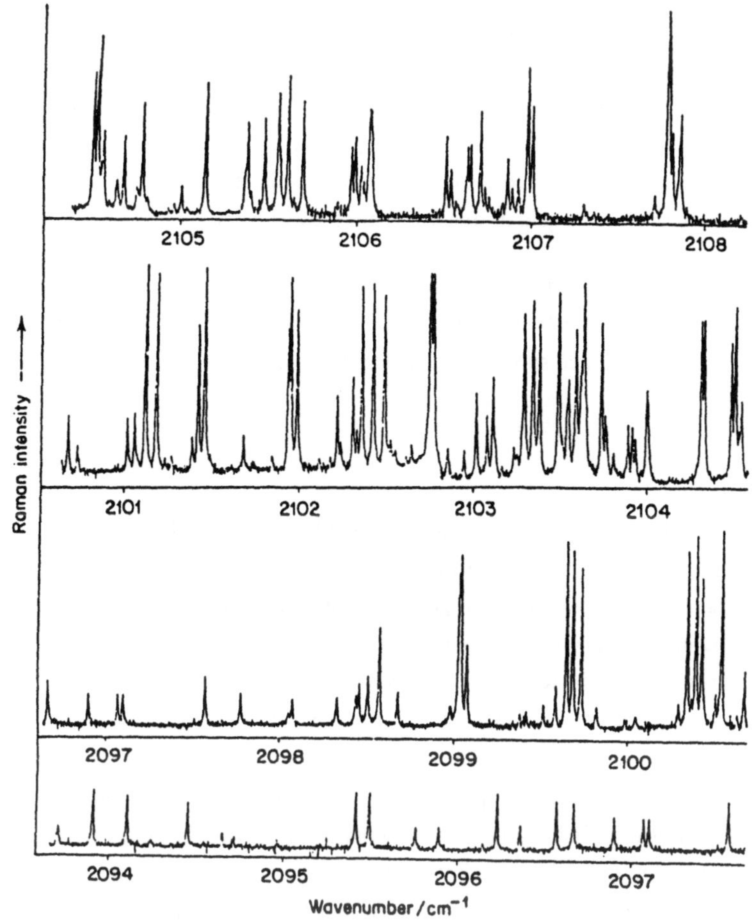

Fig.6 Part of the Raman spectrum of the ν_1 and $\nu_2 + \nu_4$ bands of $^{13}CD_4$

$\nu_2+\nu_4$ and ν_1 bands. Many lines of $\nu_2+\nu_4$ appear through the strong second-order Coriolis interaction between ν_1 and the F_1 level of $\nu_2+\nu_4$. The weak bands $2\nu_4$ (A_1) and $2\nu_2$ (A_1) have also been investigated for the first time in a methane-like molecule by coherent Raman spectroscopy. The intensity of these bands is enhanced by the strong Fermi interaction with ν_1.

We measured more than 300 Raman line positions with an accuracy better than 10^{-3} cm^{-1} in most cases. These have been combined with infrared data, in a simultaneous analysis of the pentad. A tensorial formalism for the Hamiltonian developed by Champion [*15*] and Champion and Pierre [*16*] has been used. The Raman data yielded important complementary information, in particular on the A_1 levels and their interactions with the other levels [*14*].

$^{13}CH_4$

Recently, we have recorded the Raman spectrum of the isopotomer $^{13}CH_4$ in the region of the ν_1 and $2\nu_2$ bands [*17*]. The ν_1 (A_1) and $2\nu_2$ (A_1) bands are forbidden in infrared and the corresponding upper energy levels were poorly determined by infrared data, particularly for low J values. We have investigated these bands in a sample containing 90 % $^{13}CH_4$. The multiple-pass cell provided a gain in signal seen for the $2\nu_2$ band.

The Raman data played an important role at the beginning of the first analysis of the pentad of $^{13}CH_4$ by the determination of some lower order Hamiltonian parameters, which helped assignment of the infrared spectra. The standard deviation for the Raman lines was less than 10^{-3} cm^{-1} [*17*].

$^{28}SiH_4$

In the case of silane, the ν_1/ν_3 system can be considered separately. Many studies have been devoted to this dyad, in both infrared and Raman spectroscopy. We have reinvestigated this molecule following two goals : accurate measurements of ν_1 Raman frequencies and combination with infrared data in a simultaneous analysis [*18*], and measurements and analysis of the Raman intensities of the ν_1/ν_3 dyad [*19*].

A low pressure spectrum was first obtained in a single pass cell (fig.7, lower trace). The assigned lines of this scan correspond to $J < 14$. Another region was then recorded in the M.P.C. at the same pressure (fig.7, upper trace). The enhancement of the S/N ratio is evidenced by a group of lines belonging to the two parts (fig.7). High J values were thus observed and assigned up to $J = 20$. The line frequencies have been extracted from the experimental spectrum by using a Voigt profile fit and by varying the line positions. Depending on the value of the standard deviation of these parameters derived from the fit, we attributed an uncertainty to each line position measurement. To simplify, only three categories of uncertainties have been considered which generally correspond to the situation of the fit : strong isolated lines, blended lines, weak lines.

The Hamiltonian parameters have been adjusted by fitting the Raman and infrared data. The latter were recorded by Fourier transform absorption spectroscopy at the Laboratoire d' Infrarouge, Orsay, France. The results are very good as indicated by the standard deviation :0.0004 cm^{-1} [*18*]. Moreover, the consistency between the two sets of data has been checked by comparison between infrared and Raman transitions with a common upper level. The ground state being very well known (accuracy better than 0.0001 cm^{-1}), we could directly compare the frequency calibration of the experiments : we found 21 common levels and an average difference of only 0.17×10^{-3} cm^{-1} ($\sigma = 0.49 \times 10^{-3}$ cm^{-1}). The calculated energy levels of the dyad of silane are drawn in fig.8 as a function of J. The mixing of these levels, resulting from the interaction, is displayed in this figure: each level is depicted by two horizontal lines, the length of each line being proportional to the

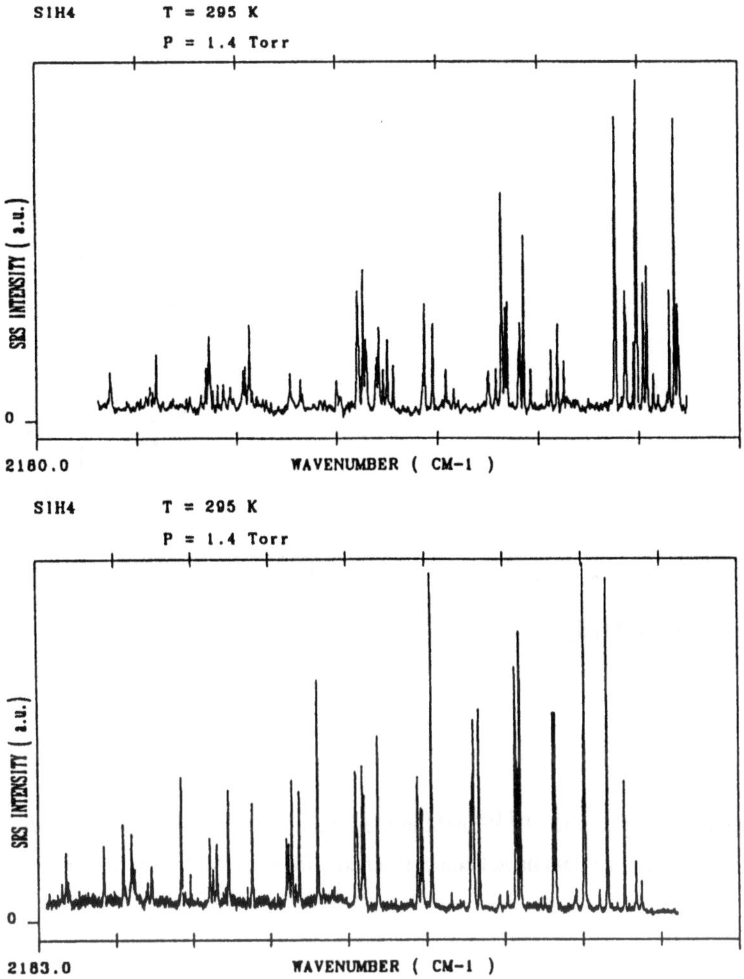

Fig.7 SRS spectrum of the ν_1 band of silane

percentage of ν_1 (left) and ν_3 (right) wavefunctions in the total wavefunction. The mixing increases with J. Furthermore, the upper levels of ν_3 present a strong mixing and are pushed away at higher energies.

The Q-branch of the ν_3 band has also been recorded. A careful calibration of Raman intensity of ν_3 with respect to ν_1 was done in order to get a set of measurements of the intensities of the dyad. These data were then analysed [*19*] by means of a model developed in the laboratory by Loete and Boutahar [*20*]. The relative precision lies in the range 5-10 %.

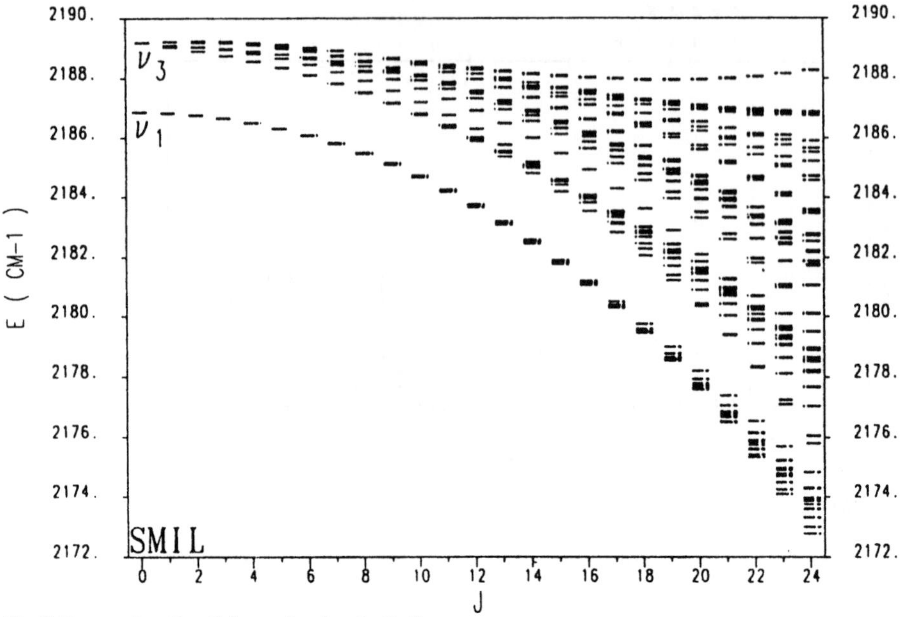

Fig.8 Energy levels of the ν_1/ν_3 dyad of silane

$^{116}SnH_4$

A recording of the Raman spectrum of $^{116}SnH_4$ in the region of ν_1 was also obtained. As in the case of silane, the Raman data have been combined with infrared data in the analysis of the dyad [*21*].

5. Symmetric Top Molecule : C_3H_6

The first high resolution Raman spectrum of cyclopropane in the region of ν_1 has been recorded with our experiment. This spectrum exhibits strong Q-branches belonging to ν_1, $2\nu_2$, and $\nu_7+\nu_9+\nu_{14}$. A nearly Doppler-limited spectrum of these Q-branches has a complicated structure. The analysis of the data led to a preliminary set of spectroscopic constants for the interacting bands mentioned above [22].

6. Conclusion

These examples demonstrate the possibility of using stimulated Raman spectroscopy to gain useful data on molecular structure of various types of molecules, generally complementary to infrared data.

Acknowledgments

The authors thank their colleagues from the SMIL Laboratory and at Orsay, Wuppertal, Munich, Cambridge, Pennsylvania, and Tomsk for their contributions to this work.

References

1. A. Owyoung, C. W. Patterson, and R. S. McDowell, Chem. Phys. Lett. **59**, 156 - 162 (1978).

2. R. S. McDowell, C. W. Patterson, and A. Owyoung, J. Chem. Phys. **72**, 1071-1076 (1980).

3. P. Esherick, A. Owyoung, and C. W. Patterson, J. Mol. Spectrosc. **86**, 250 - 257 (1981).

4. A. Owyoung, P. Esherick, A. G. Robiette, and R. S. McDowell, J. Mol. Spectrosc. **86**, 209 - 215 (1981).

5. P. Esherick, and A. Owyoung, J. Mol. Spectrosc. **92**, 162 (1982).

6. P. Esherick, A. Owyoung, and C. W. Patterson, J. Phys. Chem. **87**, 602 (1983).

7. S. Q. Mao, R. Saint-Loup, A. Aboumajd, P. Lepage, H. Berger, and A. G. Robiette, J. Raman Spectrosc. **13**, 257 - 261 (1982).

8. B. Lavorel, R. Saint-Loup, G. Pierre, and H. Berger, J.Physique Lett. Paris **45**, L295-300 (1984).

9. R. Chaux, C. Milan, G. Millot, B. Lavorel, R. Saint-Loup, and J. Moret-Bailly, J.Optics (Paris) **19**, 3 - 14 (1988).

10. R. Saint-Loup, B. Lavorel, G. Millot, C. Wenger, and H. Berger, J. Raman Spectrosc. **21**, 77-83 (1990).

11. G. Millot, B. Lavorel, and H. Berger, International symposium on coherent Raman spectroscopy, Samarkand, 18-20 September 1990, USSR, these proceedings

12. G. Millot, B. Lavorel, H. Berger, R. Saint-Loup, J. Santos, R. Chaux, C. Wenger, J. Bonamy, L. Bonamy, and D. Robert, European CARS Workshop, Dijon, 19-20 March 1990, France

13. A. Tabyaoui, B. Lavorel, G. Millot, R. Saint-Loup, R. Chaux, and H. Berger, J. Raman Spectrosc. , in press

14. G. Millot, B. Lavorel, R. Chaux, R. Saint-Loup, G. Pierre, H. Berger, J. I. Steinfeld, and B. Foy, J.Mol.Spectrosc. **127**, 156-177 (1988).

15. J. P. Champion, Canad. J. Phys. **55**, 1802-1828 (1977).

16. J. P. Champion, and G. Pierre, J. Mol. Spectrosc. **79**, 255-280 (1980).

17. J. M. Jouvard, B. Lavorel, J. P. Champion, and L. R. Brown, 45th Symposium on high resolution molecular spectroscopy, Columbus, 1990, USA (1990).

18. B. Lavorel, G. Millot, Q. L. Kou, G. Guelachvili, K. Bouzouba, P. Lepage, V. G. Tyuterev, and G. Pierre, J.Mol.Spectrosc. **143**, 35-49 (1990).

19. A. Boutahar, G. Millot, M. Loete, B. Lavorel, and C. Wenger, in preparation

20. A. Boutahar, and M. Loete, Can. J. Phys., in press (1991).

21. A. Tabyaoui, B. Lavorel, G. Pierre, and H. Burger, XIth international conference on high resolution infrared spectroscopy, Prague, 1990, Czechoslovakia

22. J. Pliva, M. Terki-Hasseine, B. Lavorel, R. Saint-Loup, J. Santos, H. W. Schrotter, and H. Berger, J. Mol. Spectrosc. **133**, 157 (1989).

Collisional Relaxation Processes Studied by Coherent Raman Spectroscopy for Major Species Present in Combustions

G. Millot, B. Lavorel, and H. Berger

Laboratoire de Spectronomie Moléculaire et Instrumentation Laser, U.R.A. CNRS n° 777, Université de Bourgogne, 6, Bd. Gabriel, F-21000 Dijon, France

The effects of collisional relaxation processes on the Q-branch profile of major species present in combustions have been studied by high resolution stimulated Raman spectroscopy. Particular interest has focused on the following collisional systems: N_2-N_2, O_2-O_2, CO_2-CO_2, O_2-N_2, N_2-CO_2 and N_2-H_2O. For each colliding pair, starting from accurate determinations of line broadening coefficients over a wide temperature range, state-to-state rates for rotational energy transfers have been deduced by using various fitting laws. Among these rate laws, special attention has been paid to the temperature dependence of the energy corrected scaling (ECS) law combined with a hybrid exponential-power law (EP) for the basis rate constants.

The improvement obtained with the ECS-EP rate law with respect to other fitting laws is discussed in detail in the case of N_2-N_2 and N_2-H_2O. A few illustrations of the ability of three fitting laws (MEG, ECS-P, ECS-EP) to reproduce the Q-branch collapse occurring at high density (up to about 600 amagat for example at 295 K for O_2-O_2 and N_2-N_2) are presented. Finally, the influence of the CO_2 and H_2O species in the SRS Q-branch of N_2 has been carefully investigated.

1. INTRODUCTION

Since the first demonstration of the ability of Coherent Anti-Stokes Raman Scattering (CARS) to measure species concentration and temperature [1,2], a great number of studies have been devoted to the use of this technique in combustion. Consequently the precision of the measurements, the spatial and temporal resolution, etc., have been substantially improved. We will focus our attention on the temperature determination. Although nitrogen is the most commonly used probe molecule due to its abundance, other molecules such as oxygen, carbon-dioxide, etc., can also be used. In each case, the temperature is determined from the spectral shape of the Q-branch, which is a strong function of pressure, temperature and foreign perturbers. So, the temperature measurement is as precise as the theoretical models describing the CARS signal. In the last decade, much progress in modelling CARS-spectra was achieved by using very accurate linewidths at low pressure, and by taking into account motional narrowing and

Springer Proceedings in Physics, Volume 63 **Coherent Raman Spectroscopy**
Eds.: G. Marowsky and V.V. Smirnov

lineshifting effects which occur at high pressure. Another nonlinear Raman spectroscopy, stimulated Raman scattering (SRS), has been proven to be very useful to provide accurate linewidths and lineshifts and to test the degree of accuracy of models for the motional narrowing phenomena.

We give some general theoretical considerations about SRS modelling and describe a few examples of the results that we have obtained.

2. BACKGROUND TO SRS MODELLING

The normalized SRS isotropic fundamental Q-branch profile may be expressed in terms of the G matrix [3,5] :

$$I(\omega) = \frac{1}{\pi} \left| <0 | \tilde{\alpha} | 1> \right|^2 \mathrm{Re} \sum_{J,J'} \rho_J \, (G^{-1}(\omega))_{J'J} \, , \tag{1}$$

where $$(G(\omega))_J = i(\omega - \omega_{1J',0J'})\, \delta_{J'J} + n\, W_{J'J} \, , \tag{2}$$

where n is the number density of perturbers, $\omega_{1J',0J'}$ denotes the isolated line position and $W_{J'J}$ is the relaxation matrix element for a transition from J to J'.

The real and imaginary parts of the diagonal matrix elements are respectively the line broadening and minus the line shifting coefficients :

$$\mathrm{Re}\, W_{JJ} = \gamma_J \, ; \; - \mathrm{Im}\, W_{JJ} = \delta_J \, . \tag{3}$$

The off-diagonal matrix elements are related to those of the scattering operator **S** by

$$W_{J'J} = \delta_{J'J} - <1J \,|\, S^+ \,|\, 1J'> \; <0J' \,|\, S \,|\, 0J> \, . \tag{4}$$

If the vibrational transfers are neglected the unitarity of the **S** operator leads to the usual sum rule:

$$- \mathrm{Re} \sum_{J' \neq J} W_{J'J} = \gamma_J - \gamma_V \, , \tag{5}$$

where γ_V is the pure vibrational phase relaxation. Let us recall that the relaxation matrix must satisfy the detailed balance principle :

$$\rho_J \, \mathrm{Re}\, W_{J'J} = \rho_{J'} \, \mathrm{Re}\, W_{JJ'} \, . \tag{6}$$

For pressures corresponding to weak overlap between lines the general expression (1) can be simplified, leading to the Rosenkranz profile [6] :

$$I(\omega) = \frac{1}{\pi} \left| <0 | \tilde{\alpha} | 1> \right|^2 \sum_J \rho_J \frac{n\gamma_J - (\omega - \omega_{1J,0J})\, n\, Y_J}{(\omega - \omega_{1J,0J})^2 + (n\, \gamma_J)^2} \, , \tag{7}$$

where the coupling parameters Y_J can be expressed as a function of the relaxation matrix elements as follows :

$$Y_J = 2 \sum_{J' \neq J} \frac{\mathrm{Re}\, W_{J'J}}{\omega_{1J',0J'} - \omega_{1J,0J}} . \tag{8}$$

At low pressure and (or) high temperature, Dicke narrowing can be taken into account by means of the Galatry profile [7]. When necessary, weak overlaps and Dicke narrowing can be simultaneously included by convoluting both effects [8].

At higher pressures the overlap increases, so that the knowledge of the full relaxation matrix is required. Fitting laws [9] are used to determine each matrix element from the line broadening coefficients. Among the various fitting laws, we have tested the modified exponential model proposed by Koszykowski and co-workers [10] and the energy corrected scaling law ECS proposed by De Pristo et al. [11].

The MEG model needs four parameters α, N, δ and β as shown by its expression :

$$- \mathrm{Re}\, W_{J'J}^{MEG} = \alpha \left(\frac{T}{T_0}\right)^{-N} \left(\frac{1 + 1.5\, E_J/kT\, \delta}{1 + 1.5\, E_J/kT}\right)^2 \exp\left(\frac{-\beta \mid E_{J'} - E_J \mid}{kT}\right) . \tag{9}$$

With the ECS law all the elements of the relaxation matrix are related to the basis rate constants W_{0L} through a spectroscopic coefficient :

$$\mathrm{Re}\, W_{J'J}^{ECS} = (2J' + 1) \exp((E_{J'} - E_{J>})/kT)\, \Omega_{J>}^2$$

$$\sum_L \begin{pmatrix} J & J' & L \\ 0 & 0 & 0 \end{pmatrix}^2 (2L + 1)\, \Omega_L^{-2}\, \mathrm{Re}\, W_{0L} , \tag{10}$$

where $J_>$ is the greater of J and J' and Ω_J is the adiabatic factor defined as

$$\Omega_J^{-1} = 1 + \omega^2_{J,J-\Delta}\, l_c^2 / 24\, \bar{v}^2 ,$$

l_c being the characteristic interaction length and $\bar{v}$ the mean relative velocity .

Let us recall that Ω_J takes into account the non-negligible rotation of the molecule during the collision. A few fitting laws have been proposed to describe the basis rate constants. In the ECS-P model the basis rates are taken as a power law :

$$- \mathrm{Re}\, W_{0L}^P = \frac{A(T)}{[L(L + 1)]^\gamma} , \tag{11}$$

where the amplitude parameter A is expressed as follows [12] :

$$A(T) = A_0 \left(\frac{T}{T_0}\right)^{-N} , \tag{12}$$

T_0 being a reference temperature.

This so-called ECS-P law is thus characterized by four parameters A_0, N, γ and the interaction length ℓ_c.

In the ECS-EP model proposed by Brunner et al. [9,13], the rapid decay of the rate constants at large $|\Delta J|$ is described by restricting the amount of angular momentum which can be transferred, by using a hybrid form between a power law and an exponential law as follows:

$$- \operatorname{Re} W_{0L}^{EP} = \frac{A}{[L(L+1)]^{\gamma}} \exp\left[\frac{-L(L+1)}{J^*(J^*+1)}\right], \tag{13}$$

introducing an additional parameter J^*.

The assumption that J^* is proportional to the thermal kinetic-energy ($J^* \propto (K_i)^{1/2}$) involves that $J^*(J^* + 1)$ is proportional to kT and because $L(L + 1)$ is proportional to the rotational energy, the ECS-EP rate law can be expressed in the following form convenient for any temperature [14] :

$$- \operatorname{Re} W_{0L}^{EP}(T) = \frac{A(T)}{[L(L+1)]^{\gamma}} \exp\left[\frac{-\beta \, E_L}{kT}\right], \tag{14}$$

where the parameter β is temperature independent.

3. DETERMINATION OF RAMAN LINE BROADENING COEFFICIENTS

Line coupling effects, responsible for motional narrowing, depend on the rotational relaxation rates $W_{J'J}$ and induce a drastic modification of the Raman spectra. So, it clearly appears that accurate rotational relaxation rates are necessary. Consequently, our goal is to define accurate rotational relaxation models for high pressure and high temperature and to test these models by stimulated Raman measurements. The method consists of starting with low pressure SRS spectra recorded over a large temperature range. Then, by using suitable profiles such as the Galatry or Rosenkranz profiles described previously, we obtain the line-broadening coefficients as a function of temperature. This set of broadening coefficients is then inverted (eqn.(5)) to obtain the state-to-state rate constants $W_{JJ'}$ using various fitting/scaling laws (eqs.9 to 14).

So, the determination of the Raman line broadening coefficients is of considerable interest. Figure 1 shows the line-broadening coefficients obtained with our high resolution stimulated Raman spectrometer for pure oxygen. A schematic diagram of the spectrometer is presented in Figure 5, and a description of it is given by B.Lavorel et al. jointly with the application to molecular structure studies. The set of line-broadening coefficients covers a large temperature range, from room temperature, 295K, to 1350K, which is necessary to determine accurate fitting parameters. The

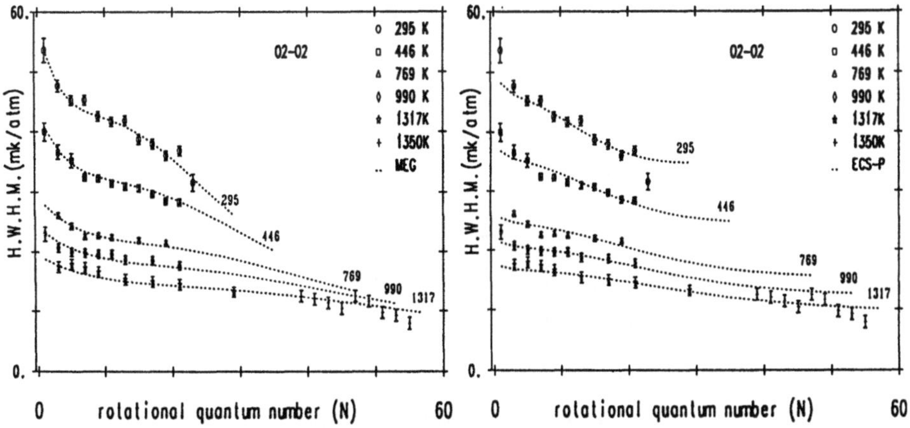

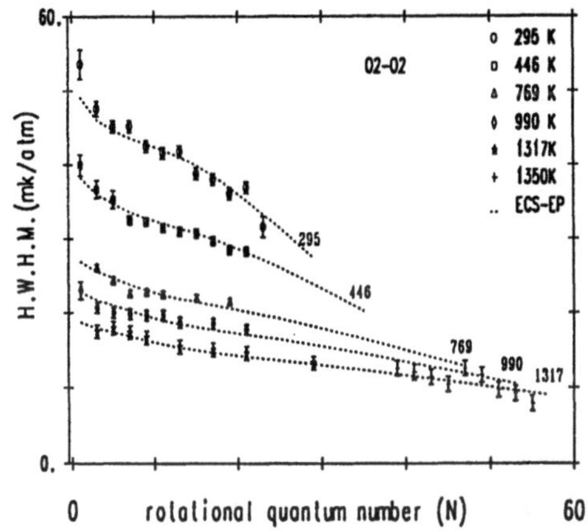

Fig. 1. Line-broadening coefficients of O_2.

experimental data are compared with the three models MEG, ECS-P and ECS-EP [15]. It clearly appears that the rotational and temperature dependences are very well reproduced with the MEG and ECS-EP models. The results are a little bit different with the ECS-P model. Indeed, the temperature dependence is quite well reproduced but at a given temperature the line broadening coefficients corresponding to high rotational quantum numbers are overestimated. So, from this comparison, we may conclude that ECS-P does not model the broadening coefficients of pure oxygen as well as the two other models. For all that, there is not sufficient evidence to reject the ECS-P law because the overestimated values correspond to nearly empty rotational levels which make only a small contribution to the collapsed Q-branch. A clear discrimination between the three models is not possible at this stage and the calculated collapsed Q-branch should be compared with experimental high density spectra for each model. Before going into more detail with the oxygen system, let us consider the case of N_2-H_2O mixture. In this case, the models are compared with accurate semi-classical

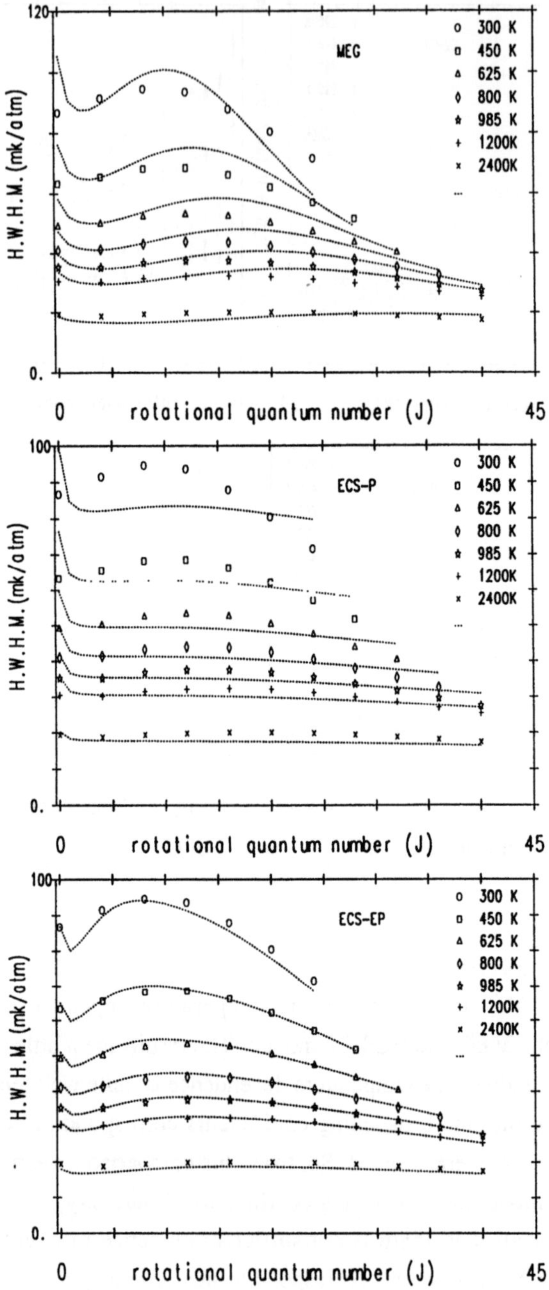

Fig. 2. Line-broadening coefficients of N_2-H_2O.

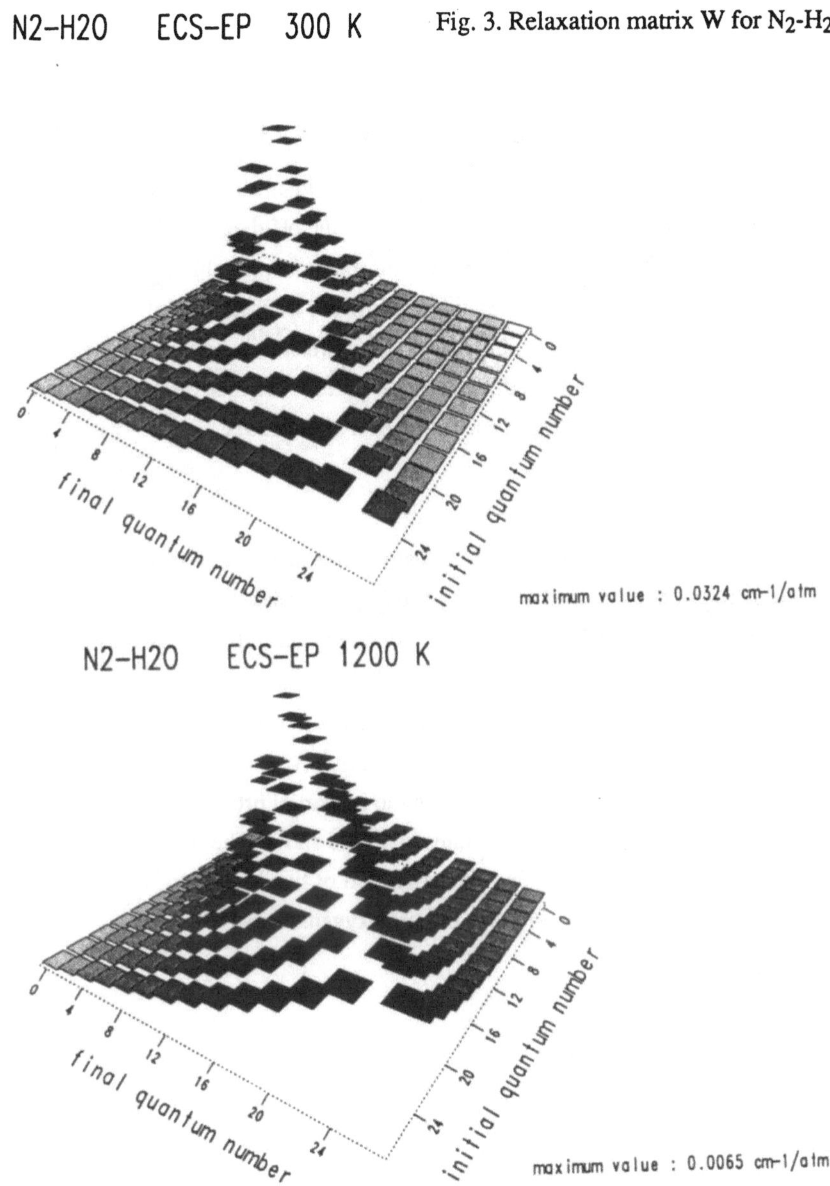

Fig. 3. Relaxation matrix W for N_2-H_2O.

calculations performed by Bonamy et al. [16]. Figure 2 shows the superiority of ECS-EP in modelling the broadening coefficients and in particular the improvement of ECS-EP with respect to ECS-P is spectacular. If the ability of a rate law to give good linewidths is not a sufficient criterion of quality, it is of course still necessary. So, the ECS-P model is rejected for the N_2-H_2O study.

When a fitting law is adequate to reproduce the line-broadening coefficients, it is then acceptable to calculate the full-relaxation matrix whatever the temperature. Figure 3 exhibits the increasing role of the excitation rates with increasing temperature for N_2-H_2O modelled by ECS-EP.

4. DETERMINATION OF RAMAN LINE SHIFTING COEFFICIENTS

It is now well known [17] that taking account of collisional shifts allows a more critical test of the relaxation model. Indeed, modification of the off-diagonal W-matrix elements has the effect of pulling Q-branch lines toward the most probable rotational quantum number, looking like an overall frequency shift. On the other hand, line-frequency shifts act as an overall frequency shift, due to their weak dependence on the rotational quantum number. In order to reduce the correlation between relaxation rates and frequency shift we performed direct measurements of lineshifts .

4.1. Measurement at low density

Direct measurements are done at low density when the lines can be considered as isolated due to the negligible effect of the off-diagonal relaxation matrix elements. We give an example for the Q(13) line at 295K for the oxygen-nitrogen mixture (Figure 4). As the lineshifts are usually very weak ($-3.10^{-3}cm^{-1}$/atm for oxygen for example), a differential method makes their measurement easier (Figure 5). Two gas cells are simultaneously used, the first one contains the mixture whereas the second one is filled with pure oxygen at a pressure corresponding to its partial pressure in the mixture. The frequency difference between the two lines gives the shift of the mixture O_2-N_2 for the corresponding pressure of nitrogen. The extraction of the lineshift of the mixture does not require a knowledge of the lineshift of pure oxygen.

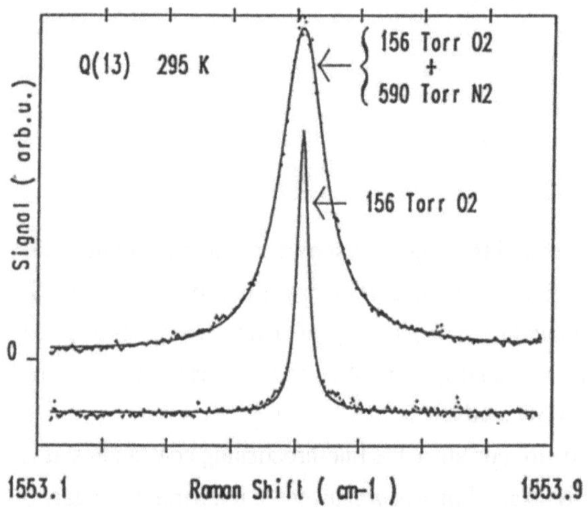

Fig. 4. Collisional shift measurement at low density for the O_2-N_2 mixture.

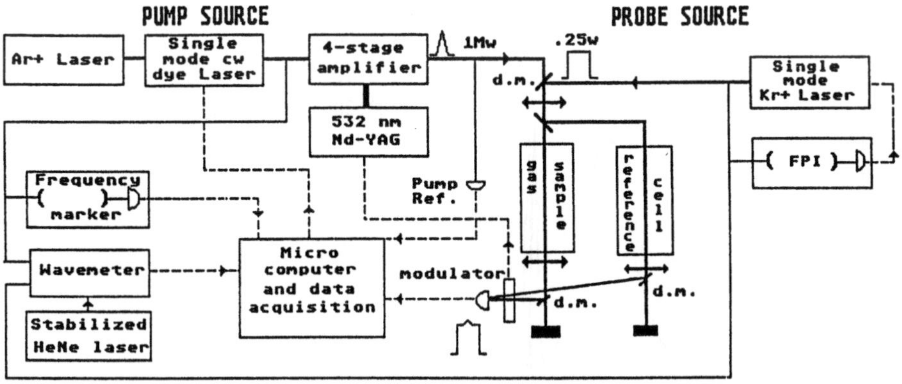

Fig. 5. Stimulated Raman spectrometer at Dijon University (Raman gain configuration).

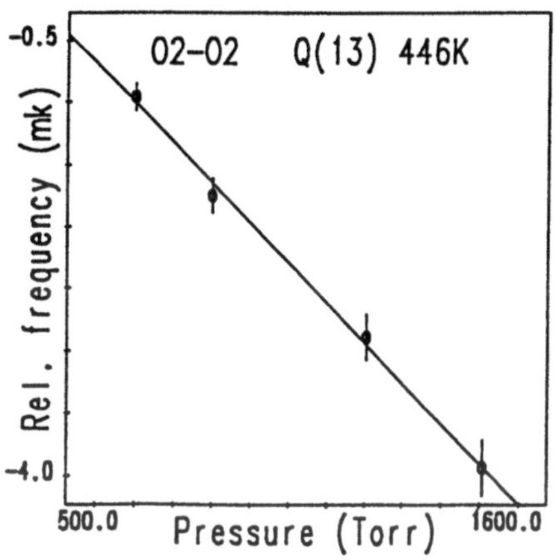

Fig. 6. Shift linearity vs pressure of O_2.

The line-shifting coefficients have been obtained for several molecular collisional systems. For pure oxygen, for example, the lineshifts have been obtained over a large temperature range for N=5 to 21. These shifts do not exibit a significant variation versus N and the mean values over N expressed in density units weakly depend on the temperature. Figure 6 illustrates the very good linear dependence of the shift versus pressure on the Q(13) line at 446K and shows the high degree of accuracy of the measurements .

4.2. Measurement at high density

A direct measurement of the collisional shift is sometimes possible at high density when collisional narrowing is nearly complete. In that case, the theoretical lineshape (Eq.(1)) weakly depends on the fitting/scaling law, and the overall frequency shift can be

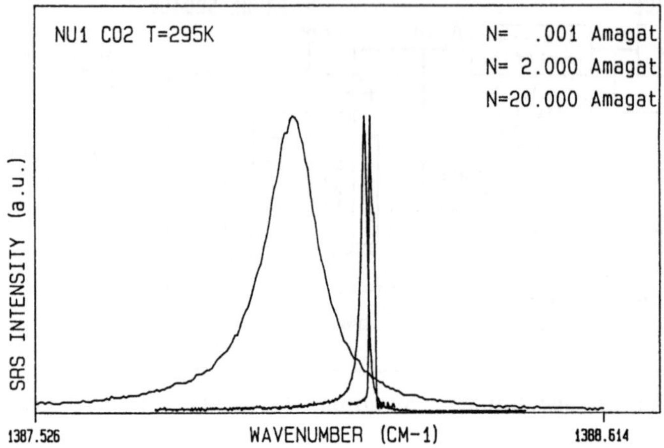

Fig. 7. Dependence of the Q-branch line shape of the ν_1 band of CO_2 as a function of density.

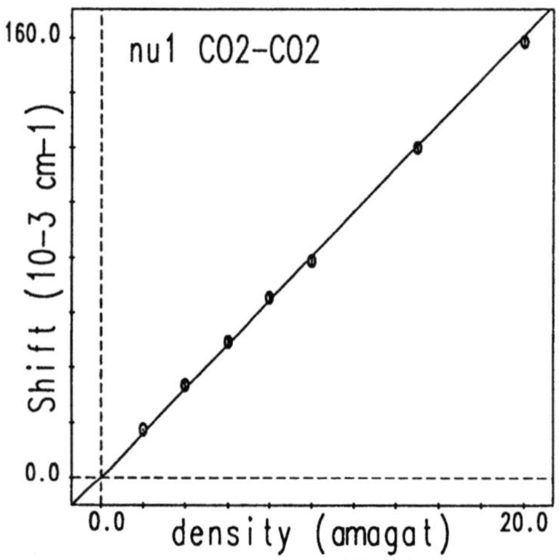

Fig. 8. Collisional shift of the ν_1 band of CO_2 vs density.

extracted independently of the relaxation rate modelling. A striking example of such a situation is given by the ν_1=1 band of CO_2 [8] (Figure 7). The line coupling mechanism is very efficient due to the very small frequency separation between J-lines, so that, at low density, motional narrowing leads to a very sharp band. When the density increases the contribution of pure vibrational dephasing becomes significant and the band broadens and shifts. The overall frequency shift of the full Q-branch is displayed in Figure 8 as a function of the density.

5. ABILITY OF THE RELAXATION RATE MODEL TO PREDICT THE COLLAPSED Q-BRANCH

In order to test the various fitting laws stimulated Raman spectra are recorded at high density. A few examples are shown (Figure 9) for pure oxygen at room temperature. The predictions of the best fit MEG, ECS-P and ECS-EP laws are compared to our observations at 19.4 amagat and 122.3 amagat. At the lower density good agreement is achieved with the three models by fitting only an overall scale factor and an overall frequency shift. On the other hand the agreement is not so good with the ECS models at higher density. Figure 10 exhibits the density dependence of the overall frequency shift obtained with each model. The solid line gives the fit of a polynomial function in powers of density. Small differences are observed. The MEG and ECS-EP models give similar results with a shift nearly linear versus density up to 400 amagat, whereas the ECS-P shift does not show such a good linear dependence versus density. By taking data below 170 amagat, each rate law gives a good linear dependence versus density, with a slope of about $-4 .10^{-3}$ cm $^{-1}$/ amagat for MEG and ECS-EP and $-4.5\ 10^{-3}$ cm^{-1} / amagat for ECS-P. These values are in relatively good agreement with the direct measurement at low density (about $-3.5\ 10^{-3}$ cm $^{-1}$/ amagat).

A similar study has been performed for pure nitrogen at room temperature as shown in Figure 11 at 136.2 amagat. Both MEG and ECS-EP give a very good spectral bandshape in contrast to the ECS-P model. It has been demonstrated by Bonamy et al. [12] that the ECS-P law properly models the line-broadening coefficients for N_2-N_2. So, the ability of a rate law to model linewidths does not prove its faculty to reproduce the lineshape of the collapsed Q-branch. Figure 12 gives the density dependence of the overall lineshift extracted from each model. The solid line is a linear extrapolation of the shift measured at low density [18]. It clearly appears that ECS-EP is the only one which gives an excellent agreement with the low density measurement. In particular the improvement obtained with ECS-EP with respect to ECS-P is spectacular. The MEG law is not adequate to give the exact collisional shift. Therefore, the ability of a rate law to model the Q-branch profile does not imply it is able to give the exact absolute collisional shift with a linear dependence versus density in the binary collisional regime. That is particularly true for the N_2-CO_2 mixture. Indeed the three models give exact Q-branch shapes, but none of the models gives the absolute shift, each of them leading to a different value. This dispersion of shift values derived from the three laws is shown in Table 1.

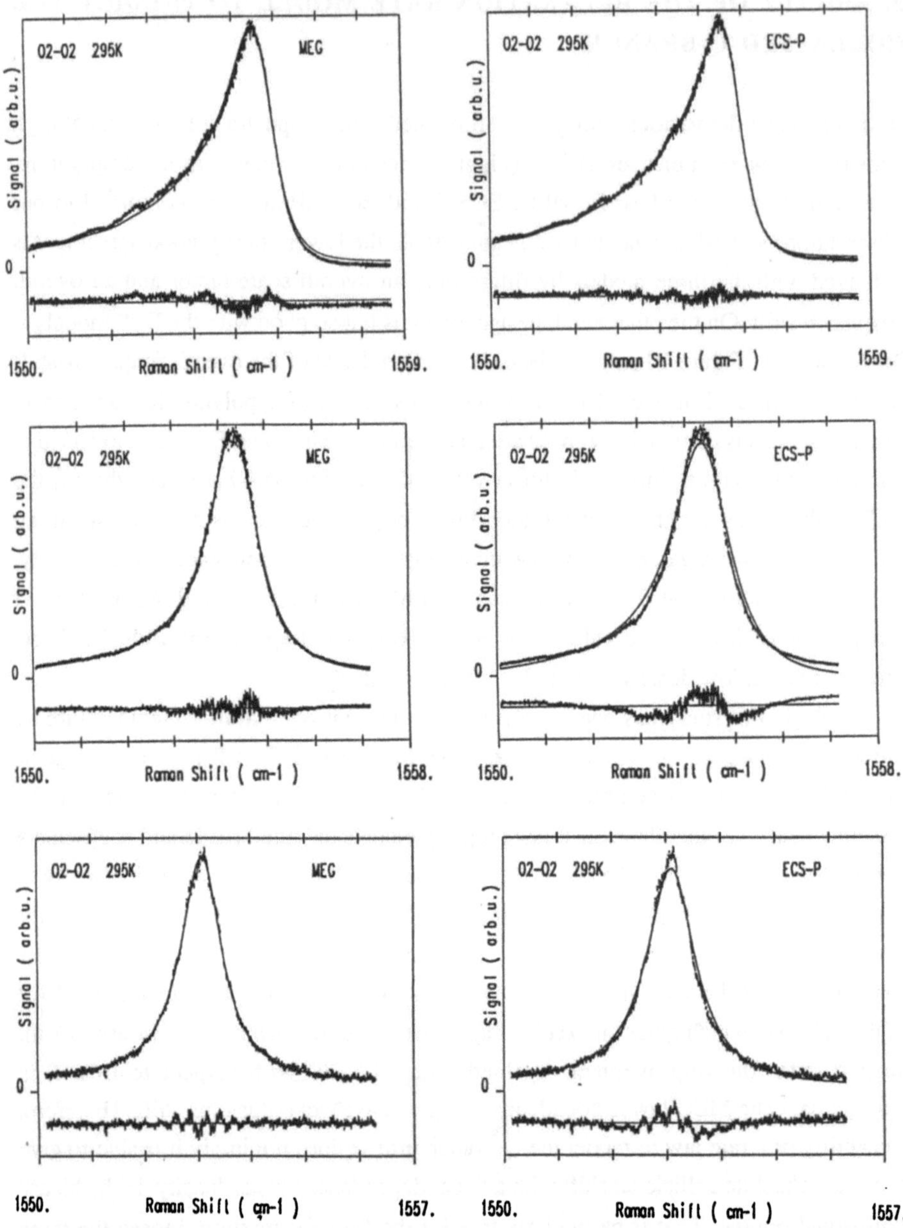

Fig. 9. Illustration of the ability of the fitting laws to provide a good spectral line shape of the Q-branch of O_2 at high density: 19.4 amagat for the left column and 122.3 amagat for the right column.

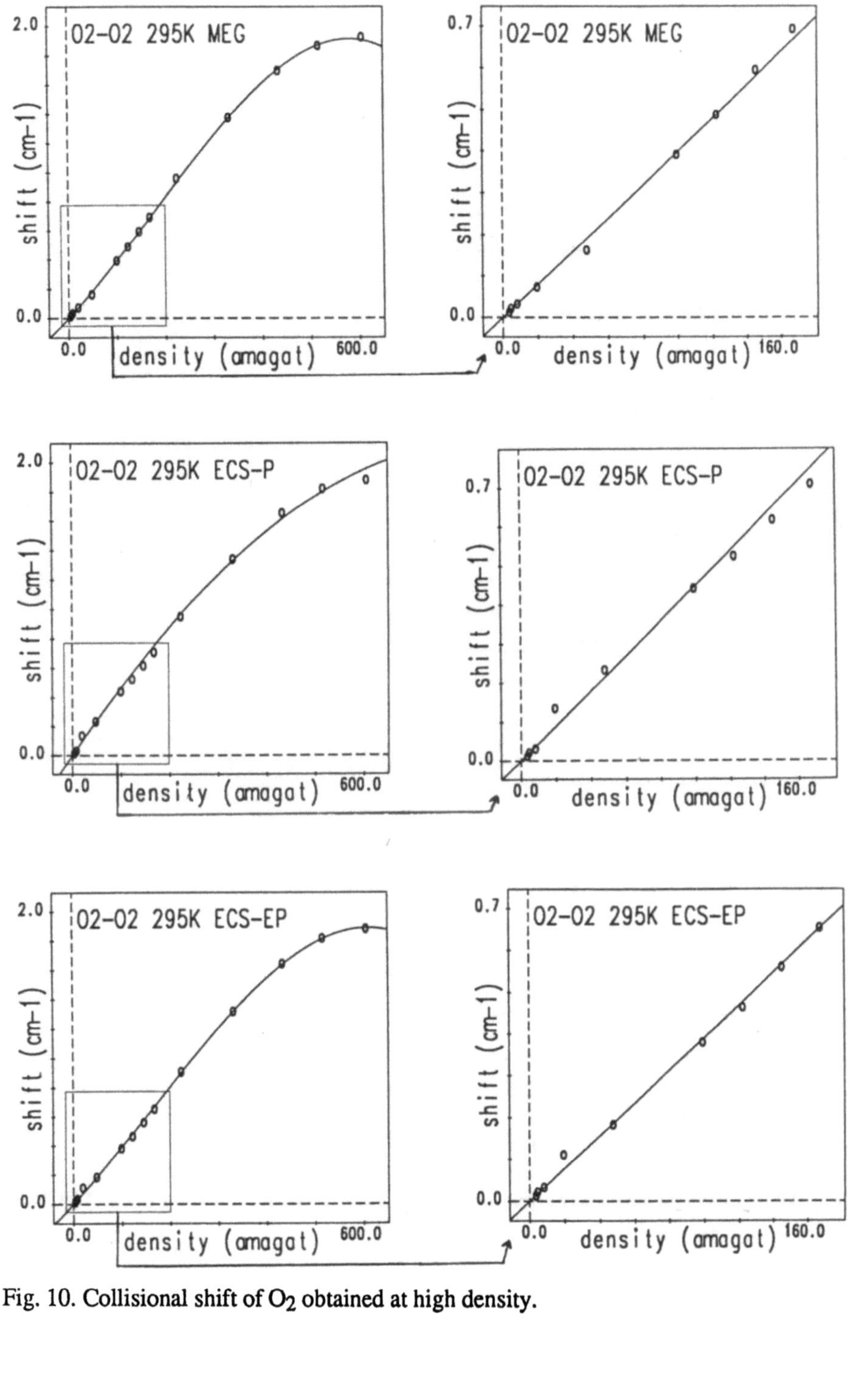

Fig. 10. Collisional shift of O_2 obtained at high density.

Fig. 11

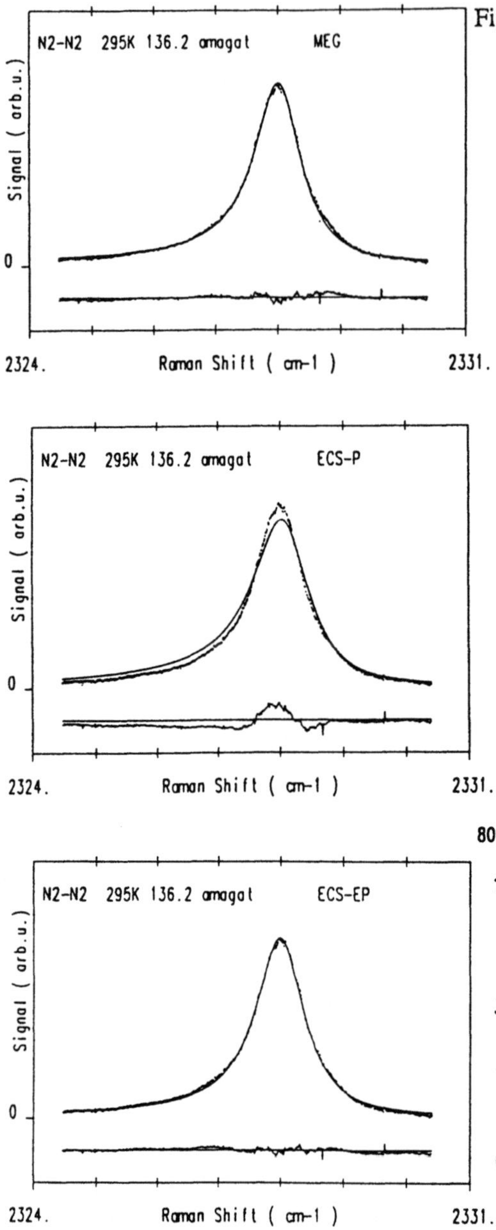

Fig. 12

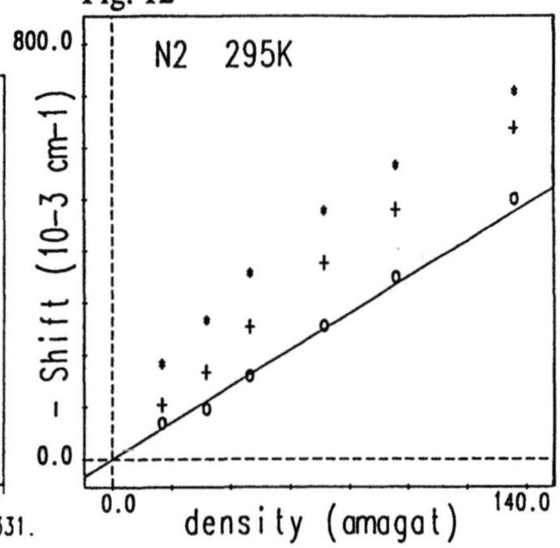

Fig. 11. Spectral line shape of N_2 at high density.

Fig. 12. Collisional shift of N_2 obtained at high density with the fitting laws: * ECS-P; + MEG; o ECS-EP.

Table 1. Collisional shift of N_2-CO_2 at high density.

Model	amagat	4.93	13.7	23.3	32.6
MEG	$-\Delta$	10.68	9.29	8.98	8.34
	$\delta\Delta$	0.25	0.10	0.07	0.06
ECS-P	$-\Delta$	16.80	13.08	11.37	9.98
	$\delta\Delta$	0.30	0.12	0.07	0.04
ECS-EP	$-\Delta$	12.90	9.25	8.25	7.42
	$\delta\Delta$	0.21	0.07	0.04	0.03

6. ROTATIONAL RELAXATION OF NITROGEN IN THE TERNARY MIXTURE N_2 - CO_2 -H_2O

In a real combustion medium, such as an atmospheric pressure flame, collisional perturbers like CO_2 and H_2O induce significant contributions to the N_2 relaxation processes. So, starting from the MEG and ECS-EP model (ECS-P does not work for N_2-H_2O even at low pressure), calculation of collisional linewidths for N_2 in the ternary

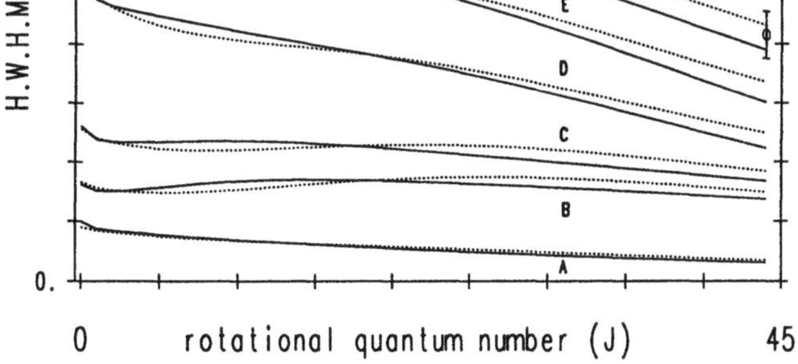

Fig. 13. Collisional widths of N_2 in the post flame gases of a methane-air flat flame at 1730K and 620 Torr. A: contribution of N_2-CO_2, B: contribution of N_2-H_2O, C: contribution of N_2-CO_2 + N_2-H_2O, D: contribution of N_2-N_2, E: results by neglecting CO_2 and H_2O, F: N_2-CO_2-H_2O

mixture N_2-CO_2-H_2O has been performed for the concentrations 74% N_2, 9% CO_2 and 17% H_2O at 1730K and 620 Torr corresponding to the experimental conditions [19] in the post flame gases of a methane-air flat flame. The results are reported in Figure 13 and give a clear indication of the significant role of the two minor species H_2O and CO_2. Both models give a similar overall agreement. It is interesting to point out that, for high J values, the contribution of CO_2 and H_2O (curves C) to the total linewidths (curves F) is about the same as the one given by N_2 (curves D). Curves E show the linewidths calculated by neglecting the influence of CO_2 and H_2O. Because the contribution to the linewidth of CO_2 and H_2O weakly depends on the rotational quantum number (curves C), the difference between curves F and E is nearly constant. That explains why Porter et al. [20] have resolved the discrepancy between CARS spectra in flame and corresponding theoretical spectra by empirically adding a dephasing component to pure nitrogen linewidths.

7 . CONCLUSION

We have demonstrated that stimulated Raman spectroscopy ia a powerful tool for studing collisional relaxation processes. In particular, SRS can provide:

- accurate line-broadening and line-shifting coefficients,
- spectra at high density,
- a critical test of rotational relaxation rate models.

So, SRS leads to significant information for CARS thermometry.

ACKNOWLEDGEMENTS

We would like to thank R.Chaux, R.Saint-Loup and C.Wenger from our laboratory and our collaborators of the Besancon University, J.Bonamy, L.Bonamy and D.Robert.

This work was supported by the Centre National de la Recherche Scientifique, Paris, and the Conseil Régional de Bourgogne.

REFERENCES

1. P.Régnier, J.P.Taran, Appl. Phys. Lett. 23, 240 (1973)
2. F.Moya, S.A.J.Druet, J.P.Taran, Opt. Commun. 13, 169 (1975)
3. U.Fano, Phys. Rev. 131, 259 (1963)
4. A.Ben Reuven , Phys. Rev. 145, 7 (1966)

5. G.J.Rosasco, W.Lempert, W.S.Hurst and A.Fein, Chem. Phys. Lett. 97, 435 (1983); in: Spectral line shapes, Vol.2, ed. W.Burnett (de Gruyter, Berlin, 1983)
6. P.W.Rosenkranz, I.E.E.E. Trans. Antennas Propag. 23, 498 (1975)
7. L.Galatry, Phys. Rev. 122, 1218 (1961)
8. B.Lavorel, G.Millot, R.Saint-Loup, H.Berger, L.bonamy, J.Bonamy and D.Robert, J. Chem. Phys. 93, 2176 (1990); 93, 2185 (1990)
9. T.A.Bruner and D.Pritchard, Dynamics of Excited States, ed.K.P.Lawley (Wiley, New York, 1982) p.589
10. M.L.Koszykowski, L.A.Rahn, R.E.Palmer and M.E.Coltrin, J. Phys. Chem. 91, 41 (1987)
11. A.E.DePristo, S.D.Augustin, R.Ramaswamy and H.Rabitz, J. Chem. Phys. 71, 850 (1979)
12. L.Bonamy, J.Bonamy, D.Robert, B.Lavorel, R.Saint-Loup, R.Chaux, J.Santos and H.Berger, J. Chem. Phys. 89, 5568 (1988)
13. S.L.Dexheimer, M.Durand, T.A.Brunner and D.E.Pritchard, J. Chem. Phys. 76, 4996 (1982)
14. G.Millot, J. Chem. Phys. 93, 8001 (1990)
15. G.Millot, R.Saint-Loup, J.Santos, R.Chaux, H.Berger and J.Bonamy, J. Chem. Phys. (to be published)
16. J.Bonamy, D.Robert, J.M.Hartmann, M.L.Gonze, R.Saint-Loup and H.Berger, J. Chem. Phys. 91, 5916 (1989)
17. B.Lavorel, G.Millot, J.Bonamy and D.Robert, Chem. Phys. 115, 69 (1987)
18.M.L.Gonze, R.Saint-Loup, J.Santos, B.Lavorel, R.Chaux, G.Millot, H.Berger, L.Bonamy, J.Bonamy and D.Robert, Chem. Phys. 148, 417 (1990)
19. L.A.Rahn, A.Owyoung, M.E.Coltrin and M.L.Koszykowski, Seventh International Conference on Raman Spectroscopy (W.F.Murphy, Ed.), p.694 (North-Holland, Amsterdam, 1980)
20. F.M.Porter, D.A.Greenhalgh, D.R.Williams, C.A.Baker, M.Woyde and W.Stricker, Proc.11th. Conf. Raman Spectrosc. London (1988)

High Resolution Inverse Raman Spectroscopy of Molecular Hydrogen

L.A. Rahn

Combustion Research Facility, Division 8354,
Sandia National Laboratories, Livermore, CA 94551, USA

Abstract. In this paper we briefly summarize our recent measurements of H_2 Raman Q-branch spectra in both pure and H_2-rare gas mixtures. This work includes a novel speed-dependent line-broadening model to explain the the H_2-Ar lineshape and broadening results.

1. Introduction

The quasi-cw inverse Raman spectroscopic (IRS) technique, first developed by Owyoung and coworkers [1], has proved to be a useful tool for the development of accurate coherent Raman spectral models for combustion-related molecules [2,3]. High spectral resolution is attained by using a pulse-amplified single-frequency cw dye laser for the pump beam and a single-frequency argon-ion laser for the probe. Quantum-limited detection noise and precise energy normalization has lead to peak-signal to RMS-noise ratios of greater than 500. The IRS technique has provided pressure-broadening and pressure-shift coefficients for many important combustion species. Measurements at higher pressures have also allowed the refinement of state-to-state rotational relaxation rate laws that are necessary for the accurate description of collisionally collapsed Q-branch spectra.

2. Results

IRS has been used to measure the dependence of the Q-branch density-shift [4] and density-broadening [5] coefficients on temperature (295 to 1000 K) and rotational quantum number ($J = 0 - 5$) in pure H_2. Using an energy-corrected-sudden scaling law analysis of the broadening coefficients, we find that resonant rotation-rotation energy transfer and pure-vibrational dephasing make important contributions to H_2 self broadening. A population-correlated J-dependence in the density-shift coefficients, previously termed the "coupling shift" from lower temperature work [6], is observed to persist up to 1000 K. These results reflect on the nature of the H_2-H_2 interaction potential, the role of resonance collisions, and the process of vibrational dephasing by collisions in pure H_2.

Springer Proceedings in Physics, Volume 63 **Coherent Raman Spectroscopy**
Eds.: G. Marowsky and V.V. Smirnov

The strong temperature dependence of the overall density-shift in H_2-Ar is found to be similar to that of pure H_2. These results suggest a strong speed dependence for the collisional interaction between H_2 and Ar and have lead to a speed-dependent inhomogeneous collisional broadening model [7] for the asymmetric vibrational line profiles observed in dilute mixtures of H_2 in argon gas. For heavy perturbers, the relative collision speed is largely determined by the laboratory speed of the light radiator. Thus, if the radiator speed persists through many collisions, inhomogeneous broadening results if the density-shift and/or density-broadening cross-sections depend significantly on relative molecular speeds. The anomalous broadening is observed under conditions normally considered to be in the impact density regime and is characterized by asymmetric line shapes and line widths that are nonlinear in the argon partial pressure. Asymmetric line shapes are also observed for H_2-Kr and H_2-N_2 mixtures but not for H_2-H_2 or H_2-He. We have found that a complete description of the line shape also requires consideration of line narrowing from speed-changing collisions. It is the high probability of speed-change during H_2-H_2 collisions and the associated narrowing of the inhomogeneity that produce the nonlinear behavior of the line width *versus* argon partial pressure.

We are currently studying the D_2-Ar system [8] with an interest in further testing of the speed-dependent model and in providing further insight into the nature of the interaction potential between hydrogen molecules and argon atoms. The line widths of the D_2 Q(1) and Q(2) transitions have also been observed, like those of H_2, to depend nonlinearly on the argon partial pressure. Line asymmetries have also been observed for very low partial pressures of D_2 in Ar. These results motivate studies of the temperature-dependent broadening and shifting coefficients in the D_2-D_2 and D_2-Ar systems. The deuterium isotope, in particular, has a much higher cross section for rotationally inelastic collisions and is expected to have a smaller speed-dependent inhomogeneity because of its increased mass.

3. Acknowledgements

This research was supported by the U. S. Department of Energy, Department of Basic Energy Research, Chemical Sciences Division.

4. References

1. A. Owyoung, in *Laser Spectroscopy IV*; edited by H. Walther and K. W. Rothe (Springer-Verlag,Berlin, 1979), pp 175-187.
2. L. A. Rahn and R. L. Farrow, in *Raman Spectroscopy: Sixty Years On*, edited by H. D. Bist, J. R. Durig, and J. .F. Sullivan (Elsevier Science, Amsterdam, 1989), pp 33-56.

3. L. Bonamy, J. Bonamy, D. Robert, B. Lavorel, R. Saint-Loup, R. Chaux, J. Santos, and H. Berger, J. Chem. Phys. **89**, 5568 (1988), and references therein.
4. L. A. Rahn and G. J. Rosasco, Phys. Rev. A **41**, 3698 (1990).
5. L. A. Rahn, R. L. Farrow, and G. J. Rosasco, Phys. Rev. A **43**, 6075 (1991).
6. A. D. May, G. Vargese, J. C. Stryland, and H. L. Welsh, Can. J. Phys. **42**, 1058 (1964).
7. R. L. Farrow, L. A. Rahn, G. O. Sitz, and G. J. Rosasco, Phys. Rev. Lett. **63**, 746 (1989).
8. L. A. Rahn, G. J. Rosasco, and R. L. Farrow, in *Proceedings of the Twelfth International Conference on Raman Spectroscopy*, eds. J. R. Durig and J. F. Sullivan (John Wiley & Sons, New York, 1990), pp. 240-241.

High Resolution CARS Spectroscopy with cw Laser Excitation

H.W. Schrötter

Sektion Physik der LMU München,
Schellingstr. 4, W-8000 München 40, Fed.Rep.of Germany

In Coherent Anti-Stokes Raman Scattering (CARS) the instrumental resolution is limited by the convoluted linewidths of the lasers used for excitation. In order to take advantage of this fact we have constructed a CARS spectrometer consisting of a stabilized c.w.single mode argon ion ring laser, an intracavity sample cell, and a tunable c.w. dye laser for excitation of high resolution CARS spectra of rovibrational bands of small molecules in the region 2000 cm^{-1} to 3500 cm^{-1}. The intracavity power of over 100 W allowed us to record CARS spectra of nitrogen, methane, hydrogen sulfide, and ammonia at pressures of about 1.5 kPa with nearly Doppler limited resolution.

1. INTRODUCTION

Up to the late 1970s, the resolution in Raman spectroscopy of gases was limited to about 0.1 cm^{-1}. The results of the investigations of pure rotational and rotation-vibrational spectra by linear Raman spectroscopy have been summarized in review articles by WEBER [1] and BRODERSEN [2]. The best resolution has been achieved by LOËTE and BERGER [3] in the Q-branch of oxygen, namely 0.05 cm^{-1}.

Through the development of the techniques of nonlinear or coherent Raman spectroscopy, where the interaction of two laser beams with the third order susceptibility of the sample creates the spectrum, a considerable improvement of the instrumental resolution was attained, because it is only determined by the convoluted linewidths of the two lasers.

In this article we want to review briefly the theory of CARS, desribe our CARS spectrometer, which is based on c.w. lasers, and give some examples of our results. More detailed information on CARS and other nonlinear Raman techniques can be found in previously published books [4 - 8], review articles [9 - 12], and conference reports [13 - 20].

The first CARS experiments in gases using the variable frequency of a dye laser were performed by REGNIER and TARAN [21]. Whereas they investigated the Q-branch of hydrogen, further experiments by other groups succeeded in resolving structure in the Q-branch of the ν_1 band of methane [22 - 24]. NITSCH and KIEFER [25] studied the Q-branches of nitrogen, oxygen and the ν_1 band of acetylene and were able to calculate the CARS spectra by computer simulation. Most of these investigations were performed by application of pulsed laser sources, where the linewidth is limited by the Fourier transform of the laser pulse.

Springer Proceedings in Physics, Volume 63 **Coherent Raman Spectroscopy**
Eds.: G. Marowsky and V.V. Smirnov © Springer-Verlag Berlin Heidelberg 1992

The full high resolution capability of the nonlinear Raman methods can be exploited by the use of single mode c.w. or quasi-c.w. lasers. BARRETT and BEGLEY [26] first demonstrated the feasibility of c.w. CARS using intracavity excitation. This technique was further developed by HIRTH and VOLLRATH [27]. The final breakthrough to nearly Doppler-limited resolution at low pressures came with the observation and assignment of rotational structure of the Q-branches of the ν_2 band of acetylene by FABELINSKY et al. [28] by c.w. CARS and of the ν_1 band of methane by OWYOUNG et al. [29] by c.w. stimulated Raman gain spectroscopy (SRGS).

2. CARS THEORY

The theory of CARS has been presented in many of the books and articles already quoted [4 - 10]. A concise summary has recently been given by NIBLER and PUBANZ [30]. They used Gaussian c.g.s. units, however, and a conversion of their main equations to S.I. units seems to be useful. Only EESLEY [4] has previously used the S.I. system.

If one starts from Maxwell's equations one arrives at the wave equation for a given frequency component ω_j, assuming plane waves (compare eq. (7) of Ref. [30])

$$\nabla \times (\nabla \times E(r,\omega_j)) - \frac{\omega_j^2}{c^2} E(r,\omega_j) = \frac{\omega_j^2}{\epsilon_0 c^2} P(r,\omega_j) \qquad (1)$$

where ϵ_0 is the electric permittivity and c the velocity of light in vacuum.

The polarization P is expanded as a power series for strong electric fields E:

$$P = \epsilon_0(\chi^{(1)} \cdot E + \chi^{(2)} : EE + \chi^{(3)} : EEE + \ldots), \qquad (2)$$

where $\chi^{(n)}$ is the susceptibility of nth order, which is a tensor of rank (n+1). EESLEY [4] has incorporated ϵ_0 into $\chi^{(n)}$, but our eq. (2) is in agreement with the definition of $\chi^{(1)}$ in most modern textbooks.

The $\chi^{(2)}$ term is responsible for frequency doubling and the (spontaneous) hyper Raman effect [31, 32] in the framework of the nonlinear polarizability theory [33] and can be neglected here. CARS and the other nonlinear Raman effects are generated by interaction of $\chi^{(3)}$ with strong laser fields. Substituting the first and third terms of eq. (2) in eq. (1) yields the wave equation

$$\nabla \times (\nabla \times E(r,\omega_j)) - n_j^2 \frac{\omega_j^2}{c^2} E(r,\omega_j) = \frac{\omega_j^2}{\epsilon_0 c^2} P^{(3)}(r,\omega_j) \qquad (3)$$

with n_j the refractive index at frequency ω_j.

Following the treatment of NIBLER and PUBANZ [30] and likewise the notation of MAKER and TERHUNE [34] one arrives at the expression for the intensity of the CARS signal (compare eq. (22) of Ref. [30])

$$I_3 = \frac{\omega_3^2}{n_1{}^2 n_2 n_3 c^4 \epsilon_0{}^2} |\chi_{CARS}|^2 \; I_1{}^2 \cdot I_2 \cdot L^2 \left(\text{sinc } \frac{\Delta kL}{2}\right)^2 \tag{4}$$

where L is the length of the interaction region and

$$\chi_{CARS} = 3\chi^{(3)}_{1111}(-\omega_3, \omega_1, \omega_1, -\omega_2) \tag{5}$$

for parallel polarization of the laser beams in isotropic media. In gases the refractive indices are $n_j \simeq 1$ and also the phase matching factor $\left(\text{sinc } \frac{\Delta kL}{2}\right)^2 \simeq 1$ and eq. (4) reduces to

$$I_3 = \frac{\omega_3^2}{c^4 \epsilon_0^2} |\chi_{CARS}|^2 \; I_1{}^2 \cdot I_2 \cdot L^2 . \tag{6}$$

Therefore, the intensity distribution in a CARS spectrum can be calculated from the frequency distribution of χ_{CARS}.

When many closely spaced transitions contribute to a CARS spectrum, as in a Q-branch of a rotation-vibration band, the approximation [35]

$$\chi_{CARS}(\omega) \propto \sum_{J,\tau} \frac{L_{J\tau} \cdot N_{J\tau}}{\omega_{J\tau} - \omega - i\Gamma_{J\tau}/2} + \chi_{NR} \tag{7}$$

can be used, where $\omega = \omega_1 - \omega_2$ is the frequency difference of the two lasers, $L_{J\tau}$ the line strength of the Raman transition, J the angular momentum quantum number, τ an additional label, $N_{J\tau}$ the population difference between lower and upper states of a transition with frequency $\omega_{J\tau}$, and $\Gamma_{J\tau}$ the linewidth (FWHM). χ_{NR} represents the nonresonant part of the CARS susceptibility.

The interference between the contributions of neighbouring lines to the real part of the complex CARS susceptibility (and with the nonresonant part) leads to asymmetry of the line profiles and slight shifts of the CARS peaks from the exact frequencies of the Raman transitions. LAANE and KIEFER [36] have demonstrated this effect by model calculations.

The approximation of eq. (7) assumes Lorentzian line profiles and therefore takes only pressure broadening into account. Doppler broadening, instrumental (laser) linewidth, Dicke narrowing, and rotational line mixing (motional narrowing) at high pressures can be taken into account by more sophisticated line shape functions [37] in order to reproduce the finer details of the spectra.

3. CARS WITH c.w. LASERS

The initial c.w. CARS experiments [23,24,26-28] demonstrated the feasibility and high resolution capability of this technique. In order to make the best possible use of the power of the available

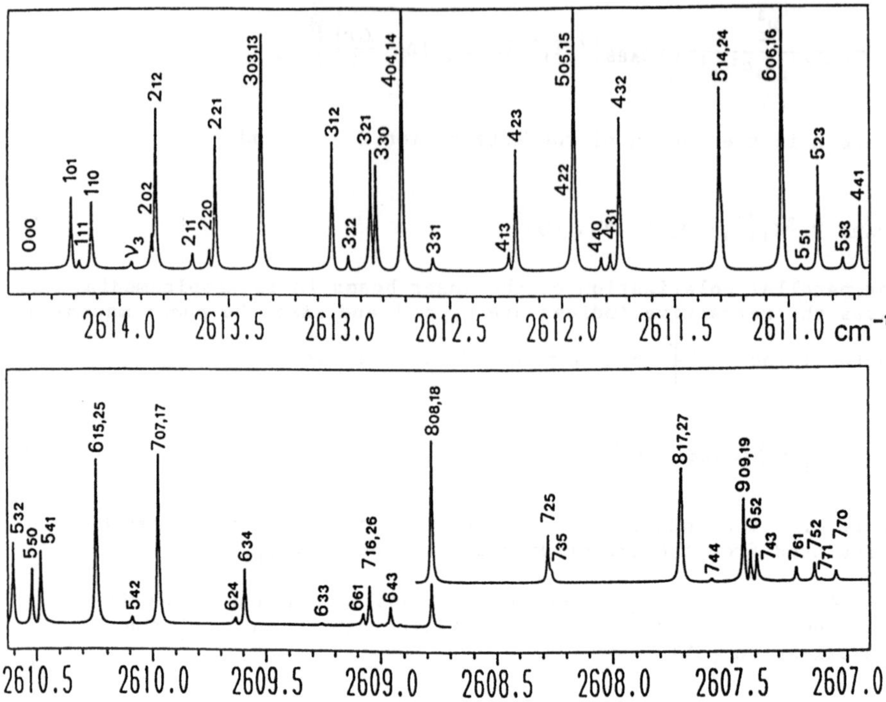

Fig. 1.

Simulated CARS spectrum of the Q-branch of the ν_1 band of hydrogen sulfide. Line assignments are given as $J_{k_a k_c}$, the quantum numbers of an asymmetric top (from Ref. 40).

lasers we decided in Munich to construct an argon ion ring laser from a commercial laser head [38]. In this way we obtained an intracavity power of more than 100 W in a single longitudinal mode at 514.5 nm. The second, tunable laser frequency is provided by a commercial dye ring laser pumped by another argon laser. CARS spectra can be generated in the region 2000 to 3500 cm^{-1}.

With this spectrometer we have investigated the CARS spectra of the Q-branches of nitrogen [38] and of the ν_1 bands of methane [39], hydrogen sulfide [40], ammonia [41] and benzene [14].

For CH_4 the assignment of about 10 energy levels in the ν_1 band has been improved [39]. The simulation of the spectra of H_2S and NH_3 on the basis of frequency data derived from infrared and microwave spectra, the theory of Raman linestrengths, and of CARS intensity profiles led to satisfactory agreement with the experimental ones.

As an example for the evaluation of a CARS spectrum we consider here the case of hydrogen sulfide. The infrared spectrum of the $2\nu_2$, ν_1 and ν_3 bands had been recorded by Fourier transform techniques and analyzed by LECHUGA-FOSSAT et al. [42]. Now the question was whether the experimental CARS spectrum of the Q-branch of the ν_1 band consisting of 54 lines [40] could be reproduced by calculation starting from the same energy levels and wave functions.

As a first step the predicted Raman line positions ν_{J_τ} and line strengths L_{J_τ} were calculated by MURPHY using his computer program [43] for asymmetric top molecules. All but one of the observed lines could thus be assigned to ν_1 transitions.

Then the CARS line profiles were calculated using a modified version of eq. (7) taking Doppler and pressure broadening into account [40]. The result is shown in Fig. 1. The agreement with the experimental spectrum was very satisfactory using a Lorentzian linewidth (FWHM) of $\Gamma_{J_\tau} = 0.006\ cm^{-1}$ for all lines at a pressure of 1.3 kPa.

4. CONCLUSION

The c.w. CARS spectrometer has been improved by adding the commercially available active stabilization to the dye laser and by the installation of a stabilization of the argon ring laser to a hyperfine component of iodine by the polarization spectroscopy technique. Presently the argon ring laser is being reconstructed on the basis of a new laser head.

ACKNOWLEDGEMENTS

I thank all my collaborators for their contributions to these results, especially H. Frunder for the construction of the CARS spectrometer. This work was supported by the Deutsche Forschungsgemeinschaft, Bonn.

REFERENCES

[1] A. Weber, "High-Resolution Rotational Raman Spectra of Gases", in "Raman Spectroscopy of Gases and Liquids" (A. Weber, Ed.), Topics in Current Physics, Vol. 11, Springer, Berlin 1979, p. 71.

[2] S. Brodersen, "High-Resolution Rotation-Vibrational Raman Spectroscopy", ibid., p. 7.

[3] M. Loëte, and H. Berger, J. Mol. Spectrosc. 68, 317 (1977).

[4] G.L. Eesley, "Coherent Raman Spectroscopy", Oxford, Pergamon, 1981.

[5] A.B. Harvey (Ed.), "Chemical Applications of Nonlinear Raman Spectroscopy", New York, Academic Press, 1981.

[6] W. Kiefer and D.A. Long (Eds.), "Non-Linear Raman Spectroscopy and Its Chemical Applications", Dordrecht, Reidel, 1982.

[7] M.D. Levenson, "Introduction to Nonlinear Laser Spectroscopy", New York, Acad. Press, 1982, revised edition 1988.

[8] R.J.H. Clark and R.E. Hester (Eds.), "Advances in Spectroscopy", Vol. 15, Chichester, Wiley 1988.

[9] J.W. Nibler and G.V. Knighten, "Coherent Anti-Stokes Raman Spectroscopy", in "Raman Spectroscopy of Gases and Liquids", (A. Weber, Ed.), Topics in Current Physics, Vol. 11, Berlin, Springer 1979, p.253.

[10] P. Esherick and A. Owyoung, "High Resolution Stimulated Raman Spectroscopy", in "Advances in Infrared and Raman Spectroscopy", (R.J.H. Clark and R.E. Hester, Eds.), Vol. 9, London, Heyden-Wiley 1982, p. 130.

[11] S.Yu. Volkov, D.N. Kozlov, A.N. Prokhorov, V.V. Smirnov, and V.I. Fabelinskiy, "Coherent Anti-Stokes Spectroscopy of Molecular Gases", in P.P. Pashinin (Ed.), Proceedings of the Institute of General Physics, Acad. Sci. USSR, Vol. 2, Nova Science Publishers, Commack N.Y. 1987.

[12] H.W. Schrötter, H. Berger, J.P. Boquillon, B. Lavorel, and G. Millot, "High-Resolution Nonlinear Raman Spectroscopy of Gases", Croatica Chimica Acta 61, 605-621 (1988).

[13] H.W. Schrötter, H. Berger, and B. Lavorel, J. Mol. Struct. 141, 195 (1986).

[14] H.W. Schrötter, and B. Lavorel, Pure and Appl. Chem. 59, 1301 (1987).

[15] H.W. Schrötter, H. Berger, J.P. Boquillon, B. Lavorel, and G. Millot, "High-Resolution Nonlinear Raman Spectroscopy of Rovibrational Bands in Gases", in: "Progress in Molecular Spectroscopy" (R. Salzer, H. Kriegsmann, and G. Werner, Eds.), Teubner-Texte zur Physik, Bd. 20, Teubner, Leipzig 1988, p. 102.

[16] H.W. Schrötter and J.P. Boquillon, "State-of-the-Art in High Resolution CARS", in: "Raman Spectroscopy: Sixty Years On" (H.D. Bist , J.R. Durig, and J.F. Sullivan, Eds.), Vibrational Spectra and Structure, Vol. 17, Elsevier, Amsterdam 1989, p. 1.

[17] L.A. Rahn and R.L. Farrow, "High-Resolution Coherent Raman Spectroscopy of Combustion-Related Molecules", in: "Raman Spectroscopy: Sixty Years On" (H.D. Bist , J.R. Durig, and J.F. Sullivan, Eds.), Vibrational Spectra and Structure, Vol. 17, Elsevier, Amsterdam 1989, p. 33.

[18] D. Robert, "Collisional Effects on Raman Q-Branch Spectra at High Temperature", in: "Raman Spectroscopy: Sixty Years On" (H.D. Bist , J.R. Durig, and J.F. Sullivan, Eds.), Vibrational Spectra and Structure, Vol. 17, Elsevier, Amsterdam 1989, p. 57.

[19] H. Berger, B. Lavorel, and G. Millot, "High Resolution Stimulated Raman Spectroscopy of Gases", in: "Recent Trends in Raman Spectroscopy" (S.B. Banerjee and S.S. Jha, Eds.), World Scientific, Singapore 1989, p. 65.

[20] H.W. Schrötter and J.P. Boquillon, "High Resolution Coherent Anti-Stokes Raman Spectroscopy of Small Molecules", in: "Recent Trends in Raman Spectroscopy" (S.B. Banerjee and S.S. Jha, Eds.), World Scientific, Singapore 1989, p. 113.

[21] P.R. Régnier and J.P.E. Taran, Appl. Phys. Lett. 23, 240 (1973).

[22] J.P. Boquillon, J. Moret-Bailly, and R. Chaux, Comptes Rend. Acad. Sci. Paris, B284, 205 (1977).

[23] M.R. Aliev, D.N. Kozlov, and V.V. Smirnov, Pis'ma ZhETF 26, 31 (1977); JETP Lett. 26, 27 (1977).

[24] A.D. May, A.D., Henesian, M.A., and Byer, R.L., Can. J. Phys. 56, 248 (1978).

[25] W. Nitsch and W. Kiefer, Opt. Commun. 23, 240 (1977).

[26] J.J. Barrett and R.F. Begley, Appl. Phys. Lett. 27, 129 (1975).

[27] A. Hirth and K. Vollrath, Opt. Commun. 18, 213 (1976).

[28] V.I. Fabelinsky, B.B. Krynetsky, L.A. Kulevsky, V.A. Mishin, A.M. Prokhorov, A.D. Savel'ev, and V.V. Smirnov, Opt. Commun. 20, 389 (1977).

[29] A. Owyoung, C.W. Patterson, and R.S. McDowell, Chem. Phys. Lett. 59, 156 (1978); 61, 636 (1979).

[30] J.W. Nibler and G.A. Pubanz, "Coherent Raman Spectroscopy of Gases", Chapter 1 in Ref. 8, p. 1.

[31] D.A. Long, "Theory of Hyper Rayleigh and Hyper Raman Scattering", in Ref. 6, p. 165.

[32] K. Altmann and G. Strey, J. Raman Spectrosc. 12, 1 (1982).

[33] S.S. Jha, and J.W.F. Woo, Nuov. Cim. 2B, 167 (1971).

[34] P.D. Maker and R.W. Terhune, Phys. Rev. A137, 801 (1965).

[35] W. Kiefer, "Application of Classical Theory of CARS to Diatomic Molecules in the Gas Phase", in Ref. 6, p. 241.

[36] J. Laane and W. Kiefer, J. Raman Spectrosc. 9, 353 (1980).

[37] B. Lavorel, G. Millot, J. Bonamy, and D. Robert, Chem. Phys. 115, 69 (1987).

[38] H. Frunder, L. Matziol, H. Finsterhölzl, A. Beckmann, and H.W. Schrötter, J. Raman Spectrosc. 17, 143 (1986).

[39] H. Frunder, D. Illig, H. Finsterhölzl, H.W. Schrötter, B. Lavorel, G. Roussel, J.C. Hilico, J.P. Champion, G. Pierre, G. Poussigue, and E. Pascaud, Chem. Phys. Lett. 100, 110 (1983).

[40] H. Frunder, R. Angstl, D. Illig, H.W. Schrötter, L. Lechuga-Fossat, J.-M. Flaud, C. Camy-Peyret, and W.F. Murphy, Can. J. Phys. 63, 1189 (1985).

[41] R. Angstl, H. Finsterhölzl, H. Frunder, D. Illig, D. Papoušek, P. Pracna, K. Narahari Rao, H.W. Schrötter, and Š. Urban, J. Mol. Spectrosc. 114, 454 (1985).

[42] L. Lechuga-Fossat, J.-M. Flaud, C. Camy-Peyret, and J.W.C. Johns, Can. J. Phys. 62, 1889 (1984).

[43] W.F. Murphy, J. Raman Spectrosc. 11, 339 (1981).

Part III

Studies of Nonstationary Processes

Vibrational Relaxation of IR-Laser-Excited SF_6 and SiF_4 Molecules Studied by CARS

S.S. Alimpiev, A.A. Mokhnatyuk, S.M. Nikiforov, B.G. Sartakov, V.V. Smirnov, and V.I. Fabelinsky

General Physics Institute, Academy of Sciences,
Vavilov St. 38, SU-117942 Moscow, USSR

Abstract. The report provides a description of a series of experiments devoted to CARS spectroscopy of CO_2-laser-excited SF_6 and SiF_4 molecules. The pump-laser induced vibrational distribution and its temporal evolution has been studied. Special attention was paid to the investigation of the channels of vibrational energy redistribution among both low-lying discrete and high-lying quasicontinuum states and to the measurements of respective rates.

1. Introduction

The wide spectrum of scientific interest together with existing practical applications of energy exchange processes involving the molecular internal degrees of freedom indicate the need to develop the methods of investigations of the spectral and kinetic characteristics of the excited vibrational and rotational states of the molecules.

The method of CARS spectroscopy of vibrationally excited molecules opens up some additional opportunities for studying vibrational relaxation due to next basic advantages:

- The high spectral, temporal and spatial resolution of CARS-spectroscopy makes it possible to obtain detailed information on the molecular distribution functions among the vibrational and rotational levels.
- Using tunable dye lasers as the biharmonic pump source it is possible to easily cover the range of frequencies of the vibrational and rotational transitions.
- For Raman active transition the selection rules permit very narrow Q-branches, their width being smaller in comparison with the anharmonical shift for many molecules. This leads to significant simplification in identification of spectra obtained from excited vibrational states.
- The square dependence of CARS intensity from number density increases the dynamic range of the measured signals when the population changes.
- Moreover, the possibility of generating significantly nonequilibrium molecular distribution among the vibrational levels in an IR-laser field [1, 2] in a combination with CARS spectroscopy as a diagnostic technique makes it possible to study the processes of molecular excitation and relaxation.

In this report we present the results of the investigation of population distribution among vibrational levels of the SF_6 and SiF_4 molecules under excitation in the resonant CO_2-laser

Eds.: G. Marowsky and V.V. Smirnov

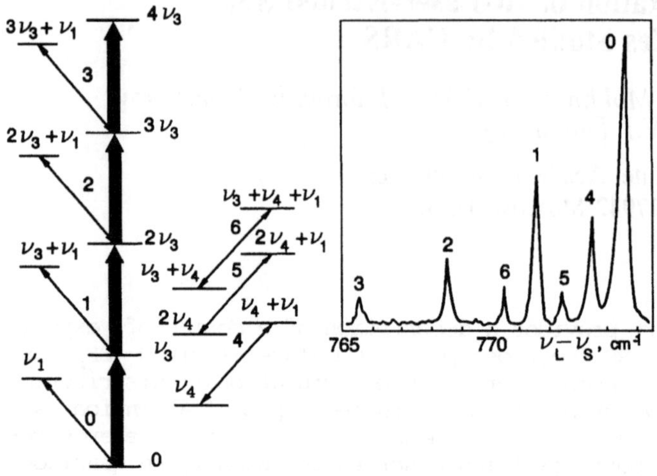

Figure 1. Transitions diagram of the experiment (a) and recorded CARS spectrum (b). Inclined arrows symbolize CARS transition. Peaks 0-3 correspond to the population of levels 0, ν_3, $2\nu_3$, $3\nu_3$ in the IR-field and are separated by the value of the anharmonism constants $x_{13} \approx 3$ cm^{-1}; peaks 4-6 correspond to the population of levels ν_4, $2\nu_4$, $\nu_3 + \nu_4$, and their spectral position is determined by the constant $x_{14} \approx 1.08$ cm^{-1}.

radiation field and the monitoring the transformation of this distributions and determination of the rates and primary channels of vibrational relaxation under collisions action.

In this study a pulsed CARS-spectrometer is used to probe the Raman-active ν_1 mode of SF_6 and SiF_4 molecules excited in a pulsed IR-laser field tuned to resonance with the IR-active ν_3 mode of these molecules (Figure 1, a). The vibrational excitation of molecules causes additional lines to appear in the CARS-spectrum (1-6 in Figure 1, b) related to scattering from the vibrational levels in the IR-laser field. The frequencies of the additional lines are shifted by $\Delta_i = x_{1i} v_i$ with respect to the frequency of the 0-ν_1 scattering line of the unexcited gas. The magnitude of the shift Δ is determined by the intermode anharmonism constant x_{1i} of the i-excited mode of the molecules with the probe mode ν_1 and the number of the excited v_i vibrational level of the i-mode. In turn the amplitudes of the scattering peaks are proportional to the squared populations of the vibrational levels. The values of the intermodal anharmonism constants x_{1i} of the SF_6 and SiF_4 molecules are rather well known. Values based on data from [3-6] are given in Table 1.

Measuring the frequency shifts and amplitudes of the additional scattering lines makes it possible to determine the distribution of energy adsorbed by the molecules in the IR-field among the vibrational levels. An investigation of the transformation of the CARS-spectra as the delay between the probe emission and the excitation IR-laser pulse increases reveals information on the kinetics and fundamental channels of collisional relaxation.

Table 1. Values of the intermodal anharmonism constants of the SF_6 and SiF_4 molecules, in cm^{-1}

x_{li}	SF_6 [3,4]	SiF_4 [5,6,7]
x_{11}	-0.9	-0.57
x_{12}	-2.44	-0.1 0.3
x_{13}	-2.93	-3.95
x_{14}	-1.02	+0.92
x_{15}	-1.15	-
x_{16}	-0.38	-

2. Experimental Arrangement

The experimental set-up consists of a pulsed CARS spectrometer (see fig. 2), line-tuned atmospheric pressure pulsed CO_2-laser and the gas cell. The CARS-spectrometer has a spectral resolution of 0.1 cm^{-1} with a continuous tuning range of ~ 200 cm^{-1} operating in a pulsed mode with the laser pulse duration 10 nsec and a repetition rate of 5-20 Hz. The CARS-probe pulses are synchronized with the excitation process, and the periodic state makes it possible to investigate the desired stage of the process repeatedly using accumulation and averaging of the recorded signal. The deviation of delay time between the CO_2-laser pulses and the CARS-spectrometer was $\pm$ 20 nsec which determined the temporal resolution of the unit.

In the experiments the molecules were excited by CO_2-laser pulses of standard shape with a peak duration of 100 nsec and a "tail" of 1 μsec, or by shortened pulses 40 nsec in duration in order to improve the temporal resolution. The standard technique based on optical breakdown in the constriction region of a

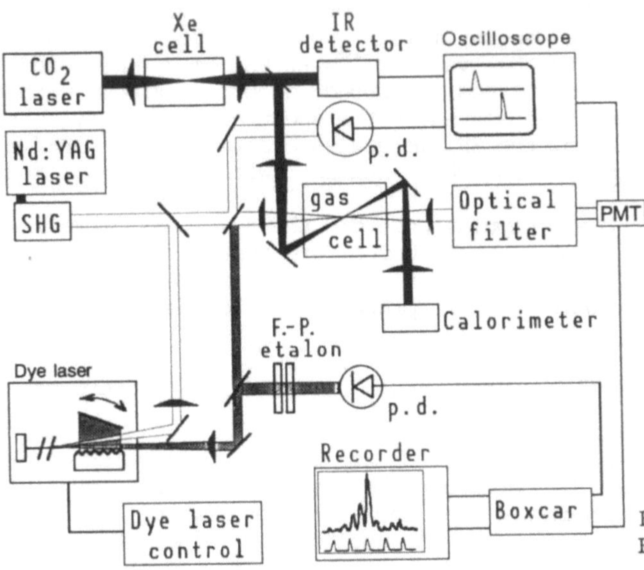

Figure 2. Experimental set-up.

short-focus telescope placed in line with the CO_2 laser beam was used to shorten the CO_2-laser pulse. The maximum energy of the shortened pulse was 15 millijoules. The CO_2-laser emission was focused by a lens with f = 1 m onto the cell containing the gas in a tube 20 mm in length and 5 mm in diameter. The area of the tube was nearly entirely filled with emission. The tube was mounted on a cold wire in the center of the cell and was cooled by liquid nitrogen to a temperature of $T \approx 150$ K. Cooling the gas to such temperatures eliminates the thermal population of the lower-lying vibrational levels, and as a result the overwhelming majority of molecules are excited in the IR-laser field from the ground vibrational state.

In order to eliminate the background CARS-signal caused by the four-wave process in air, the windows of the cell and the other optical elements, the beams of the CARS-spectrometer lasers were combined directly in the gas cell by means of a dichroic mirror.

The symmetry of the SF_6 and SiF_4 molecules (symmetry groups O_h and T_d) leads to the fact that only the Q-branch will be found in the CARS-spectrum of the ν_1 mode; the rotational structure of the Q-branches will not be resolved due to the small vibrational-rotational interaction constant of the ν_1 band (α =1.1 . 10^{-4} cm^{-1} for SF_6 [4] and 1.4 . 10^{-4} cm^{-1} for SiF_4 [6, 7].

CARS-spectrum of excited SF_6 molecules obtained at a CO_2-laser pulse energy density of Φ= 0.3 joules/cm^2 and a probe signal delay of τ_D = 50 nsec is shown in Figure 3, a. The lines corresponding to the scattering Q-branches at the $v\nu_3 \rightarrow v\nu_3 + \nu_1$ (v= 1-4) transitions and shifted by vx_{13} with respect to the frequency of the fundamental peak $0 - \nu_1$ (774.5 cm^{-1}) are clearly visible in the spectrum. The value of the anharmonism constant x_{13} determined from the spectrum is x_{13} = -2.99 ± 0.03 cm^{-1} and is in good agreement with the value x_{13}= -2.93 cm^{-1} measured in [4]. Aside from these lines a broad smooth maximum with an unresolved structure is observed in the spectra; this is evidently related to scattering by the SF_6 molecules excited in the IR-field to the upper-lying vibrational states with a mixed modal composition.

To determine the populations of the excited vibrational levels of the ν_3 mode we used a procedure that is based on the fact that the amplitudes of the scattering peaks A_v (v = 0-4) at the $v\nu_3 \rightarrow v\nu_3 + \nu_1$ transitions are proportional to the squared total populations N_v of the levels $v\nu_3$: $A_v \sim N_v^2 = n_v^2\ g_v^2$, where $g_v = (v + 1)(v + 2)/2$ are the degeneration factors of the vibrational levels of the triply degenerate mode ν_3, while $n_v = N_v/g_v$ is the population of the nondegenerate state. This procedure is based on the fact that the $v\nu_3$ levels are populated in the IR-laser field, while the population of the $v\nu_3+\nu_1$ levels is negligible, since these states are not in the resonance with IR pumping field and they will be

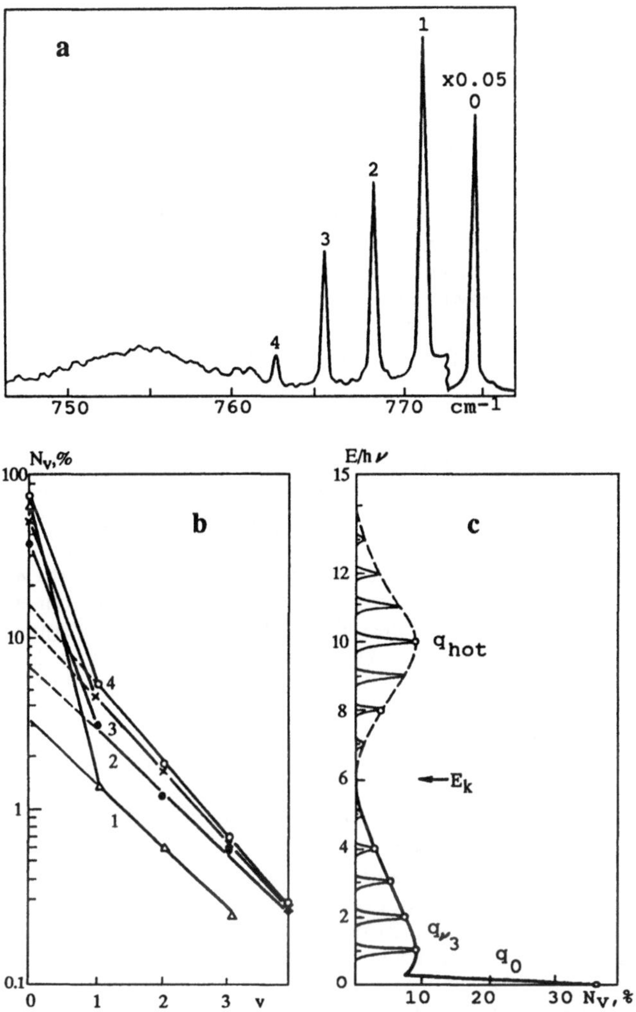

Figure 3. The CARS-spectrum of SF_6 molecules excited in the IR-field (a); the initial population distributions of the states of the ν_3 mode (v =0-4) at various IR-field energy densities and frequencies (b); the distribution function of SF_6 molecules among the vibrational levels and the breakdown of the molecules into three ensembles (c) (pressure = 0.1 torr, T = 150 K).

a - ν = 947.7 cm^{-1}; peaks 0-4 are related to the population of levels 0, ν_3, $2\nu_3$, $3\nu_3$, $4\nu_3$; b - 1 - Φ = 0.3 joules/cm^2, ν = 944.2 cm^{-1}; 2-4 - ν = 947.7 cm^{-1}, Φ = 0.3, 0.15 and 0.1 joules/cm^2; c - ν = 947.7 cm^{-1}.

populated only over extended times from excitation due to the complete thermalization of the vibrational molecular energy by collisions.

3. Laser Induced Vibrational Population Distribution

With the procedure outlined above we determined the populations of the vibrational levels immediately after IR-excitation (see Fig.3), and we drafted the dependencies of level populations on their number or energy $E = vh\nu$. Such a dependence for the spectrum shown in Figure 3,a is provided in Figure 3,b (curve 2). Since the spectrum is measured at an SF_6 pressure of $p = 0.1$ torr and a probe delay of $\tau_D = 50$ nsec ($p\tau_D = 5$ nsec . torr) we may ignore the influence of collisional processes with characteristic times $p > p_D = 5$ nsec . torr on the distribution form.

Similar population distributions of the $v\nu_3$ levels ($v = 0,1,\ldots$) were obtained from the CARS-spectra measured at various excitation energy densities Φ=0.1-0.3 joules/cm^2 for two different IR-field frequencies: 947.7 cm^{-1} (10P16) and 944.2 cm^{-1} (10P20). These results allowed to carry out a quantitative decomposition of the molecules after excitation into three ensembles (Figure 3,c) since, as we see from the curves in Figure 3,b, the molecular distribution among the excited levels $v = 1$-4 of the ν_3 mode is near a Boltzmann distribution. One exception is the population of the ground state which significantly exceeds the Boltzmann value for "cold" ensemble ($q_0 = n_0 - n_0^{Boltz}$) that may be obtained by extrapolation of the curves in Figure 3, to a value $v = 0$.

The second ensemble of molecules called the "warm" ensemble consistent with the terminology used here consists of molecules that have populated the discrete levels of the resonance adsorbing mode ν_3 in the IR-laser field. The portion of these molecules q_{ν_3} may easily be determined from the distribution in Figure 3,b, by simple summation of the populations of the levels $v = 0, 1, 2, \ldots$ of the mode accounting for their degeneration factor g_v : $q_{\nu_3} = \sum_{v=0} n_v g_v$. Moreover the average energy constant v_{ν_3} in this molecular ensemble is also of interest; this may be determined from the slope of the curves in Figure 3,b, or by simple summation $v_{\nu_3} = \sum_{v=0} v n_v g_v q$. The values of q_o, q_{ν_3} and v_{ν_3} determined from the distributions in Figure 3,b, for the various excitation levels of SF_6 (Φ_{exc}, ν_{exc}) are given in Table 2. It is clear from an analysis of these values that at low IR-excitation levels ($\Phi_{exc} < 0.1$ joules/cm^2) the total portion of molecules in the two ensembles $q_o + q_{\nu_3} = 1$ is close to unity, i.e., all molecules are concentrated on the discrete levels of the resonance absorbing mode and excitation is mode-selective. However when Φ_{exc} is increased the quantity $q_o + q_{\nu_3}$ begins to decay monotonically which, obviously, is related to the excitation of the molecules to the vibrational quasicontinuum and the formation of the so-called "hot" molecular ensemble [8]. The portion of molecules in this ensemble q_{hot} may be determined from the relation

Table 2. Values of q_o, q_{ν_3} and $\bar{v}_{\nu_3}$ for various excitation conditions.

Φ [J/cm^2]	q_o[%]	q_{ν_3}[%]	q_{hot}[%]	$\bar{v}_{\nu_3}$, [quanta/ molecule]	$\bar{v}_{hot}$ [quanta/ molecule]	$\bar{v}$ [quanta/ molecule] [10]
			ν=947.7 cm^{-1}			
0.06	50	50	0	1.5	-	0.8
0.1	45	50	5	1.6	-	1.4
0.15	40	40	20	1.7	8	2.3
0.3	30	35	35	2.2	9	4
			ν=944.2 cm^{-1}			
0.3	55	15	30	2.2	19	6

Note. The values of q_o and q_{ν_3} are determined accurate to $\pm 5\%$.

$$q_o + q_{\nu_3} + q_{hot} = 1 . \quad (1)$$

Moreover, we may determine the average excitation level $\bar{v}_{hot}$ of this ensemble of molecules from the energy balance

$$0q_o + \bar{v}_{\nu_3}q \quad + v_{hot}q_{hot} = \bar{v} , \quad (2)$$

where $\bar{v}$ is the average energy adsorbed by the molecules measured in a number of studies [2, 9, 10] on multiquantum adsorption of SF_6 in an IR-laser field.

4. Investigation of Resonance V-V Exchange and Interisotope Vibrational Exchange of Excited SF_6 Molecules

The collisional relaxation of vibrational energy plays an important role in research on the excitation of the high-lying vibrational levels of polyatomic molecules. The vibrational relaxation of polyatomic molecules is generally broken up into the relatively fast process of v-v relaxation, i.e., the relaxation within the individual modes, and the v-v' relaxation to an equilibrium between different vibrational degrees of freedom of the molecule. While v-v' relaxation of polyatomic molecules has been studied quite thoroughly during both single-photon and many-photon vibrational excitation, we do not yet have a clear picture of the rates of the resonant intramode v-v exchange in polyatomic molecules. The reason for this situation is the difficulty in directly observing a redistribution of the populations of the vibrational levels of a mode by the conventional methods.

We investigated the transformation of the CARS-spectra of the excited molecules with an increase in the delay τ_D of the probe pulse with respect to the IR-laser pulse with various pressures of the resonance absorption gas and when buffer gases

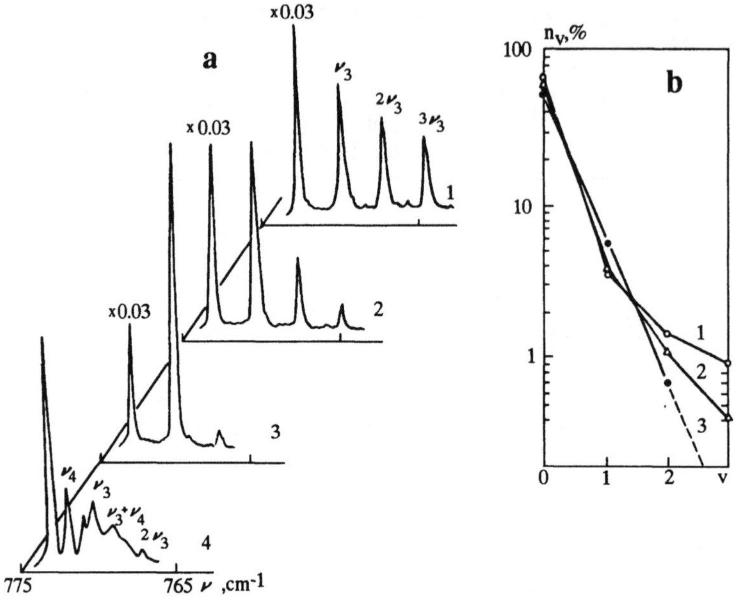

Figure 4. Transformation of the CARS-spectra (a) and SF_6 molecular distribution among the vibrational states of the ν_3 mode (b) under collisional action with an increase in the delay of the probe pulse (P_{SF_6} = 0.05 torr).
τ_D: 1 - 50 nsec; 2 - 250 nsec; 3 - 1 μsec; 4 - 10 μsec.

were added. We showed in the preceding section that a reduction in the energy density of the IR-laser field significantly simplified the form of the initial distribution function due to the rapid drop in the portion of molecules q_{hot} excited to the vibrational quasicontinuum. Hence the majority of relaxational measurements given below were performed at lower energy densities of the IR-field where the contribution of the "hot" ensemble molecules to the relaxational processes may be ignored. Fig.4, a gives the CARS-spectra of excited molecules with various probe signal delays. The equidistant peaks observed in the spectrum correspond to the scattering Q-branches from the ground state and the three excited states of the ν_3 mode. The molecular distributions among the levels of the ν_3 mode obtained from these spectra are shown in Figure 4,b. An analysis of the curves shows that immediately after the IR-laser pulse is applied ($p\tau_D$ = 5 nsec . torr) the molecular distribution among the levels of the ν_3 mode differs from the Boltzmann distribution primarily due to the existence of the "cold" molecular ensemble. However a distribution is established due to collision action by the probe time τ_D = 500 nsec ($p\tau_D$ = 50 nsec . torr) that is accurately approximated by a Boltzmann distribution (including the ground state) with a modal energy content of $\langle v_{\nu_3} \rangle \approx$ 0.55 quanta/molecule.

We derived characteristic relaxation times dependent on the number of the excited level from an analysis of the approximation of the populations of the levels to the values corresponding to a Boltzmann distribution. Analogous measurements carried out for three different SF_6 pressures (p = 0.05; 0.1; 0.2 torr) and with xenon buffer gas have revealed that while preserving the form of the initial (immediately following excitation) distribution function the characteristic times to equilibrium populations are reduced in proportion to the SF_6 pressure and are independent of the Xe pressure. The characteristic times for the populations to assume quasistationary values adjusted for SF_6 pressure were: 30 ± 5 nsec . torr for the ν_3 level, 17 ± 5 nsec . torr for the $2\nu_3$ level and 10 ± 5 nsec . torr for the $3\nu_3$ level.

Also fast relaxation process was observed in the measurements of the characteristic time of vibrational energy exchange in collisional of the $^{32}SF_6$ and $^{34}SF_6$ molecules. In these experiments the IR-laser field was applied to a specially prepared mixture of $^{32}SF_6$-$^{34}SF_6$ molecules in a 1:4 ratio. The $^{32}SF_6$ molecules were subjected to resonance excitation at the IR-field frequency of 947.7 cm^{-1} while the $^{34}SF_6$ molecules remained unexcited as a result of the significant detuning of the resonance frequency due to the isotopic shift of the mode ν_3 ($\Delta_3 \approx$ 17.5 cm^{-1}), and functioned as an artificially-created "cold" ensemble. The spectral resolution of the CARS-assembly used in this case made it impossible to directly differentiate the intraisotope and interisotope v-v exchange processes, since the isotopic frequency shift of the Raman-active mode Δ_1 is only 0.057 cm^{-1} [5]. Hence in order to investigate the interisotope vibrational relaxation process we employed an isotope mixture with a high percent concentration of the $^{34}SF_6$ component that is not excited in the laser field. Obviously in this case the primary contribution to the relaxational process comes from inter-isotope vibrational energy exchange. Figure 5 provides the population distributions for the ν_3 mode taken from the CARS-spectra at different delay times τ_D . Immediately after the excitation of the IR-laser pulse a significant fraction of the molecules (~80%) remain in the ground state. This fraction nearly coincides with the proportion of $^{34}SF_6$ nonresonant molecules. As the delay increases the collisions cause a decrease in the population of the ground state and increase of the $^{34}SF_6$ the population of the first vibrational level. This process is the result of collisional transfer of vibrational energy from the $^{32}SF_6$ molecules excited in the laser field to the $^{34}SF_6$ unexcited molecules.

As we see from Figure 5 at a total mixture pressure of 0.1 torr a near-Boltzmann distribution is established by time τ_D = 1 μs with an energy content in the ν_3 mode of about 0.25 quanta/mol. We note that when exciting the $^{32}SF_6$ monoisotope gas in the same conditions the energy content in the ν_3 mode is $\bar{V}_{\nu_3} \approx$ 1.2 quanta/ mol. It is obvious that the lower temperature of the mode in the isotope mixture is due to the transfer of vibrational energy to the $^{34}SF_6$ nonresonant molecules whose number significantly exceeds the number of excited molecules. The characteristic time of this transfer process was determined from an analysis in the ap-

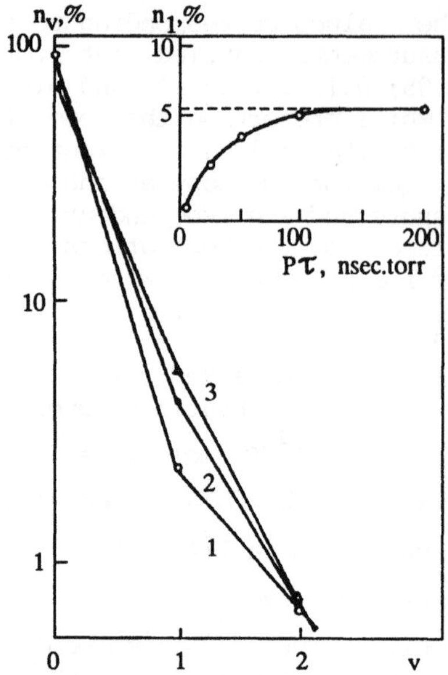

Figure 5. Transformation of the population distribution of the vibrational levels of an isotopic mixture of $^{32}SF_6$-$^{34}SF_6$ = 1 : 4 molecules (P = 0.1 torr) under excitation in an IR-field $^{32}SF_6$ (ν = 947.7 cm^{-1}, Φ = 0.3 joules/cm^2).
τ_D: 1 - 50 nsec; 2 - 500 nsec; 3 - 2 μsec.

proximation of the population n_1^{st} with an increase in τ_D for the two different SF_6 pressures: p = 0.1 and 0.2 torr (Figure 5,b). The dependence of (n_1 - n_1^{st}) on p τ_D is accurately approximated by an exponential law with a "characteristic time" of pτ = 50 nsec.torr.

The measurements show that in spite of the significant isotope frequency shift of the ν_3 mode excited by the IR-laser, the rate of vibrational exchange between the molecules of the various isotope compositions is comparable to the rate of resonance v-v-exchange in the SF_6 monoisotope gas. This result is important from the viewpoint of establishing the nature of the interaction potential of the molecules in collisions leading to v-v-exchange.

In analyzing these results we note above all that even in the simplest model of the v-v-relaxation of a single-dimensional harmonic oscillator in order to calculate the cross-section of the elementary act σ_{01}^{10} of resonance v-v-exchange it is necessary in accounting for the initial molecular distribution among the vibrational states, to solve a system of kinetic equations for the populations of a large number of levels.

The highly specific nature of v-v-exchange on polyatomic molecules is related to the degeneration of the vibrational levels. Thus, due to the triple degeneration of the ν_3 mode of the SF_6 molecule the degeneration factor g_v of the vibrational levels increases with an increase in their number v as g_v = (v

+ 1)(v + 2)/2. This causes a rapid increase in the number of possible v-v-exchange channels. It is obvious here that from the viewpoint of analyzing the nature of the intermolecular interaction potential producing v-v-exchange, the determinant factor is the partial cross-section of a single channel, at the same time that the rate of change in the population of the vibrational levels observed in experiment is determined by the total contribution of all allowed v-v-exchange channels.

In order to estimate this cross-section it is necessary to describe in detail the v-v-exchange process and to bear in mind the collisional transitions between all sublevels, i.e., account for the multichannel nature of v-v-exchange related to the degeneration of the ν_3 mode. Study [3] derived a solution of kinetic equations for the total populations of the energy levels of the ν_3 mode within the scope of a model of the single-quantum v-v-exchange of a harmonic oscillator accounting for the

$$N_v(t) = \sum_{k=0}^{\infty} B_{vk} \exp\left(-t \frac{K}{\tau_{vv}}\right), \qquad (3)$$

where $N_v(t)$ is the population of the E_v level; $\tau_{vv} \approx (3nK_{01}^{10})^{-1}$ is the characteristic time of the v-v-relaxation; N is the molecular concentration per unit of volume; K_{01}^{10} is the rate constants of v-v-exchange of a single channel at the 0-1 transition.

The values of B_{vk} are expressed through the initial population values (we note that $B_{v1} \equiv 0$):

$$B_{vk} = (1-\beta)^3 L(v,k,3,\beta) \sum_{v=0}^{\infty} N_v(0)\, L(k,v,3,\beta); \qquad (4)$$

here $\beta = v/(v + 3)$; v is the average quantum content in mode ν_3 calculated per single molecule;

$$L(v,k,3,\beta) = C_{k+2}^{2} F(-k, v + 3, 3, 1 - \beta);$$

F is the hypergeometric function. Modeling the time dependences of the populations of the vibrational levels observed in the experiment $N_v(t)$ using formula (4) made it possible to determine the value of $K_{01}^{10} \approx 5.10^{-11}$ $\sec^{-1}.\text{cm}^3$ for the ν_3 mode of the SF_6 molecule and then to estimate the characteristic exchange cross-section of a single quantum through one channel at the 0-1 transition:

$$\sigma_{01}^{10} = \sqrt{\frac{\pi m}{16kT}}\, K_{01}^{10} \approx 24\ \text{Å}^2.$$

The fact that the magnitude of this cross-section is less than the cross-section of gas kinetic collisions of SF_6 ($\sigma_{gk} \approx 70$ Å^2), leads to the conclusion that the experimentally observed high rates of change in the populations of the vibrational levels (10-50 nsec.torr with a free path time $p\tau_{gk} \approx 180$ nsec.torr) are related both to the multichannel nature of v-v-exchange for the triply degenerate mode and to the significant deviation of the initial distribution function in the IR-laser field from a Boltzmann distribution. We note that the relatively low value of σ_{01}^{10} makes it necessary to assume that the v-v-exchange cross-section is determined primarily by the short-range part of the intermolecular interaction potential.

Indeed, the experimentally-observed weak sensitivity of the v-v-exchange rate in SF_6 to the resonance defect ($\Delta_3 \approx 17.5$ cm^{-1}) with energy exchange between molecules with a different isotope composition is a serious argument against the long-term nature of the potential [4]. Thus, the estimate of the dipole-dipole interaction value $V \approx d^2/R^3$ of the two SF_6 molecules with vibrational momenta $d \approx 0.3$D [5] shows that it is possible to compensate the resonance defect at 17.5 cm^{-1} with molecular transits over a distance of 3 Å, i.e., less than their geometric dimensions and, consequently, beyond the applicability of the dipole-dipole approximation. Simple estimates show that within the scope of the dipole-dipole approximation this resonance defect cannot be compensated by transit time "broadening".

5. Collisional Vibrational Energy Exchange with Molecules Excited to the Quasicontinuum Range

As noted above, the high spectral resolution of the CARS-probe method makes, it possible to identify the fundamental channels and investigate the process of vibrational energy transfer between the various modes of a polyatomic molecule.

From the transformation of the CARS spectra (Fig.6,a and b) recorded for SF_6 and SiF_4 molecules when the probing is delayed $\tau > 500$ ns with respect to the excitation (at a gas pressure of 0.3 torr and a temperature T=150 K) we see that in addition to the populating of the $1\nu_3$, $2\nu_3$, and $3\nu_3$ states in the field, there are some other lines, which result from the populating of discrete states of the ν_4 mode in the course of collisions.

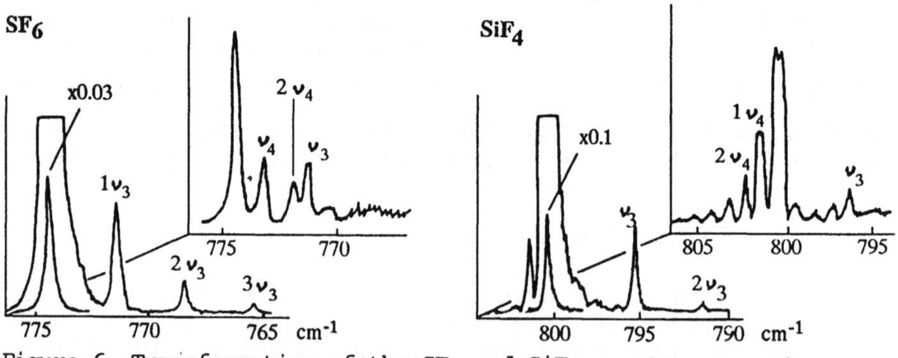

Figure 6. Transformation of the SF_6 and SiF_4 spectra over time.
SF_6: τexc=40 ns; Φexc=0.3 J/cm^2, νexc= 944.2 cm^{-1}
SiF_4: τexc=40 ns; Φexc=1.2 J/cm^2, νexc=1024 cm^{-1}.

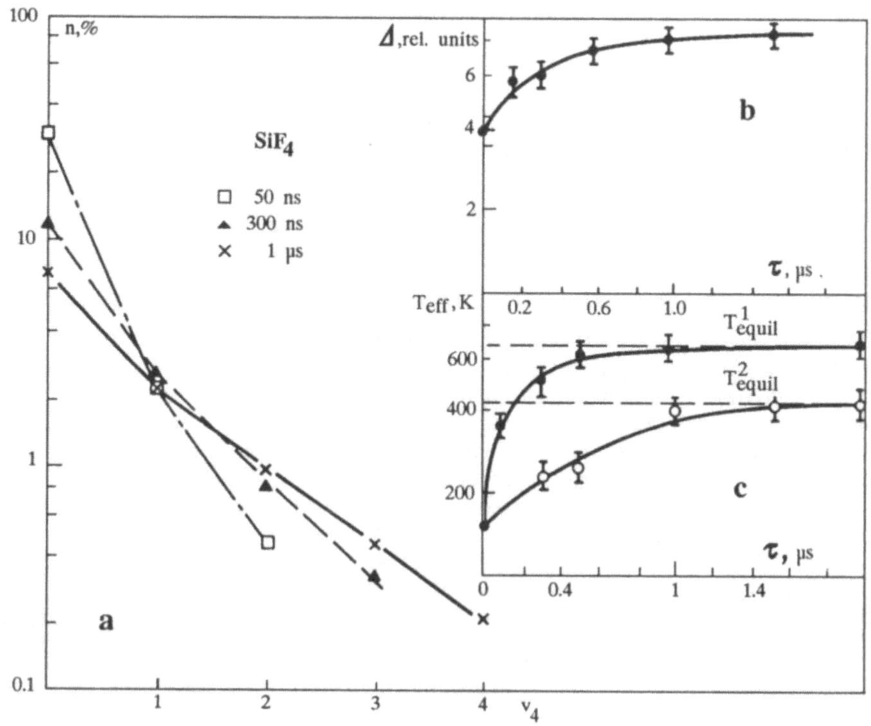

Figure 7. a: Distribution of the populations of the sublevels of the ν_4 mode (Φexc=1.2 J/cm^2); b. Time evolution of the half width of the fundamental transition band Δ; c. Time evolution of the temperature T_{eff} of the ν_4 mode. T^1_{equil} corresponds to Φ exc=1.2 J/cm^2, T^2_{equil} - to Φexc=0.4 J/cm^2.

The distribution of the populations of the $1\nu_4$, $2\nu_4$, and $3\nu_4$ levels reconstructed from the spectra is a near-Boltzmann distribution with a vibrational temperature determined by the intensity of the exciting field and the delay time τ.

Analysis of the time evolution of the temperature of the ν_4 mode and of its approach to a quasisteady value (Fig.7 for SiF_4 molecule) shows that an increase in the energy density of the IR field from 0.4 to 1.2 J/cm^2 (ν_{exc} = 1028 cm^{-1}) leads to an increase in the rate population of the vibrational states of the ν_4 mode $5.0\pm0.1 \times 10^6$ s^{-1} Torr^{-1} to $(1.2\pm0.4 \times 10^7$ s^{-1} Torr^{-1} in the case of the SiF_4 molecules, while for the SF_6 molecules an increase in the energy density to 0.4 J/cm^2 (ν_{exc} =944.4 cm^{-1}) leads to an increase in this rate to $(1.0\pm0.3 \times 10^7$s^{-1} Torr^{-1} (the characteristic rate of gas-kinetics collisions).

The broadening of the $0-\nu_1$ transition lines found experimentally is evidence that energy is also transferred to discrete levels of low-frequency modes, ν_6 in the SF_6 molecule and ν_2 in the SiF_4 molecule. The typical rates involved here are those found for the ν_4 mode. The reason for the observed rapid collisional population of the vibrational states of the ν_4 and ν_6 modes of SF_6 and the ν_4 and ν_2 modes of SiF_4 may be either a resonance of the states ν_3 and $\nu_4 + \nu_6$ in SF_6 and of ν_3 and $2\nu_4 + \nu_2$ in SiF_4 (the energy defect between these states is 14 cm^{-1} in SF_6 and 10 cm^{-1} in SiF_4) or a vibrational exchange between molecules excited into a quasicontinuun with molecules in low-lying discrete vibrational states. A characteristic feature of this exchange is its resonant nature, i.e., the ability of molecules in randomized states of the quasicontinuum to undergo a collisional exchange of a vibrational quantum of any mode. In an effort to test these suggestions, we selected excitation conditions for SF_6 molecules which resulted in an effective population of states of the quasicontinuum, while the populations at the discrete levels of the excited mode remained below the sensitivity of the apparatus. The CARS spectra recorded at a gas pressure of 0.1 Torr and a delay time τ_D = 300 ns are evidence that there is no significant population of the low-lying discrete vibrational states. When gaseous $^{32}SF_6$ in a 1:10 mixture with $^{34}SF_6$ was excited under similar conditions, at a total pressure of 0.1 Torr, with the same delay times, the CARS spectra clearly reveal lines responsible for transitions from low-lying excited vibrational states of the ν_3 and ν_4 modes. This fact is convincing evidence of an effective vibrational exchange between the deliberately prepared ensembles of $^{32}SF_6$ molecules in highly excited vibrational states and unexcited $^{34}SF_6$ molecules in the vibrational ground state.

The behavior of the CARS spectra over long delay times $1 < \tau_D < 10$ µs reflects a redistribution of the adsorbed energy in the system of vibrational states of all the modes of the molecules. Since there are many such states, the lines corresponding to transitions from them overlap markedly and form a structureless spectrum with a maximum shifted from the line of the $0-\nu_1$ transition by an amount equal to an anharmonic frequency deviation determined by the amount of energy in the vibrational degrees of freedom., The transformation of this spectrum over time and its arrival at a steady shape in a time 6 µs (Fig.8) make it possible to determine the relaxation rate of a common vibrational temperature in the system of vibrational states of the modes: $2.5 \times 10^6 s^{-1}$ $Torr^{-1}$.

This study yields a picture of the collisional redistribution of vibrational energy of polyatomic molecules excited into the vibrational quasicontinuum. A distinctive feature of this process and of the establishment of a common vibrational temperature of

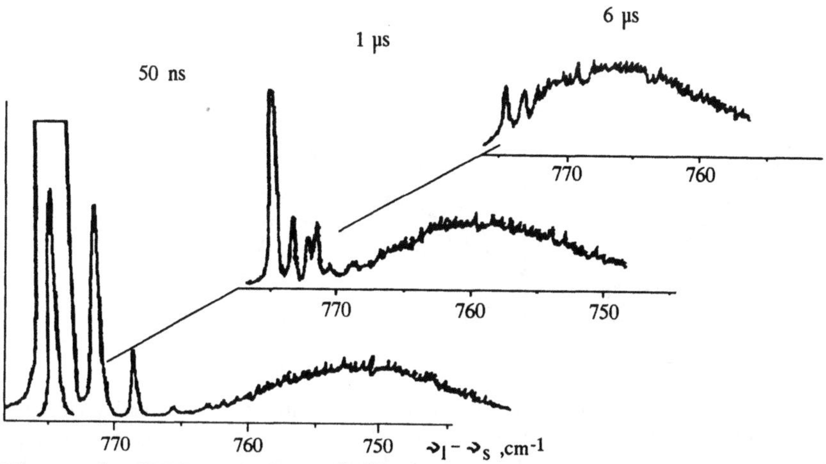

Figure 8. $CARS_2$ spectra of SF_6.
Φexc=0.4 J/cm^2, νexc=944.2 cm^{-1}, τexc=1 µs, P_{SF_6}=0.3 torr.

the molecules is the high rate of collisional energy transfer, which is a consequence of the resonant nature of the collisional exchange of the quanta of arbitrary modes along a chain: (molecule in a discrete state of mode i) - (molecule in the quasicontinuum) - (molecule in a discrete state of mode j). We observe a high rate of relaxation to a complete vibrational equilibrium ($2.5x10^6$ $s^{-1}.Torr^{-1}$) in the presence of leading channels of vibrational exchange (the ν_3 and ν_4 modes).

The results given above show that analysis of the CARS-spectra of the excited SF_6 molecules makes it possible to determine the form of the vibrational level distribution of the mode-selectively excited SF_6 molecules, where the portion of the molecules q_{ν_3} and the average energy content $\bar{v}_{\nu_3}$ in the resonance-adsorbing mode are highly dependent on the energy density Φ of IR-field pulse. It is obvious that an investigation of the q_{ν_3} and v_{ν_3} and relations with an increase in energy density Φ and,consequently, in increase in the average molecular excitation level $\bar{v}$, makes it possible to determine the energy boundary of loss of modal selectivity of excitation and, therefore, the quasicontinuum boundary of the SF_6 molecule.

The experiments described below were carried out by exciting SF_6 molecules with a standard IR-laser pulse. This made it possible to operate at the higher excitation energies of the SF_6 molecule by using the higher energy content of the unshortened pulse. Moreover, by exciting the molecules with the "long" pulse the collisional v-v-exchange process during pulse action causes the distribution of the molecules among the vibrational levels of the ν_3 mode to assume a Boltzmann form, which simplifies the interpretation of results.

Figure 9,a, gives the distribution of the SF_6 molecules among the levels of the ν_3 mode for various energy densities of the

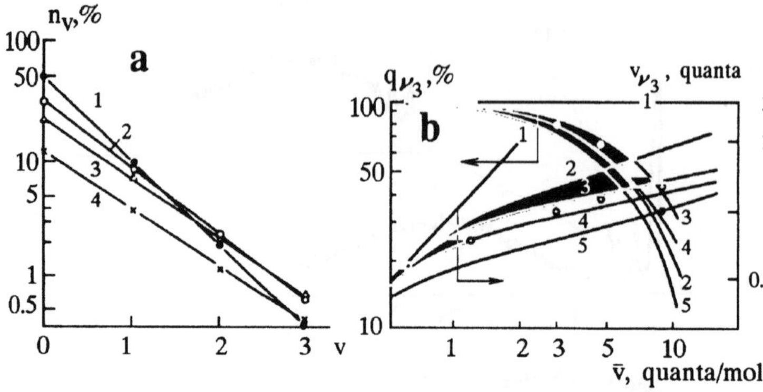

Figure 9. The distribution of SF_6 molecules among the levels of the ν_3 mode under excitation by a "long" IR-pulse of various energy densities (a) together with the percentage of molecules q_{ν_3} (solid circles) and the average vibrational energy content $\bar{v}_{\nu_3}$ in the ν_3 mode (open circles) obtained from the distributions in Figure 6,a, plotted as a function of the total energy adsorbed by the molecules $\bar{v}$ (Φ) [8] and the calculated q_{ν_3} ($\bar{v}$) and $\bar{v}_{\nu_3}$ ($\bar{v}$) curves for the various values of the adjustment parameters: E_k is the energy boundary of the quasicontinuum and δ is the portion of energy in the ν_3 mode in the quasicontinuum region $E > E_k$ (ν = 947.7 cm^{-1}, P_{SF_6} = 0.2 torr, τ_D = 500 nsec).
a: 1 - Φ = 0.04; 2 - 0.1; 3 - 0.5; 4 - 1.2 joules/cm^2;
b: 1 - $E_k = \infty$, δ = 1; 2 - E_k = 6hν, δ = 0.2; 3 - E_k = 7hν, δ = 0.1; 4 - E_k = 6hν, δ= 0.1; 5 - E_k = 5hν, δ = 0.1.

IR-field Φ = 0.04-1.2 joules/cm^2 and at a fixed probe delay τ = = 500 nsec equal to the total duration of the IR-pulse, obtained from the CARS-spectra. It is clear from these curves that by exciting the molecules with a "long" pulse and at an SF_6 gas pressure of P = 0.2 torr there is no "cold" molecular ensemble by the probe time ($P\tau_D$ = 100 nsec.torr), and their distribution among the levels of the mode may be accurately described by a Boltzmann distribution with a mode temperature dependent on the energy density of the IR-laser field. The average energy content $\bar{v}_{\nu_3}$ may be determined from the slope of the curves while summation of the populations of the levels yielded the value of the portion of molecules q_{ν_3} in the mode.

The derived values of $\bar{v}_{\nu_3}$, q_{ν_3}, plotted as a function of the total average adsorbed energy $\bar{v}$ from independent experiments [11, 12] are shown in Figure 9,b. It is clear that in the case of mode-selective excitation the portion of molecules in the mode will be conserved and will be q_{ν_3} = 1, while the average energy content will increase monotonically with growth of Φ and will be equal to the average molecular adsorbed energy $\bar{v}$ (the curves in Figure 9,b). However, as we see from Figure 9,b, an increase in Φ causes a rapid drop in the portion of molecules q_{ν_3} in the resonance adsorbing mode, while the increase in the average energy content $\bar{v}_{\nu_3}$ in the mode lags clearly behind the increase in total adsorbed energy $\bar{v}$.

Such behavior of the $q_{\nu_3}(\bar{v})$, $\bar{v}_{\nu_3}(\bar{v})$ relations, obviously, is related to the fact that beginning at a certain critical value of the vibrational energy $E_k = v_k h\nu$ (see Figure 3,c), a quasicontinuum of the vibrational states is formed. These states do not have a specific modal composition. The state density in the energy spectrum of the quasicontinuum ($E > E_k$) is significantly greater than the state density of the ν_3 mode in the $E < E_k$ range and hence with a sufficiently large energy content ($\bar{v} > E_k/h\nu$) the majority of molecules will be concentrated in the quasicontinuum, and as a result the portion of molecules q_{ν_3} in the ν_3 mode and the level of quanta $\bar{v}_{\nu_3}$ will be less than 1 and $\bar{v}$, respectively.

The following simple model based on the derived quantitative data makes it possible to determine the value of E_k. We will assume that the v-v-quantal exchange of the ν_3 mode occurs earlier ($P\tau_{\nu_3} \approx 30$ nsec.torr) than the mode-nonselective energy transfer from the molecules in the quasicontinuum to the molecules at discrete levels ($P\tau \approx 100$-700 nsec.torr). Further we will assume that due to the quantal exchange of the ν_3 mode a quasi-equilibrium distribution is established from which the values of q_{ν_3} and $\bar{v}_{\nu_3}$ are determined. The feature of this distribution is that the energy levels populated by the IR-field $E_v = vh\nu$ ($h\nu$ is the IR-field quantum, $\nu \approx \nu_3$) beginning at energy E_k no longer have a specific modal composition, i.e., the energy E_v when $E_v > E_k$ no longer is concentrated entirely in the ν_3 mode and only a portion of this energy E_v is in the ν_3 mode, while the rest of the energy - $(1-\delta)E$ - is distributed among the remaining modes of the molecule. The quasi-stationary population distribution $N(E)$ is determined by equality of the output and downward flows: $E_v \rightleftarrows E_{v+1}$, with quantal exchange of the ν_3 mode. Using the results from [13] to calculate the v-v-exchange rates in polyatomic molecules we obtain

$$\frac{N(E_v)}{N(E_{v+1})} = \frac{E_{\nu_3}(E_{v+1})}{3h_3 + E_{\nu_3}(E_v)} \exp\frac{h\nu_3}{kT_3}, \qquad (5)$$

where $h\nu_3/kT_3$ is the Boltzmann factor of the $_3$ mode; $E_{\nu_3}(E_v)$ is the energy concentrated in the ν_3 mode on the $E_v = vh\nu_3$ energy level (when $E_v < E_k$ $E_{\nu_3}(E_v) \equiv E_v$, when $E_v > E_k$ $E_{\nu_3}(E_v) = \delta E_v$). Expression (5) determines through E_k, and the Boltzmann factor $\exp(h\nu_3/kT_3)$ the form of the $N(E_v)$ distribution. The quantities E_k and δ are the adjustment parameters of the model, and the value of $\exp(h\nu_3/kT_3)$ is known from experiment. Using relation (5) we were able to calculate the $N(E)$ distribution and to obtain the dependencies of q_{ν_3} and $\bar{v}_{\nu_3}$ on $\bar{v}$ for select values of the adjustment parameters E_k and in Figure 9,b (curves 1-5). We see from Figure 9,b that for SF_6 the best agreement with experiment (curves 3, 4) is achieved when $6 < E_k/(h\nu_3) < 7$, $\delta = 0.1$ and, consequently, the quasicontinuum boundary of the molecule lies in the $6h\nu_3 - 7h\nu_3$ energy range. The best correlation between the experimental points and the calculated curves with a parameter value $\delta = 0.1 \ll 1$ indicates that in the $E > E_k$ range there is a redistribution of vibrational energy among the many vibrational degrees of freedom and only a small portion of the vibrational energy E_v stored by the molecules is concentrated in the ν_3 mode due to its high frequency ($\nu_3 > \nu_1$, ν_2, ν_4, ν_5, ν_6).

Thus, the derived results make it possible not only to determine the energy position of the boundary E_k of the loss of modal selectivity of SF_6 molecular excitation but also it indicates the sharp or threshold activation of intermodal interactions above the critical energy level E_k.

6. Conclusion

The analysis of the CARS-spectra of molecules excited in a powerful IR-laser field has made it possible to establish the initial form of the SF_6 molecular distribution among the vibrational levels and the degrees of freedom as well as to determine the portion of molecules and the average energy contained in the various molecular ensembles as well as to investigate the transformation of the initial molecular distribution with changes in energy density and frequency of excitation IR-field.

The temporal transformation of CARS spectra of the excited molecules gives also the extensive information about the channels and rates of allowed vibrational energy redistribution among as discrete levels as quasicontinuum states.

References

[1] Akulin V.M, Alimpiev S.S, Karlov N.V. et al. Tr. FIAN, (1979), Vol.114, pp.107-138.
[2] Bargatashvili V.N., Letokhov V.S., Makarov A.A., Ryabov E.A. Moscow: VINITI, (1980), Itogi nauki i tekhniki, Vol.2.
[3] Dowell R.S., Aldridge J.B., Holland R.F. J.Phys.Chem., (1976), Vol.80, pp.1203-1207.
[4] Aboumajd A., Berger H., Saint-Loup R. J.Mol.Spectrosc., (1979), Vol.78, pp.486-492.
[5] McDowell R.S., Reisteld M.J., Patterson C.W., Krohn B.J., Vasque M.C., Laguna G.A. J.Chem.Phys., (1982), Vol.77, No.9, p.4337-4343.
[6] Alimpiev S.S., Mokhnatyuk A.A., Odabashyan G.L., Sartakov B.G., Smirnov V.V., Fabelinsky V.I. Preprint of the General Physics Institute, No.266, (1987).
[7] Vereschagin K.A., Volkov S.Yu., Kozlov D.N., Mokhnatyuk A.A., Sadovsky D.A., Smirnov V.V., Fabelinsky V.I. Proc. of the XX-th Spectroscopy Cogress, Kiev, p.1, (1988), p.272.
[8] Bagratashvili V.N., Dolzhikov V.S., Letokhov V.S. ZhETF, (1979), Vol.76, p.18.
[9] Lyman J.L., Quigley G.P., Judd O.P. Los Alamos, (1979), 157 pages. Los Alamos Lab. Rep.; No.79-2605).

[10] Alimpiev S.S., Zikrin B.O., Sartakov B.G., Khokhlov E.M. Preprint of the Lebedev Physics Institute, No.205, Moscow, (1983).
[11] Alimpiev S.S. Izv. AN SSSR. Ser. fiz., (1985), Vol.49, No.3, pp.490-495.
[12] Jensen C.C., Andersen T.G., Reiser C., Steinfield J.I. J.Chem.Phys., (1979), Vol.71, pp.36-48.
[13] Alimpiev S.S., Sartakov B.G. Preprint of the Institute of General Physics of the Academy of Sciences, No.63, Moscow, (1985), 50 pages.
[14] Mkrtchan M.M., Platonenko V.G. Kvantovaya elektronika, (1978), Vol.5, pp.2104-2109.
[15] Alimpiev S.S., Karlov N.V. ZhETF, (1974), Vol.66, pp.542-547.
[16] Gordietz B.F., Osipov A.I., Shelepin L.A. Moscow: Nauka, (1980), 512 pages.
[17] Bates R.D., Knudson J.T., Flynn G.W., Ronn A.M. Chem. Phys.Lett., (1971), Vol.8, pp.103-110.
[18] Shuryak E.V. ZhETF, (1976), Vol.71, pp.2039-2056.
[19] Akulin V.M., Alimpiev S.S., Karlov N.V., Sartakov B.G. ZhETF, (1978), Vol.74, pp.490-500.

Nonlinear Transient Spectroscopy Using Four-Wave Mixing with Broad-Bandwidth Laser Beams

P.A. Apanasevich, V.P. Kozich, A.I. Vodchitz, and B.L. Kontsevoy

Institute of Physics, BSSR Academy of Sciences,
Lenin Avenue 70, SU-220602 Minsk, USSR

Abstract. The degenerate four-wave mixing ($\omega_s=\omega_1+\omega-\omega$; $\omega_1=2\omega$) process using incoherent Gaussian exciting beams has been realized for one- and two-component media with the predominant orientational nonlinearity. The picosecond molecular orientational relaxation times of organic liquids and dye solutions have been measured using nanosecond exciting beams.

1. Introduction

Ultrafast phenomena spectroscopy has developed rapidly over the last two decades. It consists in exciting and probing the short-lived states of matter with ultrashort pulses. In such a technique, the time resolution is determined by the pulse width of the light sources. To have high temporal resolution in this case one needs to apply very short light pulses. This leads to a number of difficulties such as the necessity to use rather complicated pico- or femtosecond lasers and the time-resolution limitation due to linear and nonlinear pulse broadening. A new approach to transient spectroscopy using broad-bandwidth incoherent (or "noisy") light beams with known statistical properties has been developed recently [1-7]. In this technique the light pulses may be long and non-transform-limited, i.e. $t_p\Delta\omega > 1$, where t_p is the pulsewidth and $\Delta\omega$ is the spectral bandwidth. In this case the time resolution is governed by the correlation time $\tau_c \sim 1/\Delta\omega$ instead of the duration of the light pulses. The correlation time of broad-bandwidth laser pulses can be sufficiently small to provide subpicosecond time resolution. This permits one to measure short relaxation times in pico- and femtosecond regions using relatively simple nanosecond lasers. Some results of incoherent light technique applications to study electronic and vibrational dephasing, the population and molecule orientation relaxation are discussed in [1-7].

In this paper the results of the investigation of the Kerr nonlinearity relaxation in liquids using degenerate four-wave mixing (three-wave mixing , TWM) of incoherent laser beams ($\omega_s = \omega_1 + \omega - \omega$; $\omega_1 = 2\omega$) are presented. The Kerr-shutter configuration for TWM in

Springer Proceedings in Physics, Volume 63 **Coherent Raman Spectroscopy**
Eds.: G. Marowsky and V.V. Smirnov

benzene, nitrobenzene and ethanol solutions of Rh6G and one of pentacarbocyanine (PCC) dyes was realized experimentally. In the fitting of the theoretical and experimental curves the Kerr nonlinearity with a maximum of two comparable components with different relaxation times was considered. Two contributions to the Kerr nonlinearity may be due to either various physical mechanisms or the presence of two kinds of molecules.

2. Theory

The technique of induced birefringence relaxation measurements using incoherent light sources consists in the following [7]. The sample under investigation with third-order Kerr nonlinearity is irradiated by two broad-bandwidth light beams with central frequencies ω_1 and ω (Fig.1). The polarization configuration of the interacting waves may be arbitrary. For the model calculations we chose the wave $E_1(t)$ to be linearly polarized along the Y axis and the polarization of the wave E(t) was oriented at 45° to the first one. The phase matching conditions for wave vectors $k_s = k_1 + k - k$ in the case of degenerate four-wave mixing $\omega_s = \omega_1 + \omega - \omega$ are fulfilled for any crossing angle of the beams if the fields $E(\omega)$ and $E^*(-\omega)$ propagate along the same direction, i.e. the process is k-degenerate. One wave with frequency ω induces a birefringence in the substance that is probed by the weak wave with frequency ω_1. In the experiment the signal transmitted through a polarizer crossed with the probe wave E_1 is detected versus the time delay τ of the probe wave relative to exciting one. The field of the signal is proportional to the nonlinear polarization of the medium, given as [8]

$$P_x(t,\tau) \sim E_{1y}(t) \int_{-\infty}^{\infty} r_{xyyx}(t-t')E_y(t')E_x^*(t')dt', \qquad (1)$$

where $E(t')$ and $E_1(t)$ are the amplitudes of the exciting and probing fields, and $r_{xyyx}(t-t')$ is the nonlinear

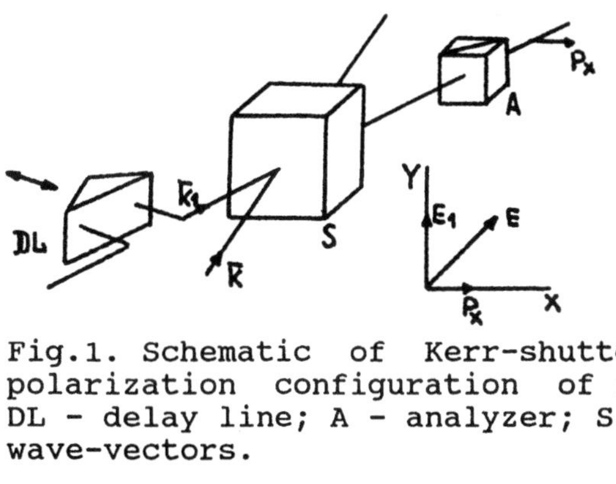

Fig.1. Schematic of Kerr-shutter experiment and polarization configuration of interacting waves. DL - delay line; A - analyzer; S - sample; k, k_1 - wave-vectors.

response function. The observed signal energy is proportional to the value

$$\varepsilon(\tau) \sim \int_{-\infty}^{\infty} <|P_x^{(3)}(t,\tau)|^2>dt, \quad (2)$$

where the symbols <> mean statistical averaging.

The experiment and fitting calculations have been made for the case when $E_1(t)$ is the second harmonic of the field E(t) modelled as a stationary Gaussian random process [3,9]:

$$E_x(t) = E_y(t) = \alpha_1(t)R(t),$$
$$E_{1y}(t) = \alpha_2(t)R(t-\tau)R(t-\tau), \quad (3)$$
$$<R^*(t)R(t-\tau)> = G(\tau) = \exp(-|\tau|/\tau_c),$$

where $\alpha_{1,2}(t)$ are slowly varying regular time functions and R(t) is a complex random function describing the Gaussian process with correlation time τ_c. The relaxation of both the components of Kerr nonlinearity is assumed to be exponential, with time constants T and T'. Then the total nonlinear response function of the medium can be expressed as

$$r_{xyyx}(t-t') = r_{xyyx}(0)\{\exp[-(t-t')/T] + \beta\exp[-(t-t')/T']\}, \quad (4)$$

i.e. as a superposition of the contributions of two components in the proportion determined by the coefficient β. The coefficient depends on the characteristics of the molecules, their concentrations and interaction as well as on the excitation conditions. As mentioned above, the contributions may be from either solvent and solute molecules or orientations of molecules as a whole and librational motions which have different relaxation times.

Substituting (1),(3),(4) into (2), factorizing the obtained eighth-order moment of a Gaussian random process with second-order moments and integrating we obtain the dependence $\varepsilon(\tau)$. This is given here in the following form because the full expression is very cumbersome:

$$\varepsilon(\tau) = A + B\exp[-2|\tau|/\tau_c], \quad \tau \le 0;$$

$$\varepsilon(\tau) = A + C\exp[-2\tau/\tau_c] + D\exp[-\tau/T] + D'\exp[-\tau/T'] + F\exp[-(1/T+2/\tau_c)\tau] + F'\exp[-(1/T'+2/\tau_c)\tau] + J\exp[-2\tau/T] + J'\exp[-2\tau/T'] + H\exp[-(1/T+1/T')\tau], \quad \tau > 0. \quad (5)$$

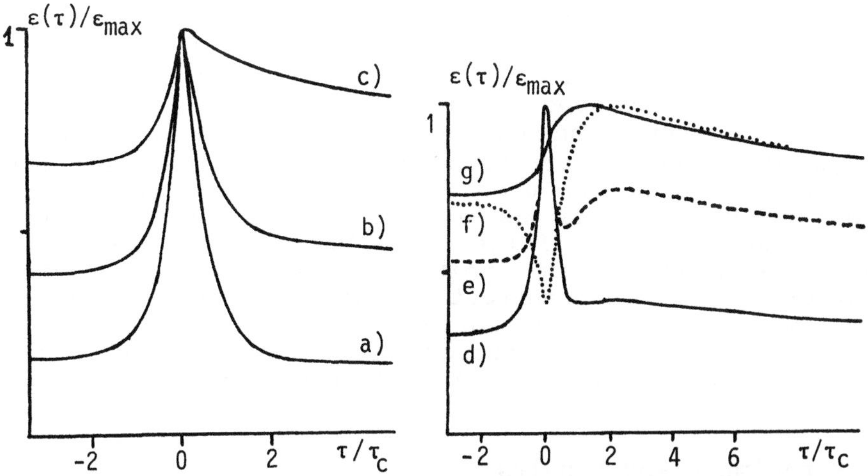

Fig.2. The calculated dependences $\varepsilon(\tau)/\varepsilon_{max}$ for $T/\tau_c = 0.1$, $T'/\tau_c = 10$ and β equal to 0.001 (a); 0.025 (b); 0.150 (c); -0.015 (d); -0.018 (e); -0.050 (f); $|\beta| \gg 1$ (g).

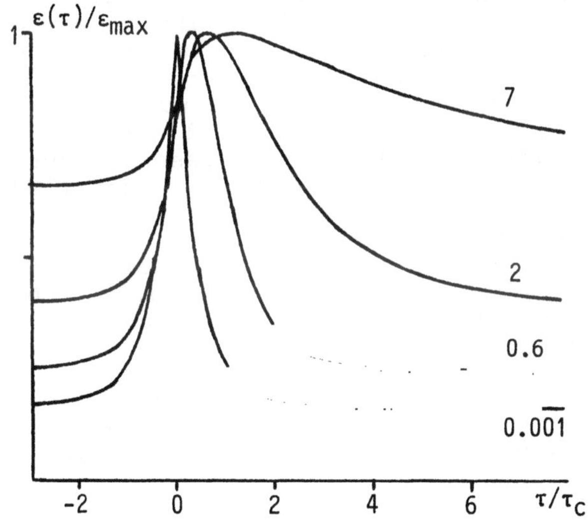

Fig.3. The signal energy $\varepsilon(\tau)$ versus the time delay between exciting beams for different orientational relaxation times T and $\beta = 0$ (theory).

Here the coefficients A,B,...,H are functions of τ_c,T,T' and β. When $\beta = 0$ the dependence $\varepsilon(\tau)$ coincides with that for the case of the single-exponential response function presented in [7]. The theoretical dependence $\varepsilon(\tau)$ was analyzed with a computer. The calculated curves $\varepsilon(\tau)$ for $T/\tau_c = 0.1$ and $T'/\tau_c = 10$ with different β are represented in Fig.2. It can be seen that the shape of curve $\varepsilon(\tau)$ is very sensitive to a variation of the contributions ratio β when the contributions to the Kerr nonlinearity have opposite signs ($\beta < 0$).

Fig.3 shows the curves $\varepsilon(\tau)$ for the case of a single-exponential Kerr response ($\beta = 0$) and different ratios T/τ_c = 0.001; 0.6; 2; 7. The finite relaxation time exhibits itself as a modification of the curve shape and contrast. If the ratio T/τ_c is increased then the curve asymmetry and their shift relative to zero are increased also. The contrast having the maximum value of $\varepsilon_{max}/\varepsilon(\infty) = 6$ for the instantaneous response of the medium ($T \ll \tau_c$) is decreased with increasing T/τ_c and becomes equal to 1.5 for $T/\tau_c = 7$, for example.

3. Experiment and Results

The fundamental frequency $E(\omega)$ and the second harmonic $E_1(2\omega)$ radiations of the YAG:Nd^{3+} laser with an unstable nonselective resonator were used as the exciting and probing beams respectively. The fundamental radiation was characterized by the correlation time τ_c = 45±5 ps and pulse width t_p = 15 ns. It should be noted that the value of τ_c is influenced substantially by the regime of laser operation, and the mutual correlation of $E(\omega)$ and $E_1(2\omega)$ fields depends on the second harmonic generation efficiency. The question of light field statistics transformation in nonlinear processes is very interesting and requires a separate detailed investigation.

The verification of the assumption of Gaussian statistics of the radiation was realized by a comparison of the dependences $\varepsilon(\tau)/\varepsilon_{max}$ for theory ($\beta = 0$, $T \to 0$) and experiment in glass, the medium with instantaneous electronic nonlinearity. The correlation time τ_c was also determined from these experimental curves.The real-time control of the optical delay line, recording, processing and graphic output of the data obtained were carried out by using a computer and the CAMAC system.

3.1. Kerr Nonlinearity Relaxation in Benzene and Nitrobenzene

The Kerr nonlinearities and their relaxation times for benzene and nitrobenzene differ by an order of magnitude. Characteristically also, the Kerr contribution to the third-order nonresonant susceptibility of these substances predominates. The experimentally obtained scattered signal energy $\varepsilon(\tau)$ versus the time delay between the fields $E_1(2\omega)$ and $E(\omega)$ for benzene and nitrobenzene are represented in Fig.4 by points. Assuming the presence of only one component in the Kerr nonlinearity and using as a sole fitting

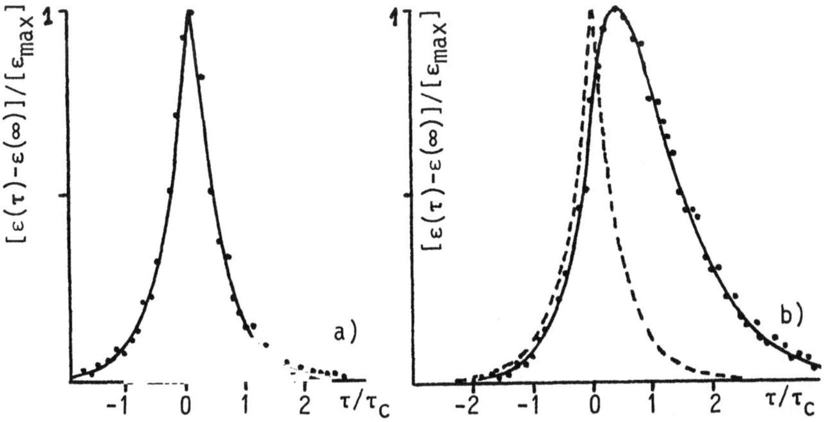

Fig.4. The normalized energy of the TWM signal versus time delay between the fields $E_1(2\omega)$ and $E(\omega)$ for benzene (a) and nitrobenzene (b). The points represent experiment and the solid line theory. The dotted line corresponds to the instantaneous response of the medium (glass).

parameter the value of T/τ_c for the case $\beta = 0$, the best agreement of the experimental and theoretical curves has been obtained for $T \simeq 4$ ps for benzene and $T \simeq 40$ ps for nitrobenzene. The theoretical curves for these values of T are shown in Fig.4 by solid lines. The dotted line corresponds to the instantaneous response of the medium, $r_{xyyx}(t-t') = r_{xyyx}(0)\delta(t-t')$. In both cases $T \leq \tau_c$ so the curves' asymmetry is small. The Kerr relaxation shows itself as a broadening of the curves and shifting of their maxima relative to the point $\tau = 0$.

The measured values of orientational relaxation of benzene and nitrobenzene molecules (the very mechanism that predominates in the Kerr nonlinearity of these liquids) agree well with the results obtained by other methods [10]. This agreement confirms the adequacy of theoretical modelling of the radiation field of the YAG:Nd^{3+} laser with a nonselective resonator (~ 200 longitudinal modes) as a random stationary Gaussian process. The last conclusion is justified because in the case of $T \ll \tau_c$ the curve shape $\varepsilon(\tau)$ also reflects the statistical properties of the exciting light. In contrast to [11] we have determined the slowest relaxation time for nitrobenzene, because we used the radiation with a larger correlation time τ_c and measured in a greater range of time delays τ.

3.2. Two-component Kerr Nonlinearity Relaxation in Rh6G and Pentacarbocyanine Dye Solutions

One can reasonably assume the presence of two contributions to the Kerr nonlinearity of the dye solution: one is from dye molecules and the other from solvent ones. This gives the possibility of testing our technique for the case of two-component Kerr nonlinearity relaxation.

The ethanol solution of Rh6G is interesting due to the fact that the second harmonic radiation is resonant with the electronic transition $S_0 \Rightarrow S_1$ of the dye molecules. Because this field was a probe one in the experiment we did not observe the resonant enhancement of the Kerr-effect signal for the ground state molecules when this field was weak. Another situation is when the probe radiation power is enough to populate the long-lived singlet state S_1 where the orienting field E(t) is resonant to the $S_1 \Rightarrow S_2$ electronic transition and the Kerr nonlinearity of excited dye molecules may be very substantial. It is possible to change the excited state population by changing the second harmonic power and as a result to vary the contribution of excited molecules to the Kerr nonlinearity (coefficient β in theoretical calculations).

Fig.5 shows experimental curves for 10^{-4} mole/liter Rh6G ethanol solution at different values of the second harmonic pulse energy. It can be seen that in the case (a) the solvent contribution predominates because the curves almost do not differ from ones for pure ethanol and in the case (b) the solute contribution predominates. Both nonlinearity components are comparable for curve (c).

Before fitting the calculated and experimental curves, we investigated pure ethanol. Comparison of the dependences $\varepsilon(\tau)$ for ethanol and glass has shown coincidence of the curves within an experimental error. It shows that the Kerr relaxation of ethanol is rather fast, in any case $T/\tau_c \leq 0.05$. The same properties of the Kerr nonlinearity were observed for such solvents as 2-propanol and ethylene glycol.

Variation of the parameters T, T′, β allowed calculated and experimental curves to be fitted. The best agreement was obtained for $T/\tau_c = 0.05$, $T'/\tau_c = 5.56$ and the values of β: curve (a) $\beta = -0.012$; (b) $|\beta| > 1$; (c) $\beta = -0.042$. As a result we have obtained $T < 2$ ps and $T' = 250 \pm 50$ ps. We have done the evaluation only for ethanol. The component is rather fast and it looks as if a librational or electronic mechanism dominates here. According to [12] the shortest orientational relaxation time for ethanol is equal to 1.6 ps and may be related to the reorientations of OH dipoles. Probably the contribution to the nonlinearity from diffusive (thermal) reorientations of molecules in this liquid is small.

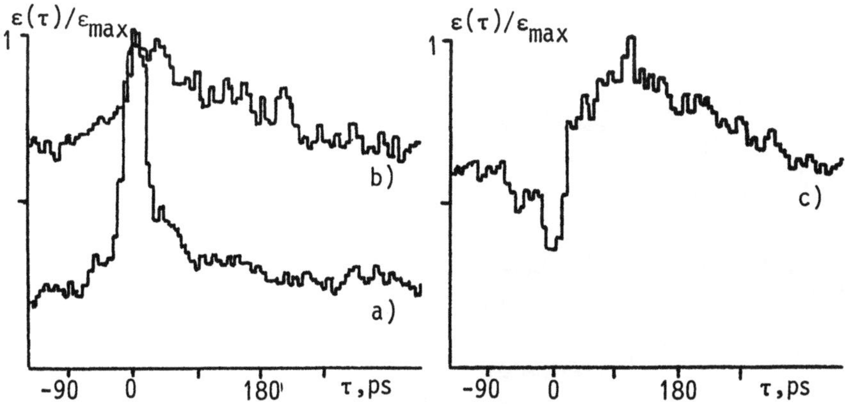

Fig.5. The experimental dependences $\varepsilon(\tau)/\varepsilon_{max}$ for the ethanol solution of Rh6G for different probe pulse energies ε_1: 0.005 mJ (a); 0.460 mJ (b); 0.250 mJ (c).

The measured time T′ characterizes an orientation relaxation of the Rh6G molecules in the excited singlet state S_1. Really, the excited state lifetime is equal to a few nanoseconds [13,14] so more fast orientation relaxation has time to occur before the molecule is returned to the ground state. The measured time agrees well with the result obtained by other methods [14]. We found that the orientation contributions to the Kerr nonlinearity of ethanol and excited Rh6G molecules had opposite signs. It should be noted that if second harmonic radiation of the $YAG:Nd^{3+}$ laser is used as an exciting beam and the weak radiation of the fundamental frequency as a probe, then the orientation relaxation of Rh6G molecules in the ground state can be studied.

The ethanol solution of PCC dye differed from the one discussed above by the excitation conditions. In this case the birefringence inducing field E(t) is resonant to the $S_0 \Rightarrow S_1$ transition [1,15]. Fig.6 shows the experimental curves $\varepsilon(\tau)$ for the different dye concentrations increasing in the ratio 1:2:3.3 (the curves a, b, c, respectively). The pump and probe pulse energies were constant and equal to $\varepsilon \simeq 4$ mJ and $\varepsilon_1 \simeq 0.46$ mJ, respectively. The characteristic feature of the dependences obtained is that here the shift of the curve maxima relative to zero delay and curves′ shape variation were not manifest so sharply as in the case of Rh6G.

The best agreement of the calculated and experimental curves was obtained for $T'/\tau_c = 13$, $T/\tau_c = 0.05$ and $\beta = 0.0015$ (a), $\beta = 0.01$ (b), $\beta = 0.047$ (c). The measured orientational relaxation time for PCC dye molecules is $T' = 590 \pm 90$ ps.

What electronic state does this time correspond to? In general, molecular reorientation will occur both in

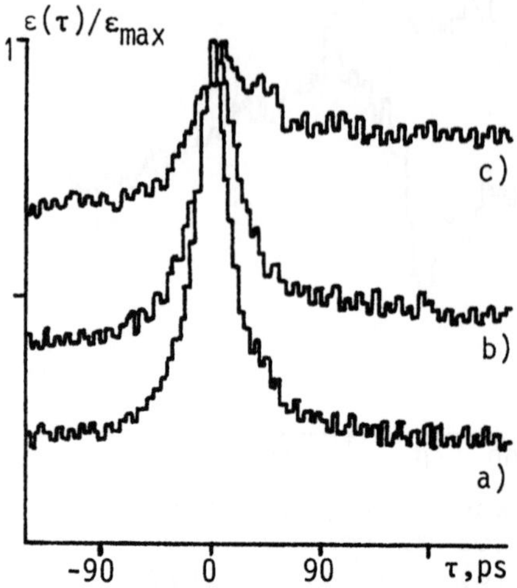

Fig.6. The experimental dependences $\varepsilon(\tau)/\varepsilon_{max}$ for the ethanol solution of PCC dye for different concentrations increasing in the ratio 1:2:3.3 (curves (a), (b), (c) respectively).

the ground and excited electronic states. In the case of different orientation relaxation times of these states the measured time will depend on the quasi-stationary population distribution between the levels $S_0 \Leftrightarrow S_1$ [13]. It is known that the lifetime of the excited electronic state of PCC dye is about 40 ps [1,15]. This time is substantially less than the relaxation constant T′ obtained. Then the measured time T′ should be related to the ground state molecules because their contribution substantially predominates.

There is no linear dependence between the coefficient β and dye molecule concentration when fitting calculated and experimental curves. Probably this is because the fraction of molecules in the excited electronic state changes with concentration, i.e. the fraction of molecules undergoing effective orientation by the resonant field E(t).

4. Conclusion

The investigations carried out demonstrate the possibility of measuring the picosecond Kerr nonlinearity relaxation by TWM with "noisy" broad-bandwidth nanosecond laser beams. To get the relaxation times, the calculated and experimental curves have been fitted. A two-exponential nonlinear response of the medium and Gaussian statistics of the radiation were assumed. Such a response function enabled us to take into account the contributions from solute and solvent molecules, when studying solutions, or ones due to different physical mechanisms (electronic, librational or orientational), when studying pure substances. Besides the relaxation characteristics of the nonlinearity components, the technique developed enabled us to determine the relative magnitude and sign of two dominant contributions.

The calculations show that in the technique discussed one can take into account the nonlinearity due to electronic state population motion and determine population relaxation. This is possible if this nonlinearity is not very small relative to the Kerr one.

The realized Kerr-shutter configuration for TWM in which birefringence inducing and probing radiations have different frequencies (ω and 2ω) offers a some advantages in comparison with frequency degenerate four-wave mixing. These are a larger signal/pedestal ratio and the possibility of excluding the contribution to the signal detected from nonlinearity due to isotropic mechanisms (electrostrictional, thermal and others).

The investigations have shown that, using radiation with known statistical properties, it is possible to measure relaxation times in the range $T < \tau_c$ up to $T/\tau_c = 0.1$ if one nonlinearity component predominates. The application of more broad-bandwidth radiation (τ_c decreasing) leads to the prospect of attaching subpicosecond scale measurements.

References

[1] M.A.Vasil'eva, V.I.Malyshev, A.V.Masalov. FIAN Short Phys.Commun. 1, (1980) 35-39.

[2] S.Asaka, H.Nakatsuka, M.Fujiwara, M.Matsuoka. Phys.Rev.A. 29, (1984) 2286-2292.

[3] N.Morita, T.Yajima. Phys.Rev.A. 30, (1984) 2525-2536.

[4] R.Beach,S.R.Hartmann. Phys.Rev.Lett. 53,(1984) 663.

[5] H.Nakatsuka, Y.Katsahima, K.Inouye,R.Yano,S.Vemura. Opt.Commun. 74, (1990) 219-222.

[6] T.Kobayashi, A.Terasaki, T.Hattori, K.Kurokawa. Appl.Phys.B. 47, (1988) 107-125.

[7] P.A.Apanasevich, V.P.Kozich, A.I.Vodschitz. J.Mod.Opt. 35, (1988) 1933-1938.

[8] R.W.Hellwarth.Prog.Quantum Electron. 5,(1977) 1-68.

[9] S.A.Akhmanov,Yu.E.D'jakov,A.S.Chirkin. Inroduction to Statistical Radiophysics and Optics. Moscow: Nauka, (1981) 640.

[10] S.L.Shapiro (editor). Ultrashort Light Pulses. Moscow: Mir, (1981) 479.

[11] K.Kurokawa, T.Hattori, T.Kobayashi. Phys.Rev.A. 36, (1987) 1298-1304.
[12] W.J.Chase,J.W.Hunt.J.Phys.Chem. 79, 26,(1975) 2835.
[13] A.B.Myers, R.M.Hochstrasser. IEEE J. Quantum Electron. QE-22, 8, (1986) 1482-1492.
[14] D.W.Phillion, D.J.Kuizenga, A.E.Siegman. Appl.Phys. Lett. 27, (1975) 85-87.
[15] M.A.Vasil'eva, V.I.Malyshev, A.V.Masalov, P.S.Antsiferov. Izv.Akad.Nauk SSSR. Ser.Fiz. 46, 6, (1982) 1203.

Application of Single-Pulse Broadband CARS to Shock-Tube Experiments

A.S. Diakov and P.L. Podvig

Moscow Institute of Physics and Technology,
Institutski 9, SU-141700 Dolgoprudny, Moscow Region, USSR

Coherent anti-Stokes Raman spectroscopy is a powerful diagnostic tool in hostile environments when measurements of temperature and species concentration are required, and when traditional methods for some reason cannot be applied. Such conditions occur in those non-stationary gas-phase chemical reactions in which nitrogen plays a significant role. In this case, CARS provides information about the most important parameters of the process with a sufficient temporal and spatial resolution. In our experiments, the broadband CARS technique is used to obtain an evolution of the vibrational distribution function of N_2 in the process of thermal dissociation of nitrogen-dioxide behind a reflected shock wave.

The process of thermal dissociation of N_2O was intensively investigated by conventional methods of infrared spectroscopy [1, 2]. Some results of this work indicate a presence of vibrational non-equilibrium in the products of dissociation and a mutual dependence of the chemical reaction itself and the vibrational relaxation process.

The thermal dissociation of N_2O at temperatures of about 2000 K is caused primarily by the following reactions:

$$N_2O + M \leftrightarrows N_2 + O + M\,,$$
$$N_2O + O \rightarrow O_2 + N_2\,,$$
$$N_2O + O \leftrightarrows NO + NO\,,$$

where M represents any molecule of the mixture. Molecules of N_2, O_2 and NO which occur in these reactions can have a substantial amount of vibrational energy. Since the process of equilibration of vibrational energy has a single-quantum mechanism, one might conclude that under some conditions a non-equilibrium in the vibrational distribution of the molecules can occur. The rate constants of the reactions above are in turn observed to depend substantially on the vibrational states of the collision partners, so the mechanism of the dissociation under non-equilibrium conditions must differ from that under equilibrium. To study this mechanism it is necessary to obtain the time dependence of the concentration of the molecules which are involved in the process, and evolution of their vibrational level distribution.

A shock tube with a 25×25 mm cross-section was used to heat up a gas mixture containing N_2O to a temperature of approximately 2500 K. The initial pressure in the low-pressure section of the shock tube was 0.05–0.1 atm, and in the high-pressure section up to 10 atm. Pressure bursting of aluminium

Springer Proceedings in Physics, Volume 63 **Coherent Raman Spectroscopy**
Eds.: G. Marowsky and V.V. Smirnov

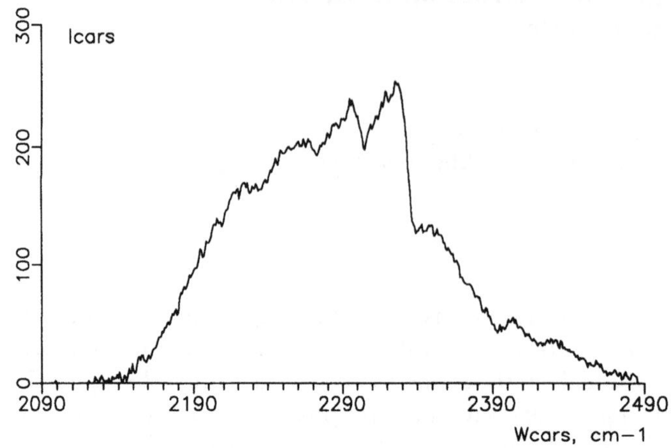

Figure 1: CARS spectrum of nitrogen dissotiated from N_2O behind the shock wave. The initial mixture is 20% N_2O + 80% Ar. The effective vibrational temperatures are T_{01} = 3500 K, T_{12} = 7000 K, $T_{23} = T_{34}$ = 10000 K. T_R = 2200 K, p = 4 atm, concentration of N_2 is 11 ± 1%.

diaphragms initiated the shock waves. The temperature of the gas just behind a shock wave can be determined from shock velocity measurements. The velocity was measured by two pressure gauges placed in the shock tube. Signals from the gauges were also used to synchronize the laser pulse with the time the shock wave passes through the test point. It was also possible to introduce an additional delay between these moments.

The experimental set-up is typical for CARS experiments with collinear phase-matching geometry. The CARS pump beam is provided by a frequency-doubled Nd:YAG laser, which has a bandwidth of 0.05 cm^{-1} and an energy of 150 mJ (532 nm — 60 mJ), with a laser pulse time of 20 ns. Part of the 532 nm radiation pumps a broadband dye laser, which has an energy of 10 mJ, and bandwidth of about 150 cm^{-1}. The CARS spectra are detected by a system containing a grating spectrograph (PGS-2), with a dispersion of 5 Å/mm, a multichannel plate intensifier, and an optical multichannel analyzer (OMA 1205). The dispersion of the whole detection system is 0.8 cm^{-1}/element.

A typical CARS spectrum of nitrogen produced through dissociation of N_2O behind a reflected shock wave is shown in Fig. 1. Experimental conditions, i.e. high temperatures of the medium (up to 2200 K) and high pressures (up to 7 atm), as well as low spectral resolution of the detector, results in the unresolved rotational structure of the Q–branch. To interpret such spectra, a special computer code has been developed. The code yields temperature, species concentration and other parameters of the spectrum by fitting the calculated spectrum to an experimental one. Because of the high pressure conditions of the experiments such effects as collisional narrowing must be taken into account [3, 4]. The modified exponential gap model is used to emulate the temperature and J dependence of N_2 linewidths [5, 6].

The main parameters of the fitting algorithm are the rotational temperature and effective temperatures of vibrational transitions. When the vibrational dis-

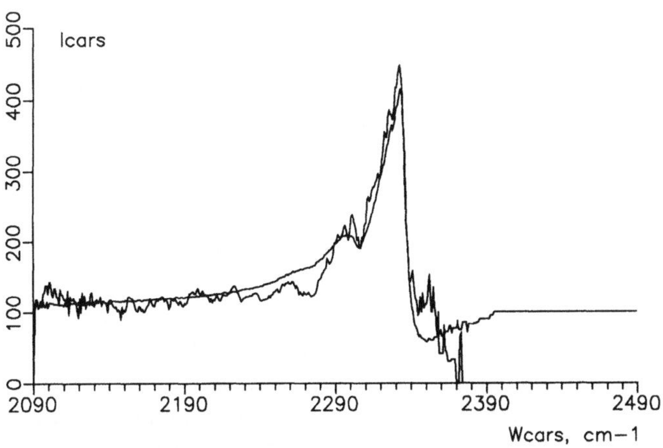

Figure 2: Normalized CARS spectrum of nitrogen in a mixture of 25% air + 75% Ar behind the reflected shock wave (normalized experimental data with the best fit curve). Temperature is 1650 ± 50 K. Best fit curve corresponds $T = 1710$ K and 22% concentration of N_2.

tribution is presumed to be nonthermal, the code makes it possible to vary these parameters independently. To take into account effects of nonresonant susceptibility we introduce a parameter which corresponds to the resonant/nonresonant susceptibility ratio. The fitting routine does not require the exact mole fraction of the species to be known. To determine species concentration of nitrogen from this parameter, it is necessary to know the local nonresonant susceptibility of the whole mixture. This value was derived from calibration experiments where spectra of N_2 in a mixture of 25% air and 75% Ar were obtained. We considered the nonresonant susceptibilities of the main components of the mixture, N_2, NO and Ar, to be equal. Other essential parameters of the algorithm are those describing the shape of the Stokes spectrum. In our experiments this shape was nearly Gaussian.

To test the whole procedure of interpreting single-pulse CARS spectra, two series of experiments were performed. First, spectra of nitrogen in a 25% air + 75% Ar mixture at 300 K and different pressures were used to determine the nonresonant susceptibility, and the shape and width of the instrumental slit function of the detector. The nonresonant susceptibility was found to be 5×10^{-37} cm^6/erg (with a Q–branch Raman cross-section of 1.1×10^{-31} cm^2) corresponding to available data. The instrumental slit function has a Gaussian form with spectral width (HWHM) of 1.9 cm^{-1}.

These data were used to determine temperature and species concentration in the second series in which a mixture of 25% air and 75% Ar was heated by a shock wave. The temperature of the gas behind the reflected shock wave can be determined independently by measuring the shock wave speed. The results of one of these experiments is presented in Fig. 2. Shock wave speed, in this case, is 1.05 km/sec, which yields $T = 1650 \pm 50$ K. The best-fit calculated spectrum, shown on the same figure, corresponds to $T = 1710$ K and a mole fraction of N_2 of 22%.

Accuracy of the fitting algorithm is within about 10 K of the considered temperature range. Nonetheless, the real uncertainty in temperature measurements is substantially greater. The main source of the uncertainty in temperature and concentration measurements is so-called "spectral noise", which results from stochastic mode intensity distribution in the spectrum of the dye laser [7]. It was determined that the present level of noise makes it impossible to measure temperature with uncertainty less than 150–200 K. At the same time, species concentration measurements are less sensitive to the spectral noise.

After having tested the whole procedure of the concentration and vibrational distribution measurements, we applied it to the study of the N_2O dissociation process. Spectra of N_2 were obtained with various time delays behind the shock wave, i.e. at different stages of dissociation. In Fig. 1 is shown the spectrum of N_2 on the final stage of the process (the time delay is 50 μs). The mole fraction of nitrogen determined from the spectra is $11 \pm 1\%$, corresponding to the fact that the net yield of N_2 in N_2O dissociation is $\approx 50\%$ (initial mixture consisted of 20% N_2O and 80% Ar). The effective temperatures of vibrational transitions of N_2 were determined to be $T_{01} = 3500$ K, $T_{12} = 7000$ K, $T_{23} = T_{34} = 10000$ K. The rotational temperature of N_2 is 2200 K. Though the problems mentioned above prevent us from determining the temperatures more precisely, one can see that the vibrational distribution is substantially nonthermal. In the first stages of the dissociation (time delays 5 μs and 10 μs) the vibrational distribution is also characterized by non-equilibrium. The mole fraction of N_2 in these cases turned out to be approximately equal to 2% which is about the sensitivity limit of our experiments.

The single-pulse broadband CARS spectra technique has been shown to be an appropriate tool for study of non-stationary gas-phase chemical reactions in spite of such unfavorable experimental conditions as high temperatures, high pressures, and the impossibility of carrying out the process in the periodic mode. It was also found that accuracy of temperature measurements is limited by the quality of the Stokes spectrum. Diminishing dye laser spectral noise, as well as stabilization of the Stokes spectra shape, might make it possible to obtain information about non-equilibrium states of nitrogen, and about the evolution of the vibrational distribution function with more accuracy. Accuracy of species concentration measurements is less sensitive to the level of spectral noise; it is determined primarily by uncertainty in available nonresonant susceptibility data.

References

1. A. P. Zuev and B. K. Tkachenko, Chemical Physics 23, 742 (1982), (in Russian).
2. N. N. Kudriavtsev, S. S. Novikov, P. B. Svetltchnyi, Journ. of Applied Mechanics and Technical Physics 5, 9 (1974), (in Russian).
3. D. A. Greenhalgh and R. J. Hall, Opt. Commun. 57, 125 (1986).
4. M. L. Koszykowsky, R. L. Farrow, and R. E. Palmer, Opt. Lett. 10, 478 (1985).

5. M. L. Koszykowsky, L. A. Rahn, R. E. Palmer, and M. E. Coltrin, J. Phys. Chem. 91, 41 (1987).
6. L. A. Rahn, R. E. Palmer, M. L. Koszykowsky, and D. A. Greenhalgh, Chem. Phys. Lett. 133, 513 (1987).
7. D. A. Greenhalgh and S. T. Whittley, Appl. Opt. 24, 907 (1985).

Pump–Probe Measurements of Rotational Transfer Rates in N_2–N_2 Collisions

*R.L. Farrow and G.O. Sitz**

Combustion Research Facility, P.O. Box 969,
Sandia National Laboratories, Livermore, CA 94551, USA
*Current address: Department of Physics, University of Texas at Austin, Austin, TX 78712, USA

1. Introduction

Rotational energy transfer is an important process affecting the spectroscopy of molecules in the gas phase. The collisional broadening observed in Raman and infrared-absorption transitions is typically dominated by rotational relaxation.[1] The saturation characteristics of N_2 coherent anti-Stokes Raman spectroscopy (CARS) spectra[2] and the collapse of N_2 Q-branch spectra at high pressures[3] are known to depend sensitively on rotational transfer. To predict N_2 Raman spectra under these conditions, many empirical models for the J-dependent transfer rates have been proposed.[3]

We have performed direct, pump-probe measurements of the state-to-state rotational energy transfer rates for collisions of N_2 ($v = 1$) with N_2 ($v = 0$) at 298 K. Stimulated Raman pumping[4,5] was used to prepare populations in a selected rotational level in the $v = 1$ vibrational state. Allowing time for collisions to transfer population to other rotational levels, we probed the rotational level populations using 2+2 resonance-enhanced multiphoton ionization (REMPI) through the $A^1\Pi_g \leftarrow X^1\Sigma_g^+$ transition.[6] A best-fit set of rotational transfer rates between even rotational levels was obtained using a fitting procedure that was not model-based and did not place any constraints on the rates other than detailed balance. We compare these rates to previously published empirical and semi-empirical models and to a detailed semi-classical scattering calculation.

2. Experiment

We used stimulated Raman pumping of Q-branch ($v=0 \rightarrow 1$) transitions to prepare a selected rotational state of N_2 in the $v = 1$ state (see Fig. 1). The pump/Stokes lasers used for the Raman excitation were single-mode and had near-transform-limited spectral bandwidths. The pump laser was an injection-seeded, frequency-doubled Nd:YAG laser with a linewidth of 0.002 cm^{-1} full-width at half maximum (FWHM) at 532 nm. The Stokes laser was a stabilized cw ring-dye laser, pulse-amplified using most of the pump laser output, with a linewidth of 0.003 cm^{-1} following amplification. We investigated the efficiency of the resulting population excitation, and found that the Raman Q-branch transitions could be

†Research supported by the U.S. Department of Energy, Office of Basic Energy Sciences, Chemical Sciences Division.

Springer Proceedings in Physics, Volume 63 Coherent Raman Spectroscopy
Eds.: G. Marowsky and V.V. Smirnov

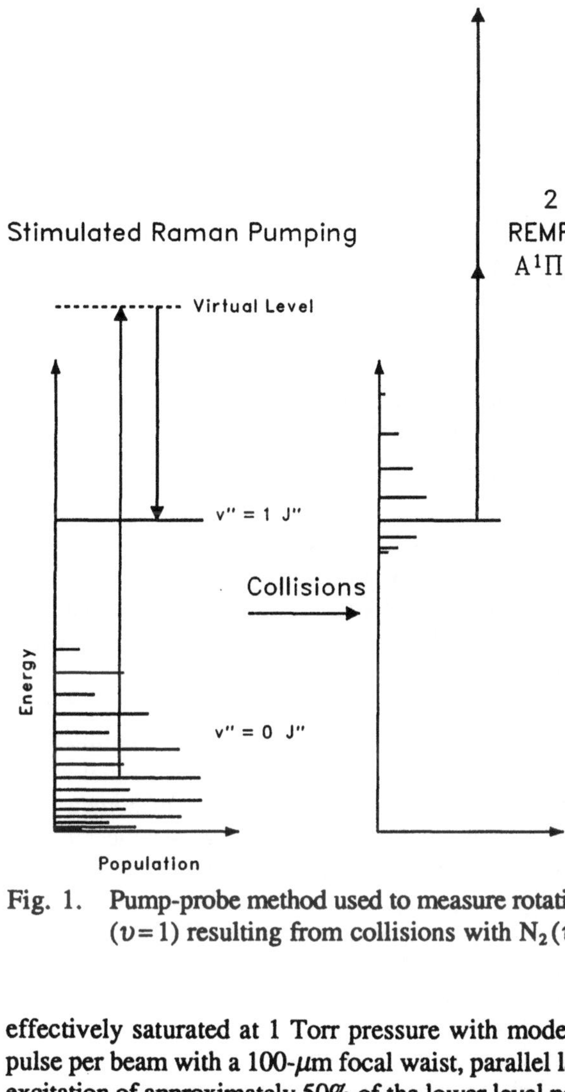

Fig. 1. Pump-probe method used to measure rotational energy transfer rates of N_2 ($v=1$) resulting from collisions with $N_2(v=0)$.

effectively saturated at 1 Torr pressure with modest laser intensities (10 mJ per pulse per beam with a 100-μm focal waist, parallel laser polarizations), resulting in excitation of approximately 50% of the lower level population.

After allowing an appropriate time interval for collisions to occur, 2+2 resonance-enhanced multiphoton ionization (REMPI) was used (through the $A^1\Pi_g$ - $X^1\Sigma_g^+$ transition) to detect the relative population of the pumped level and other levels to which rotational energy transfer had occurred. The (2,1) band was chosen for excitation because of its moderate Franck-Condon factor and good overlap with the gain curve of an efficient laser dye (Rhodamine 6G). The tunable light used to excite this two-photon transition in the wavelength range of 286 nm was generated by frequency doubling the output of a dye laser (Lambda-Physik FL2002). The dye laser was pumped with the second harmonic of a second single-mode Nd:YAG laser and is operated with an etalon in the oscillator cavity, providing a bandwidth of approximately 0.04 cm^{-1} in the visible. Typical energies were 4-5 mJ per pulse in the ultraviolet, with a pulse duration of 8 ns. The UV beam counterpropagated

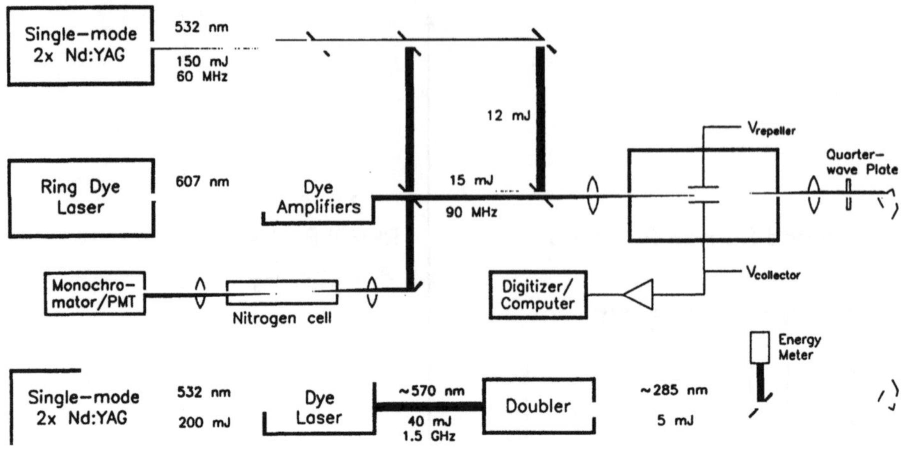

Fig. 2. Schematic diagram of the apparatus.

with respect to the Raman pumping beams and was focused at the same spot. The UV light was circularly polarized by means of a quarter-wave plate. The use of circular polarization increased the signal[7] by 50% and resulted in a significant narrowing of the MPI lines. The narrowing effect was most likely due to a decrease of dynamic Stark broadening of the line.[8]

Figure 2 shows a diagram of the experimental setup. The experiments were performed in a stainless-steel cell. N_2 gas was allowed to flow continuously through the cell at approximately 1 l/min. The cell pressure was measured with a thermocouple-type pressure gauge calibrated against a capacitance manometer and ranged from 1-3 Torr. Electrons from the multiphoton process were collected with a parallel-plate-type arrangement, typically with a 30-volt bias across a 2-cm spacing. This voltage gradient was purposely kept low to reduce ion cascading. We split off part of the Raman excitation beams to generate CARS signal in a separate cell containing N_2 at 5-10 Torr pressure. The CARS signal served as a convenient monitor of the excitation process and facilitated tuning the pump/Stokes frequency difference into resonance with the Raman transitions.

We performed a series of measurements in which a single even rotational level (J_i = 0-14) was excited and the time-dependent level populations were recorded at three or more delay times, typically 20-80 ns. (Examples of REMPI spectra obtained with various delays after pumping J_i = 0 are shown in Fig. 3.) The data were analyzed by integrating the area under each peak, and correcting for the two-photon transition probability using known line strengths.[7] This procedure yields the relative populations of the various rotational quantum states. [We were able to verify the accuracy of the relative populations by performing measurements at sufficiently large pressure-time products (400 ns-Torr) that the $v = 1$ rotational state distribution had equilibrated among the even J levels.] The entire data set, consisting of time-dependent populations for different time delays and initially excited levels, was globally fit to determine the best-fit set of state-to-state rate constants describing the observed populations. The predicted time-dependent populations for a given rate matrix were obtained by solving the coupled set of

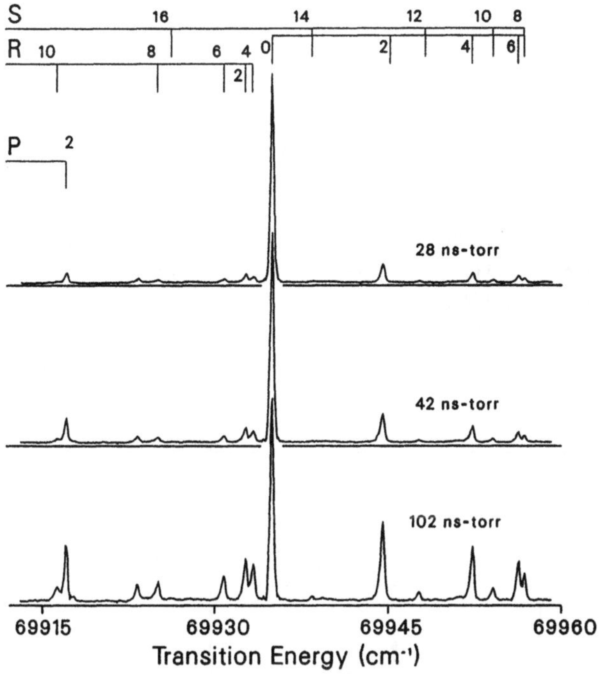

Fig. 3. Time-dependent MPI spectra of the (2,1) band of the $A^1\Pi_g \leftarrow X^1\Sigma_g^+$ transition taken at three different pressure-time products following excitation of the *Q*(0) Raman transition. The growth of population in the satellite levels results from rotational energy transfer.

differential rate equations describing the even-level populations for $J = 0\text{-}18$ via a numerical matrix diagonalization.[9] The fitting procedure did not place constraints such as an energy gap law on the *J*- or energy-dependence of the rates.

The presence of nuclear spin in the N_2 nucleus forbids transitions with odd ΔJ. It is worth noting that no evidence is seen in any of our spectra of vibration-vibration (v-v) inelastic processes.

3. Results and Discussion

The results of the fit are given in Table I along with error estimates. These rates can be compared to other experiments, in particular to the high-resolution line broadening measurements of Rosasco *et al.*[10] In the absence of vibrational relaxation and elastic vibrational dephasing, the line broadening coefficient is related to one-half the sum of rotationally inelastic rates that transfer population out of the upper and lower vibrational levels.[11,12] If the rates are assumed to have no vibrational level dependence (reasonable for N_2 due to its relatively low vibrational amplitude), the line broadening coefficient is then given by

$$\gamma_j = \sum_{j \neq k} \gamma_{jk} \quad . \tag{1}$$

Table I. State-to-state rotational rate constants for N_2 (υ=1). Units are μs^{-1} $Torr^{-1}$ (divide by 247.85 for cm^{-1} atm^{-1}).

J_f / J_i	0	2	4	6	8	10	12	14
0		6.64 (118)	3.76 (83)	2.73 (61)	0.86 (51)	0.64 (12)	0.29 (6)	0.22 (4)
2	1.40 (25)		5.13 (58)	2.40 (44)	1.52 (34)	0.97 (22)	0.28 (14)	0.12 (4)
4	0.50 (11)	3.26 (37)		4.7 (6)	2.2 (4)	1.4 (2)	0.54 (13)	0.21 (4)
6	0.31 (7)	1.30 (24)	4.03 (54)		3.37 (47)	2.22 (29)	0.71 (14)	0.23 (5)
8	0.10 (6)	0.84 (19)	1.92 (31)	3.43 (48)		2.52 (41)	1.12 (22)	0.29 (7)
10	0.087 (17)	0.62 (14)	1.41 (24)	2.63 (34)	2.93 (48)		2.68 (43)	1.04 (13)
12	0.052 (10)	0.23 (11)	0.71 (17)	1.10 (22)	1.69 (33)	3.49 (56)		1.83 (26)
14		0.056 (11)	0.15 (5)	0.393 (85)	0.51 (12)	0.62 (16)	1.96 (25)	2.64 (38)

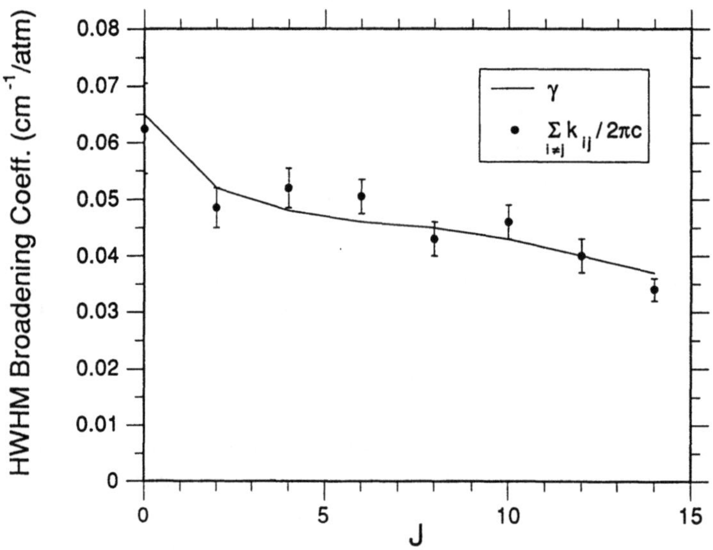

Fig. 4. Comparison of Raman Q-branch line broadening coefficients from Ref. 10 with experimental rotational transfer rates depopulating a given J level. The line broadening coefficients are connected by straight lines to guide the eye, and have uncertainties ranging from ±5 to ±10%. The excellent agreement observed is consistent with negligible linewidth contributions by elastic vibrational dephasing,[11] and supports the validity of the rate measurements.

Here, γ_{jk} is the transfer rate from level J_i to J_j and γ_j is the collisional broadening rate. Figure 4 gives a comparison between the line broadening coefficient for N_2 measured by Rosasco *et al.*[10] with that calculated as a sum of our experimental rate constants. The agreement is seen to be excellent, indicating that rotational relaxation accounts for all of the Raman linewidth. In other words, we observe no significant elastic vibrational dephasing contributions associated with any of the J

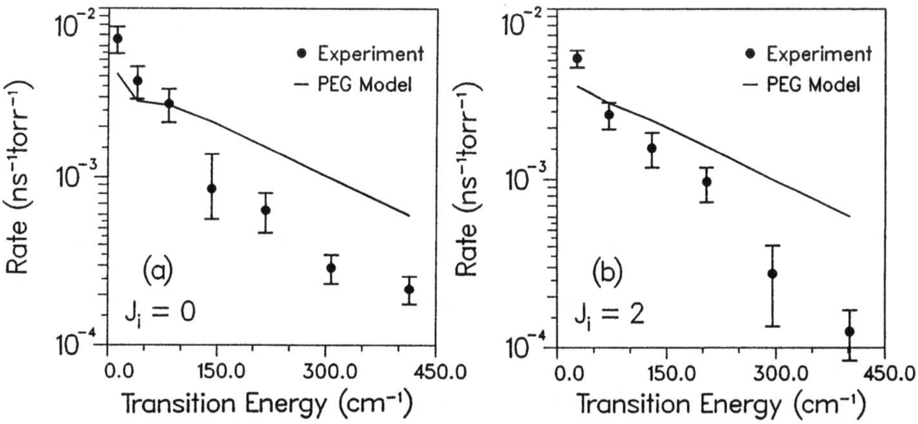

Fig. 5. Comparison of state-to-state rate constants determined from data such as that shown in Fig. 3 with predictions of the PEG model. Panel (a) is for $J_i = 0$ and (b) for $J_i = 2$.

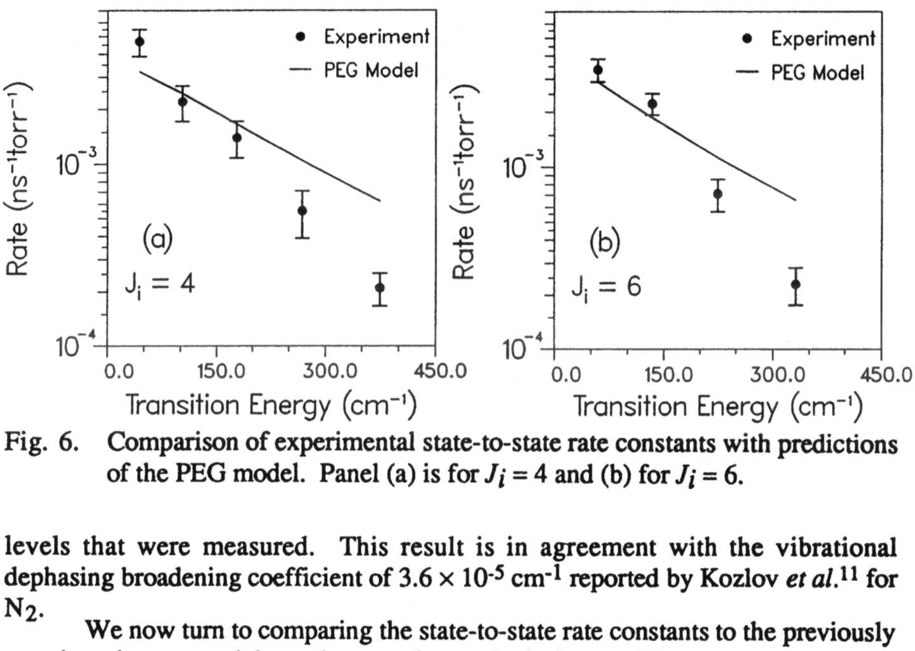

Fig. 6. Comparison of experimental state-to-state rate constants with predictions of the PEG model. Panel (a) is for $J_i = 4$ and (b) for $J_i = 6$.

levels that were measured. This result is in agreement with the vibrational dephasing broadening coefficient of 3.6×10^{-5} cm^{-1} reported by Kozlov *et al.*[11] for N_2.

We now turn to comparing the state-to-state rate constants to the previously mentioned rate models and scattering calculations. We compare with the predictions of four models that were developed using high-quality Raman linewidth measurements measured with inverse Raman scattering (IRS). The models are (1) a polynomial energy-gap model (PEG) proposed by Greenhalgh *et al.*,[13] (2) the MEG model of Koszykowski *et al.*,[14] (3) a statistical polynomial energy-gap model[15] (SPEG) based on the functional form proposed by Sanctuary,[16] and (4) an energy-corrected sudden[17] model (ECS) proposed by Bonamy *et al.*[18]

Figures 5 and 6 show a comparison of the rates from $J_i = 0$, 2, 4, and 6 with the predictions of the PEG model.[13] This model was derived by fitting to

Raman collisional linewidths at different temperatures but did not consider predictions of the Q-branch spectral rotational collapse at high pressures.[19] The model has the form

$$\gamma_{jk} = \rho_j \sum_{m=0}^{n} c_m \, \xi_{jk}{}^{m} , \tag{2}$$

where $\xi_{jk} = |E_j - E_k|^{-1}$, i.e., the inverse of the energy gap of the levels coupled by the collision, ρ_j is the fractional population of the lower level, and the c_m are coefficients determined by fitting. The predictions of the model are in relatively poor agreement with the experimental rates. We see from Figs. 5 and 6 that the model underestimates rates with small ΔJ and overestimates those with large ΔJ. It should be noted that the PEG model was later compared with experimental Raman Q-branch spectra and was shown to predict insufficient rotational collapse.[20] From further comparisons to the scattering calculations of Koszykowski *et al.*,[14] the authors reached essentially the same conclusion as above concerning the predictions of this model.

One of the first successful applications of an exponential gap-type model to the inelastic N_2 rates was developed by Koszykowski *et al.*[14] These authors used an energy-dependent pre-exponential factor, resulting in a modified exponential gap (MEG) model. The parameters of the model were derived by fitting to Raman Q-branch collisional linewidths[21] but were further tested by comparing the spectral rotational collapse predicted by the model with experimental spectra. Looney and Rosasco[15] similarly obtained the SPEG model parameters by fitting Q-branch linewidths measured with IRS[10] and by comparing to experimental spectra. These two models are considered statistically based[22] in that the fitting parameters bear no rigorous physical significance. The expressions used for the excitation rates (initial $J_i <$ final J_j) of the models are[14,16]

$$\gamma_{ij} = \alpha p f(E_i, E_j) \exp(-\beta |\Delta E_{ij}| / kT), \tag{3}$$

where, for the MEG model,

$$f(E_i, E_j) = \left(\frac{1 + 1.5E_i / kT\delta}{1 + 1.5E_i / kT} \right)^2 , \tag{4}$$

and for the SPEG model,

$$f(E_i, E_j) = \left(\frac{|\Delta E_{ij}|}{B_0} \right)^{-\delta} . \tag{5}$$

The de-excitation rates were determined by microscopic reversibility. Here α, β, and δ are fitting parameters (a different set is used for each model, given in Refs. 15 and 20), E_i is the term energy of level J_i, $\Delta E_{ij} = E_i - E_j$, p is the pressure, and B_0 is the rotational constant. Note that for a given initial level the MEG model is a simple exponential in ΔE_{ij}, but the SPEG model includes an additional power law dependence in the $f(E_i, E_j)$ term.

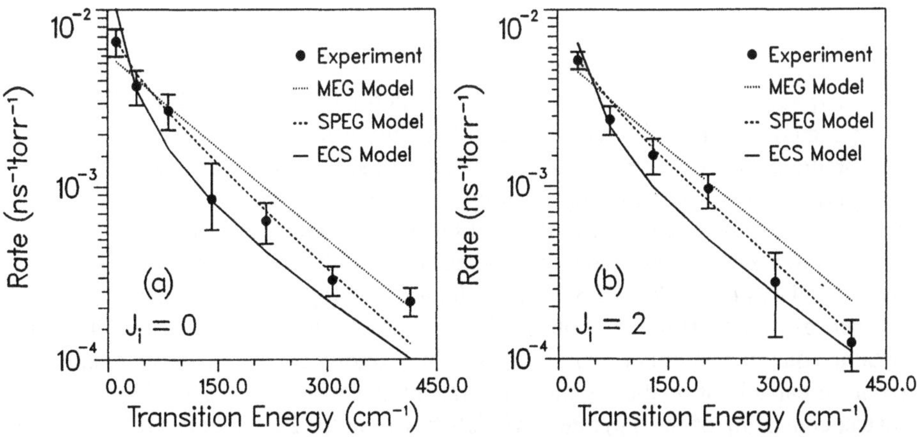

Fig. 7. **Comparison of experimental state-to-state rate constants with predictions of the MEG, SPEG, and ECS models. Panel (a) is for $J_i = 0$ and (b) for $J_i = 2$.**

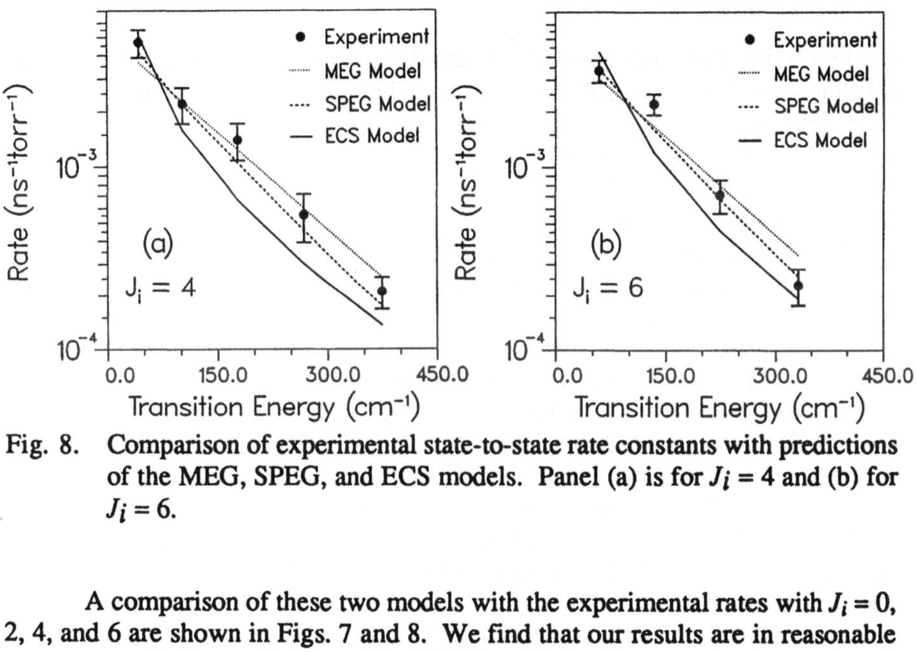

Fig. 8. **Comparison of experimental state-to-state rate constants with predictions of the MEG, SPEG, and ECS models. Panel (a) is for $J_i = 4$ and (b) for $J_i = 6$.**

A comparison of these two models with the experimental rates with $J_i = 0$, 2, 4, and 6 are shown in Figs. 7 and 8. We find that our results are in reasonable agreement with the predictions of the MEG model, but in somewhat better overall agreement with the SPEG model. In particular, the SPEG model reproduces the steeper dropoff observed for rates with $\Delta J = 2$ and $\Delta J = 4$. In other words, the logarithm of the excitation rates from a given initial level appear to exhibit a curvature with ΔE which is not predicted by an essentially exponential-gap law such as MEG. (The MEG model rates out of a given level are purely exponential in ΔE, resulting in a straight line in Figs. 7 and 8.) The SPEG model has the ability to predict this curvature as a result of its power-law term in ΔE [Eq.(5)]. This trend is particularly evident in the data for rates out of $J = 2$ and 4, as shown in Fig. 7a and

7b: the MEG model consistently underpredicts the $\Delta J = 2$ rate and overpredicts the rates for $\Delta J > 2$. The differences between the two gap models and the data lessen at higher J_i values. This is seen in Fig. 8a and 8b for $J_i = 4$ and 6 which shows that the gap models agree with the data almost equally well, although the SPEG model still predicts the $\Delta J = 2$ rate more accurately. It should be emphasized that these comparisons were made using existing values of the constants in the MEG and SPEG models; no attempt was made to fit these constants to our data.

We now compare the experimental rates with more fundamentally based theoretical results. The ECS scaling theory developed by DePristo *et al.*[17] is derived by considering exact physical properties of the S matrix from which are obtained quantum-number scaling relationships among the state-to-state rates. The theory is capable of predicting the entire rate matrix when given a single row of the matrix consisting of the γ_{k0} rates. Models based on the ECS theory are developed by considering different models for the basis rates. Bonamy *et al.*[18] have advanced the following form for the basis rates:

$$\gamma_{k0} = \frac{A}{[J_k\,(J_k + 1)]^{\alpha}}\,, \tag{6}$$

based on fitting Raman Q-branch linewidth data, where A and α were determined by fitting. We show a comparison to this model in Figs. 7 and 8. We find that this model generally predicts a drop-off in the rates with ΔE that is steeper than what is observed. That is, ECS values for the $\Delta J = 2$ rates are higher than the observed values and values for $\Delta J > 2$ are lower than the observed values. These trends with ΔJ can be seen more clearly in Fig. 9 where we have plotted the rates as function of J_i for $\Delta J = 2$ and $= 4$. The two exponential gap models are also plotted in Fig. 9; the consistent slight underestimation of $\Delta J = 2$ rates by the MEG model is evident. The SPEG model, however, is in excellent agreement with these rates.

The poorer agreement obtained between the experiment and the ECS model of Bonamy *et al.*[18] could be a result of their assumed form for the basis rates [Eq. (6)] and may not represent a failing of the scaling capability of the ECS theory. To test the ECS scaling relationships more directly we have used our measured rates for $J_i = L \rightarrow 0$ (column 1 of Table 1) as the basis rates and calculated the rest of the rate matrix using the ECS expressions as given in Ref. 18 and the value of Bonamy *et al.*[18] for l_c. Results are shown in Fig. 10 for $J_i = 2$ and $J_i = 4$, where it is seen that *when an accurate set of basis rates are used the ECS model predictions agree with the measured rates to within experimental error.* The overall level of agreement between the ECS model and the fit is then approximately equivalent to that obtained with the SPEG model, which is the best of the statistically based laws. Thus, we conclude that the ECS theory successfully describes the scaling of the inelastic rates of N_2 at room temperature.

Comparisons with the semiclassical trajectory calculations of Koszykowski *et al.*[14] are limited since only the state-to-state rates reported for $J_i = 4$ can be compared with our data. Figure 11 shows that the agreement obtained is well within the uncertainties of the data (shown by error bars) and the calculation (estimated[23] to be ±10%). We cannot presently determine whether the small underestimation of the rates for small ΔJ by the theory is significant or the result of

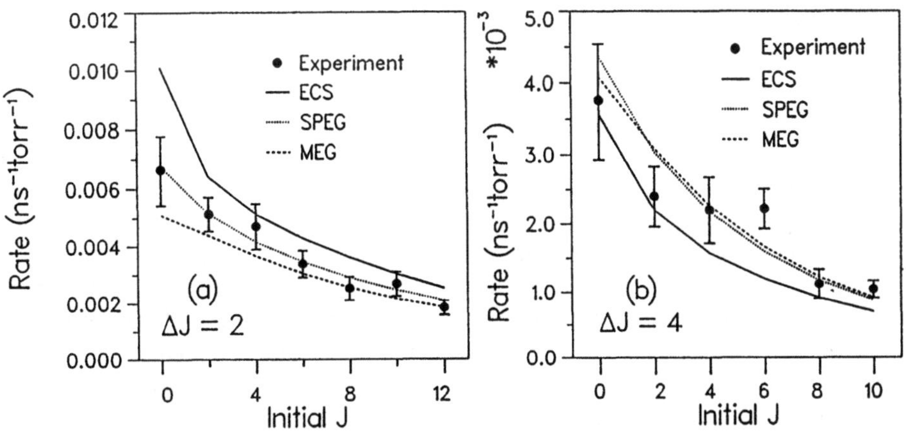

Fig. 9. Comparison of experimental state-to-state rate constants and the predictions of the ECS, SPEG and MEG models. The rates are plotted as a function of J_i for $\Delta J = 2$ and $\Delta J = 4$.

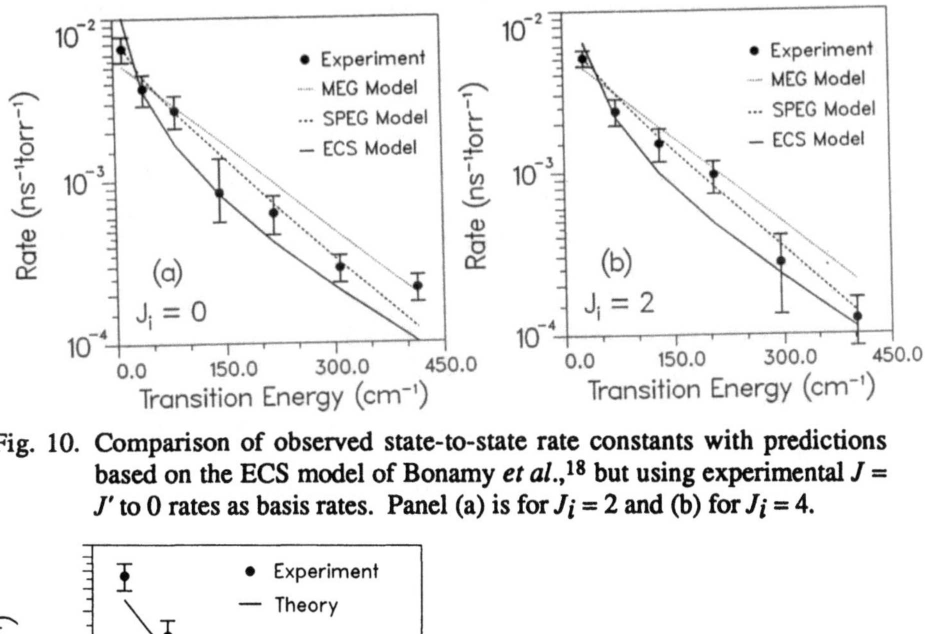

Fig. 10. Comparison of observed state-to-state rate constants with predictions based on the ECS model of Bonamy *et al.*,[18] but using experimental $J = J'$ to 0 rates as basis rates. Panel (a) is for $J_i = 2$ and (b) for $J_i = 4$.

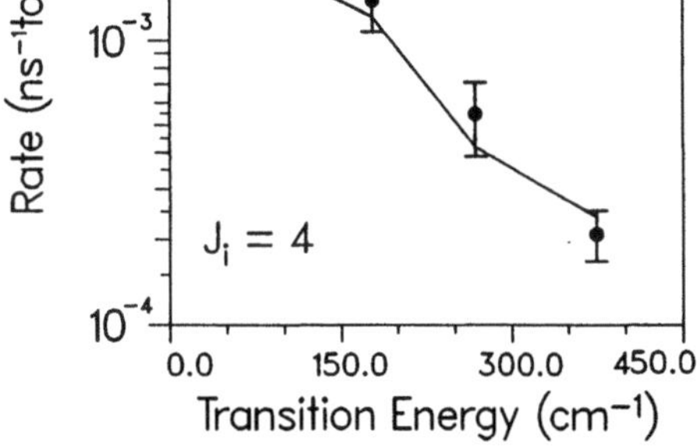

Fig. 11. Comparison of observed state-to-state rate constants with results of a semiclassical scattering calculation by Koszykowski *et al.*[14]

systematic errors in the calculation or experiment. Nevertheless, we consider the agreement to be excellent.

4. Summary and Conclusions

Of the phenomenological models, the modified exponential gap (MEG) model and a statistical polynomial exponential gap (SPEG) model exhibit superior agreement with our measured rates. A polynomial energy gap (PEG) model based only on comparison with Raman linewidth data exhibited the poorest agreement. We attribute this result to the lack of sufficient information in the linewidth data alone to determine the entire rate matrix: the linewidths give information predominantly on the *sum* of the off-diagonal rates in the matrix and not on the distribution of the rates. Models which were developed by also examining spectral rotational collapse were generally in better agreement with our results. An energy-corrected sudden (ECS) model had unsatisfactory agreement, but was significantly improved by the use of our experimentally-determined γ_{k0} rates as basis rates. This result supports the validity of the ECS scaling relationships. We also found excellent agreement between our results and those predicted by detailed semiclassical calculations.[14]

References

1. J. Van Kranendonk, Can. J. Phys. **41**, 433 (1963).
2. R. P. Lucht and R. L. Farrow, J. Opt. Soc. Am. B **5**, 1243 (1988).
3. See, for example, D. A. Greenhalgh, "Quantitative CARS Spectroscopy," Chapter 5 of *Advances in Nonlinear Spectroscopy*, R. Clarke and R. Hester, Eds., (Wiley, New York, 1987).
4. F. de Martini and J. Ducuing, Phys. Rev. Lett. **17**, 117 (1966).
5. R. Frey, J. Lukasik, and J. Ducuing, Chem. Phys. Lett. **14**, 514 (1972).
6. K. L. Carleton, K. H. Welge, and S. R. Leone, Chem. Phys. Lett. **115**, 492 (1985).
7. A. C. Kummel, G. O. Sitz, and R. N. Zare, J. Chem. Phys. **85**, 6874 (1986).
8. W. M. Ho, K. P. Gross and R. L. McKenzie, Phys. Rev. Lett. **54**, 1012 (1985).
9. D. W. Chandler and R. L. Farrow, J. Chem. Phys. **85**, 810 (1986).
10. G. J. Rosasco, W. Lempert, W. S. Hurst, and A. Fein, Chem. Phys. Lett. **97**, 435 (1983).
11. D. N. Kozlov, V., V. Smirnov, and S. Yu. Volkov, Appl. Phys. B **48**, 273 (1989).
12. A. Schenzle and R. G. Brewer, Phys. Rev. A **14**, 1756 (1976).
13. D. A. Greenhalgh, F. M. Porter, and S. A. Barton, J. Quant. Radiat. Transfer **34**, 95 (1985).
14. M. L. Koszykowski, L. A. Rahn, R. E. Palmer and M E. Coltrin, J. Phys. Chem. **91**, 41 (1987).
15. P. J. Looney and G. J. Rosasco, private communication.
16. B. C. Sanctuary, Chem. Phys. Lett. **62**, 378 (1979).

17. A. E. DePristo, J. J. BelBruno, J. Gelfand, and H. Rabitz, J. Chem. Phys. **74**, 5031 (1981).
18. L. Bonamy, J. Bonamy, D. Robert, B. Lavorel, R. Saint-Loup, R. Chaux, J. Santos, and H. Berger, J. Chem. Phys. **89**, 5568 (1988).
19. V. Alekseyev, A. Grasiuk, V. Ragulsky, I. Sobel'man, and F. Faizulov, IEEE J. Quantum Electron **QE-4**, 654 (1968).
20. L. A. Rahn, R. E. Palmer, M. L. Koszykowski, and D. A. Greenhalgh, Chem. Phys. Lett. **133**, 513 (1987).
21. L. A. Rahn and R. E. Palmer, J. Opt. Soc. Am. B **3**, 1164 (1986).
22. T. A. Brunner and D. Pritchard, "Fitting Laws for Rotationally Inelastic Collisions," in *Dynamics of the Excited State*, ed. K. P. Lawley, (John Wiley & Sons, New York, 1982), pp. 589-641.
23. M. L. Koszykowski, private communication, Sandia National Laboratories, Livermore, CA, 94551-0969.

Dicke Effect Manifestation in Nonstationary CARS Spectroscopy

F. Ganikhanov, I. Konovalov, V. Kuliasov, V. Morozov, and V. Tunkin

R.V. Khokhlov Nonlinear Optics Laboratory, International Laser Center, Moscow State University, SU-119899 Moscow, USSR

Abstract. Previously the Dicke effect was studied almost entirely by frequency-domain spectroscopy methods. Time-domain spectroscopy methods give additional opportunities for the observation of this effect. The experimental evidence supporting this statement was obtained here in nonstationary CARS experiments with the $4F_{7/2}$- $4F_{5/2}$ transition of thulium atoms.

1. Introduction

In his pioneering work Dicke considered the radiation of an atom moving between two walls [1]. The atom radiation is frequency modulated by the Doppler effect. Dicke showed that the Fourier spectrum of this radiation contains a frequency unshifted component if the distance between the walls is smaller than the radiation wavelength.

Such experimental conditions can be realized in magnetic resonance. In [2], for example, the rubidium atoms were enclosed in a cell with dimensions smaller than the wavelength of the transition between ground level hyperfine components. The walls of the cell were covered with a special material which did not dephase the excitation during atomic collisions with the walls. In accordance with Dicke's predictions, a narrow spike on the top of a much broader base was observed.

But usually the collisions between atoms or molecules serve as a movable wall. For example, Dicke and Wittke [3] mixed atomic and molecular hydrogen, due to collisions with molecules the spectral lines of atomic hydrogen were narrowed and the hyperfine splitting of the ground level was redetermined with better accuracy.

For the most part the Dicke effect was observed rather in molecular than in atomic species. Different spectroscopic methods were applied to record it for the molecular hydrogen or deuterium Q-lines: stimulated Raman scattering [4], spontaneous Raman scattering [5], CARS with pulsed lasers [6,7], CARS with continuous wave lasers [8] and Raman amplification [9]. Using spontaneous Raman scattering [10] and diode laser spectroscopy [11] the effect was recorded for pure rotational transitions in hydrogen. Hydrogen and deuterium are used most frequently in such experiments, this follows from the fact that collisional broadening, which prevents the observation of the Dicke effect, is anomalously small in the case of hydrogen (deuterium) transitions.

Methods of frequency-domain spectroscopy were employed in all of the above-mentioned experiments. In our experiment performed a decade ago [12] we used one of the time-domain

Springer Proceedings in Physics, Volume 63 **Coherent Raman Spectroscopy**
Eds.: G. Marowsky and V.V. Smirnov

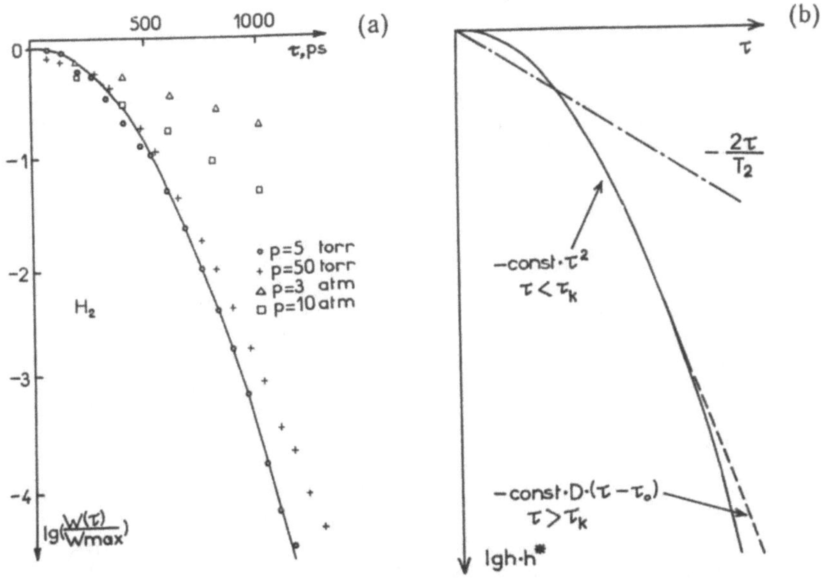

Fig.1. (a) Anti-Stokes signal in pure hydrogen versus delay time [12]; solid line-calculated signal in the Doppler limit. (b) Transformation of the pulse response with pressure p [13]: diffusion coefficient D~1/p, dephasing time T_2 ~ 1/p, correlation time for velocity τ_k ~ 1/p, $\tau_0 \approx \tau_k$.

spectroscopy methods, nonstationary CARS, to record the Dicke effect for one of the components of the hydrogen Q-line. The results of this experiment are reproduced in Fig.1. In the Doppler limit the anti-Stokes signal plotted on a logarithmic scale versus delay time is close to a parabolic dependence. The solid curve in this figure is just the Doppler parabola for room temperature. With increasing pressure, the trailing part of the anti-Stokes signal dependence increases. For a hydrogen pressure of 3atm the whole anti-Stokes signal dependence is straight and the dephasing is slowest. For pressures higher than 3 atm collisional broadening makes the dephasing faster.

In 1983 Djakov theoretically showed [13] that some rise of the trailing part of the pulse response occurs for every transition. Djakov derived the expression for the pulse response, taking into account two processes: the process of collisional phase distortions, characterized by dephasing time T_2, and the process of collisional velocity distortions, characterized by correlation function $B(\Theta)=\langle v(t)v(t+\Theta)\rangle$. If these two processes are statistically independent then the pulse response is given by the following expression:

$$h(\tau) = \exp\left[-\frac{\tau}{T_2} - k^2\cdot\int_0^{\tau}(\tau-\Theta)B(\Theta)d\Theta\right] \qquad (1)$$

where k is the wave vector for the frequency of the transition under investigation. The behaviour of this pulse response with pressure can be represented by Fig.1. In the Doppler limit ($T_2 = \infty$, the correlation time for velocity $\tau_k = \infty$) the pulse response on a logarithmic scale is parabolic. With increasing pressure the pulse response deviates from parabolic in two ways. First, the trailing part ($\tau > \tau_k$) due to the Dicke effect becomes straight and dephases slower than in the Doppler limit. Second, the initial part due to phase distortions also becomes straight, but dephases faster. Thus the Dicke effect and phase distortions manifest themselves, for small enough pressures, in different parts of the pulse response. So, in time-domain spectroscopy, at least in principle, an opportunity exists to observe these two effects separately. This can be done in real experiment if the dynamic range of pulse response measurements is sufficiently large.

2. Experimental

The scheme of our picosecond nonstationary CARS spectrometer is shown in Fig.2. The fundamental passively mode-locked Nd-YAG oscillator is feedback controlled using a high current rapid photomultiplier and electrooptic modulator. The weak pulse is split off by a beam splitter placed inside the cavity (Fig.3). The output of the photomultiplier is applied to the electrooptic modulator with rather low halfwavelength voltage (400V). Negative feedback functions until a negative electrical pulse formed by the avalanche transistors is applied to one of the photomultiplier dynodes approximately 30 μs after the beginning of the oscillations. During this period picosecond pulses with small energy but stable parameters are produced. After the photomultiplier is switched off a short train of stable picosecond pulses is formed.

To determine the stability of the pulse parameters, the central pulse from the train was selected, frequency doubled

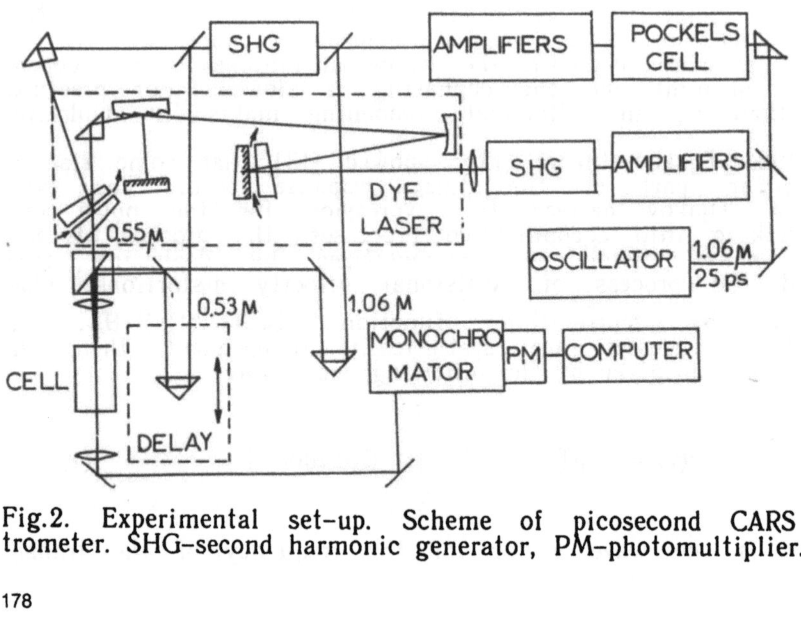

Fig.2. Experimental set-up. Scheme of picosecond CARS spectrometer. SHG-second harmonic generator, PM-photomultiplier.

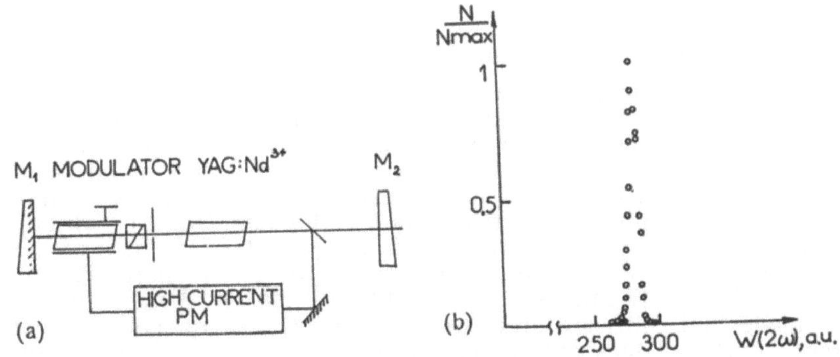

Fig.3. Feedback controlled passively mode-locked Nd-YAG oscillator. (a) M-mirrors, PM-photomultiplier. (b) The statistics of the second harmonic energy, generated with 1% efficiency by single pulse selected from the train.

with 1% efficiency and the statistics of the second harmonic energy were measured (Fig.3). The statistics were accumulated for 20 minutes at a repetition rate of 1.5 Hz. The energy of the second harmonic is stable within ±2%. The duration of the pulses is 30 ps. The output pulse train energy is 1.5 mJ.

The output pulse train is split into two parts. One part is amplified, frequency doubled and serves as a pump for the synchronously mode-locked dye laser. Another part goes through the pulse selector. The selected pulse is amplified and then is also split into two parts. One single pulse is amplified, frequency doubled and serves as a pump for the dye amplifier. Another single pulse, after amplification, is in its turn split into two parts: one part is used as an exciting pulse, the second part is frequency doubled, delayed in the variable delay line and used as a probing pulse. The energies of the exciting pulses are on the fundamental wavelength 5 mJ and on the dye laser wavelength 0.5 mJ; the energy of the probing pulse is 3 mJ.

The exciting and probe beams were combined using a dichroic mirror and a Glan polarizer and then focused by a 50 cm focal length lens into a heat-pipe with the atomic vapour under investigation. The temperature of the heat-pipe central zone, where the exciting and probe beams were focused, was controlled either by a thermopile thermometer or by a pyrometer.

The anti-Stokes radiation was collected by a lens, focused on the slit of the monochromator and then detected by a photomultiplier. The radiation energy is proportional to the electrical charge at the output of the photomultiplier. This charge is transformed into two voltages by two charge-voltage converters, working in parallel, with different amplifications 1 and 100. These voltages are digitized by analog-digital converters with a dynamic range of 1000 and then processed by computer. The computer selects the signal from the channal with amplification 100, if this signal is in linear range, or from another channel, if this is not the case. Thus a dynamic range of 10^5 is obtained in measurements of the anti-Stokes radiation energy.

3. Results and discussion

In our previous experiments with the $6P_{1/2}$-$6P_{3/2}$ transition of thallium (Tl) atoms it was shown that the character of the hyperfine components quantum beats depends on the polarizations of the exciting and probe beams [14]. Quantum beats are most pronounced for this transition when the polarizations allow the generation of the anisotropic scattering signal. Since the isotropic scattering is forbidden ($\Delta J=1$) this is accomplished simply by using parallel polarizations of the exciting beams [15]. It is interesting to show here the results of the anti-Stokes signal versus delay time measurements made not only in the Doppler limit, as in [14], but for different buffer gas pressures. Figure 4 shows the results for buffer gas (neon) pressures of 10, 360 and 800 Torr in the case of pure anisotropic scattering. Computer simulations are presented by solid lines. The minima of the quantum beats did not change their positions in delay time, i.e. the frequency interval between components also does not change. This fact is in agreement with theoretical work [16], where it was shown that the collisional shifts of hyperfine components are in general different, but for a small number of transi tions (including the Tl transition under investigation $J=1/2 \rightarrow J=3/2$) they are equal.

The following expression for pulse response $h(\tau)$ was used in the computer simulation:

$$h(\tau) = \exp\left[-\frac{\tau}{T} - \frac{\tau^2 k_B T k_\Omega^2}{2m} \right] \sum_{F_b, F_c} [F_b F_c \kappa] \cdot \exp(i\omega_{F_b F_c} \tau) , \quad (2)$$

where $[F_b F_c \kappa] = (2F_b + 1)(2F_c + 1) \cdot \begin{Bmatrix} J_b & J_c & \kappa \\ F_c & F_b & I \end{Bmatrix}^2$ are the amplitudes of

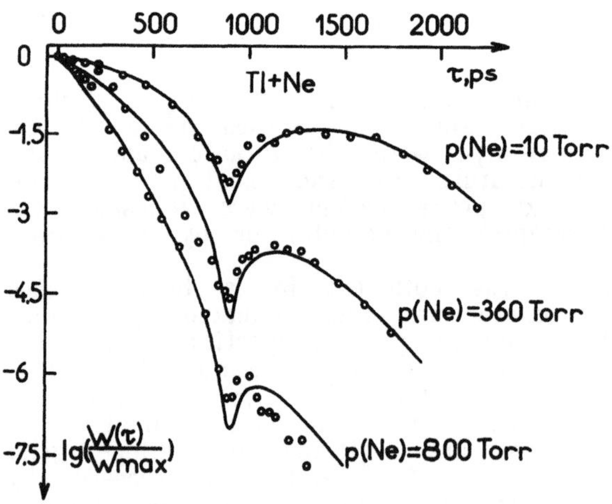

Fig.4. Anti-Stokes signal versus delay time for the $6P_{1/2}$-$6P_{3/2}$ transition of Tl with Ne as a buffer gas. Solid lines computer simulation.

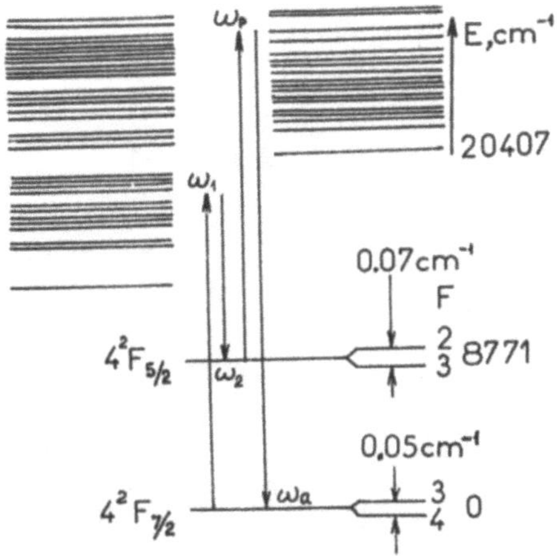

Fig.5. Quantum diagram of excitation and probing for the $4F_{7/2}$–$4F_{5/2}$ transition of Tm. 0.07 cm^{-1} and 0.05 cm^{-1} hyperfine splittings of upper and ground levels; F full atomic moments; ω_1 and ω_2 frequencies of exciting pulses, ω_p frequency of probing pulse, ω_a frequency of anti-Stokes pulse.

the hyperfine CARS-spectrum components for the multipole scattering of rank κ [15], { } is the 6j symbol, J_b and J_c are electronic moments of the ground and excited states, respectively, F_b and F_c are the full atomic moments, I is the nuclear spin, k_B the Boltzmann constant, T the temperature, k_Ω the wavevector, m the atomic mass and $\omega_{F_bF_c}$ the frequency of the hyperfine transition. For a collinear geometry: $k_\Omega = k_1 - k_2$, where k_1 nd k_2 are the wavevectors of the exciting beams. We assume here that multipole scattering of single rank is recorded. Assuming then Gaussian amplitudes for the exciting and probing pulses with durations of 30 ps and using expression (2) for $h(\tau)$, the energy of the anti-Stokes pulse versus delay time is found, and presented in Fig.4.

The transformation with pressure of the anti-Stokes signal versus delay time for the $4F_{7/2}$–$4F_{5/2}$ thulium (Tm) transition is different. A quantum diagram of excitation and probing and also the hyperfine structure of the corresponding levels are shown in Fig.5. Due to the resonance conditions for excitation and probing, the intensity of antisymmetric scattering is greater than the intensity of anisotropic scattering (isotropic scattering is forbidden). The relative intensities of antisymmetric and anisotropic scatterings were found by rotating the analyzer in the anti-Stokes beam intothe following polarization situation: polarizations of the exciting beams are perpendicular to each other and form a 45° angle with the polarization of the probe beam (Fig.6). In this case the polarizations of the anisotropic scattering and of the probe beam coincide; the polarizations of the probe beam and of the

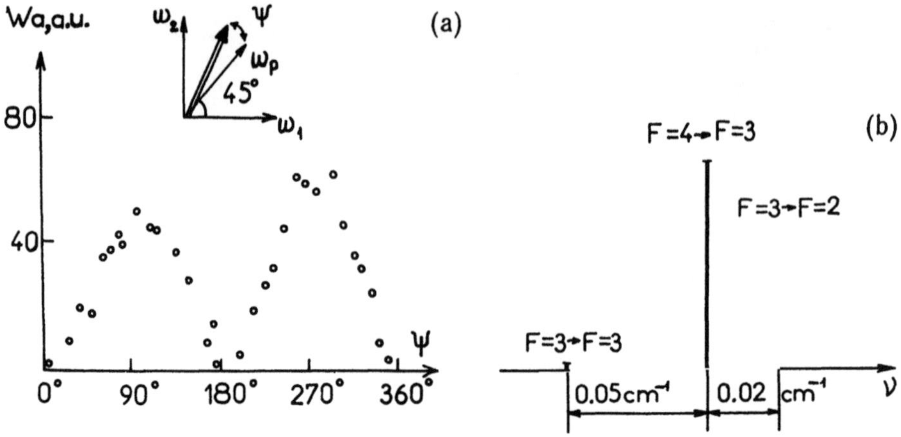

Fig.6. (a) Polarization of the anti-Stokes beam, Ψ is the angle between the analyzer axis and the polarization of the probe beam. (b) Hyperfine structure of the CARS spectrum of the $4F_{7/2}$-$4F_{5/2}$ Tm transition for antisymmetric scattering; one component F=4 → F=2 is forbidden; the ratio of amplitudes of other components (calculated) is 1:27:20.

antisymmetric scattering are always perpendicular to each other. The intensity of the antisymmetric scattering is 30 times greater than the intensity of the anisotropic one. For this reason, in order to favour the generation of the antisymmetric scattering, the mutually perpendicular polarizations of the exciting beams were used. For the same reason computer simulations were made for the antisymmetric scattering.

Fig.6 shows the hyperfine structure of the CARS spectrum of the $4F_{7/2}$-$4F_{5/2}$ Tm transition for the antisymmetric scattering. One component F=4 → F=2 is forbidden; the ratio of amplitudes of other components (calculated) is 1:27:20.

The transition under investigation is the transition between orbits which are screened by electrons on external orbits. This diminishes the collisional broadening and allows the observation of the Dicke effect manifestation with the anti-Stokes signal dynamic range reached in our experiments.

The anti-Stokes signals versus delay time for the $4F_{7/2}$-$4F_{5/2}$ Tm transition and different buffer gas (neon) pressures are shown in Fig.7. The temperature in the excitation and probing volume was 1450 K. Each experimental point was averaged over 20 laser shots. The solid line is the theoretical curve for the Doppler limit. For comparison reasons, the same curve is reproduced in each figure. With increasing pressure the experimental dependences of the anti-Stokes signal on delay time are modified in two ways. First, the minimum of the hyperfine components' quantum beats is shifted to longer delay times. Second, the signal for delay times more than 1.5 ns decays more slowly; the slowest decay is at a neon pressure of 7 atm. This is an obvious manifestation of the Dicke effect.

Thus, the decay slowing down does not reach the initial part of the anti-Stokes signal dependences, as in hydrogen,

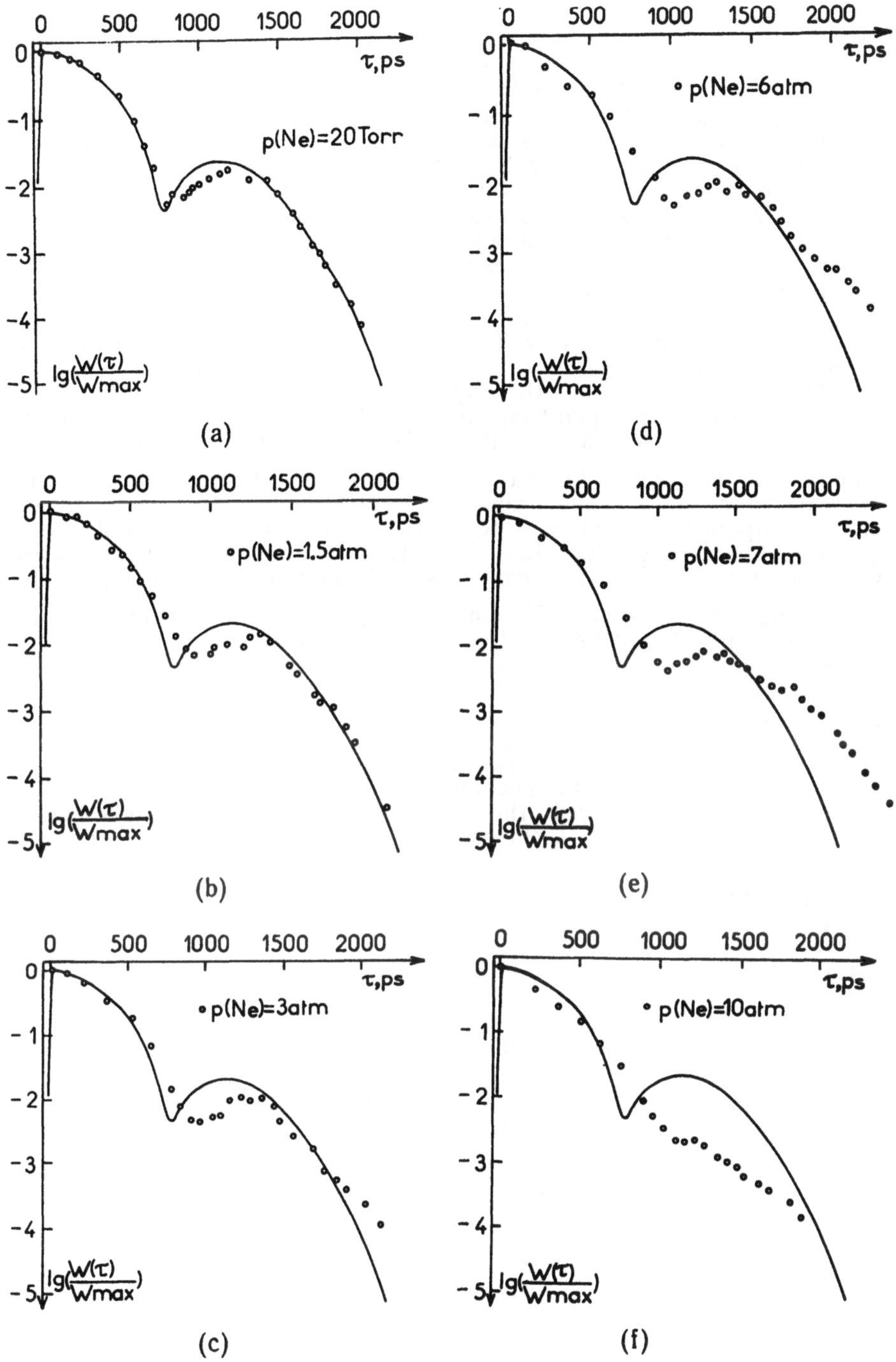

Fig.7a–f. Anti-Stokes signal versus delay time for the $4F_{7/2}$–$4F_{5/2}$ Tm transition. Solid line: computer simulation for the Doppler limit.

and, hence, the spectral line narrowing is quite small. The CARS spectra calculations based on the experimental pulse responses give only 5% narrowing. At the same time, the manifestation of the Dicke effect in time-domain spectroscopy is quite significant: the anti-Stokes signal for a delay time of 2.5 ns and a neon pressure of 7 atm is two orders of magnitude more intense than in the Doppler case.

The decay slowing down as the manifestation of the Dicke effect was also observed with helium as a buffer gas, but in this case the effect is less pronounced. The CARS spectra, calculated using experimental pulse responses, show no line narrowing in this case.

The quantum beat's minimum is shifted with helium as a buffer gas in approximately the same manner as with neon. Two reasons can be given for this shift. First, different ordinary collisional shifts of the hyperfine components which manifest themselves in quantum beats. If we assume this reason then we obtain for a relative collisional shift of these components: $2 \cdot 10^{-3}$ cm^{-1}/Amagat. According to the results of theoretical work [16] the absolute collisional shift of hyperfine components should be still greater. The collisional spectral broadening for the Tm transition under investigation with helium as a buffer gas was determined experimentally and was found to be ~ $1 \cdot 10^{-3}$ cm^{-3}/Amagat [17]. But the collisional broadening should be greater than collisional shift.

The second reason is the effect analogous to molecular Q-line "motional narrowing". The computer simulation of the anti-Stokes signal based on this last assumption for different buffer gas pressures gives closer agreement with experimental results. Further experimental investigations are under way to determine the real mechanism of the shift.

So, we have observed the manifestation of the Dicke effect in nonstationary CARS-spectroscopy of the atomic Tm transition $4F_{7/2}$- $4F_{5/2}$ and shown that methods of time-domain spectroscopy are more adequate for the observation of this effect.

References

[1] R.H.Dicke: Phys.Rev. **89**, 472 (1953)
[2] J.C.Camparo, R.P.Frueholz, H.G.Robinson: Phys.Rev. **A40**, 2351 (1989)
[3] J.P.Wittke, R.H.Dicke: Phys.Rev. **103**, 620 (1956)
[4] P.Lallemand, P.Simova, G.Bret: Phys.Rev.Lett. **17**, 1239 (1966)
[5] A.Javan, J.R.Murray: J.Mol.Spect. **29**, 502 (1969)
[6] F.DeMartini, G.P.Giuliani, B.Santamato: Opt.Comm. **5**, 126 (1972)
[7] M.R.Malikov, A.D.Savel'ev, V.V.Smirnov: Pis'ma Zh. Eksp. Teor. Fiz. **39**, 527 (1984)
[8] M.A.Henesian, L.Kulevskii, R.L.Byer, R.L.Herbst: Opt. Comm., **18**, 225 (1976)
[9] A.Owyoung: Opt.Lett. **2**, 91 (1978)
[10] V.G.Cooper, A.D.May, E.H.Hara, H.F.P.Knapp: Can.J.Phys. **46**, 2019 (1968)
[11] J.Reid, A.R.W.McKellar: Phys.Rev. **A18**, 224 (1978)
[12] S.A.Magnitskii, V.G.Tunkin: Kvantovaja Elektronika **8**, 2008 (1981)

[13] Yu.E.Djakov: Pis'ma v JETP **37**, 14 (1983)
[14] V.D.Vedenin, F.Sh.Ganikhanov, S.Dinev, N.I.Koroteev, V.N. Kuliasov, V.B.Morozov, V.G.Tunkin: Opt.Lett. **14**, 113 (1989)
[15] A.I.Alekseev, V.N.Beloborodov, O.V.Zhemerdeev: J.Phys.B. **22**, 143 (1989)
[16] V.N.Rebane: Optika i Spektroskopija **41**, 372 (1976)
[17] N.I.Agladze, V.N.Kuliasov: Optika i Spektroskopija **63**, 12 (1987)

Picosecond Coherent Raman Spectroscopy of Excited Electronic States of Polyene Chromophores

N.I. Koroteev, A.P. Shkurinov, and B.N. Toleutaev

R.V. Khokhlov Nonlinear Optics Laboratory, International Laser Center, Moscow State University, SU-119899 Moscow, USSR

Abstract. The results are presented of a detailed analysis of the vibrational structure of polyene molecules in the excited electronic states. The molecules in a solution were excited by picosecond UV pulses and then probed by means of resonant coherent anti-Stokes and Stokes Raman scattering (CARS and CSRS).

The polarization sensitive CARS spectra of *trans*-stilbene in the S_1 state were measured with high resolution and signal-to-noise ratio. The technique of coherent ellipsometry was applied to determine the complex components of the third-order susceptibility tensor. The extremely high value of the resonant electronic hyperpolarizability of S_1 *trans*-stilbene detected was the result of significant π-electron delocalization in the polyene chain under the electronic excitation. The observed phases of different vibrational resonances measured relative to the phase of electronic susceptibility are considered as new molecular characteristics which reflect the features of the electron-vibronic coupling in the excited molecule.

The conformational and vibrational relaxation in excited *trans*-stilbene was studied in the presence of optical depletion of the S_1 state. The depletion of the ground state of β-carotene under the laser excitation and subsequent recovering kinetics was detected by time-resolved CARS.

1. Introduction

Recently, there has been considerable interest in the time-resolved vibrational spectroscopy of excited electronic states [1, 2]. The transient picosecond resonance Raman technique (linear or nonlinear) provides the means for investigating the changes in molecular vibrations under the electronic excitation and gives more direct and detailed information about the structure of polyatomic molecules in liquids, which is inaccessible to absorption or fluorescence spectroscopy [3, 4].

The advantages of the resonant coherent anti-Stokes and Stokes Raman scattering (CARS and CSRS) methods over spontaneous Raman spectroscopy in the study of short-lived excited electronic states are well established [5-9]. When measuring the spectra of excited states with lifetimes in the picosecond range a significant problem arises in detecting extremely weak spontaneous Raman signals, which are generally overwhelmed by the intense fluorescence from the excited molecules.

1) In CARS/CSRS, the signal level is orders of magnitude higher, thus providing good accuracy and wide dynamic range in measurements. The better

Springer Proceedings in Physics, Volume 63 **Coherent Raman Spectroscopy**
Eds.: G. Marowsky and V.V. Smirnov

spectral resolution, limited only by the laser bandwidths, gives the possibilities of more detailed Raman line-shape analysis. The beam characteristics of the coherent Raman signal allows the spectral, polarization and spatial rejection of the luminescent background [5, 6].

2) Using the methods of nonlinear optics one can realize the full recording of amplitude, polarization and phase characteristics of the coherent scattered light field [6]. Here, as in optical holography, the recording of spectral information is carried out by actively controlled interference of coherent resonant components (Raman, electronic, etc.) with the non-resonant background from the solvent, the latter playing the role of a reference wave. In other words, the coherent Raman scattering technique can be considered as an analog of the optical heterodyne scheme, having the potential to obtain novel information on the molecule.

Excited electronic states of polyatomic molecules play an important role in many ultrafast photochemical and biological processes. There is now a series of publications on the transient coherent Raman spectroscopy of short-lived excited states, which has demonstrated the advantages of this technique in research into the dynamics of large organic molecules, including biological chromophores [3, 7-12].

In the present contribution, we show the potential of the nonlinear-optical methods to obtain new spectroscopic characteristics, which are derived from the measurement of third-order susceptibilities in excited states. These new parameters can provide detailed information about the structure and dynamics of polyatomic molecules in solutions.

2. Experimental

The principle of the "pump-probe" resonant CARS and CSRS spectroscopy of excited electronic states is illustrated in the energy diagram of the stilbene molecule, Fig. 1. A picosecond UV pulse prepares transient excited molecules in the S_1 state which are then probed by a pair of visible pulses ($\omega_1 > \omega_2$), appropriately delayed with respect to the excitation pulse. Both frequencies ω_1 and ω_2 are in resonance with $S_n \leftarrow S_1$ transition so that the generated CARS ($\omega_a = 2\omega_1 - \omega_2$) or CSRS ($\omega_s = 2\omega_2 - \omega_1$) signal is resonantly enhanced.

A picosecond laser system for CARS/CSRS spectrochronography is based on a *cw*-pumped, acousto-optically mode locked and *Q*-switched Nd:YAG laser, operating at 5 kHz repetition rate [13, 14]. The Nd:YAG second and fourth harmonics, along with a tunable dye-laser, yield picosecond pulses with high peak intensities at low average power, which cover a wide spectral range in the visible and UV regions. The high repetition rate and the use of time-correlated single-photon counting enable the reliable detection of Raman scattering signals with the highest possible sensitivity and good signal-to-noise ratio.

The optical scheme of the picosecond CARS/CSRS spectrometer is depicted in Fig. 2. The trains of Nd:YAG-laser pulses (λ = 1064 nm) consist of 30 pulses each of about 80 ps duration separated by 10 ns. After frequency doubling in a $LiIO_3$ crystal, which yields about 30% in the second harmonic (λ = 532 nm), the residual fundamental radiation is separated by a dichroic

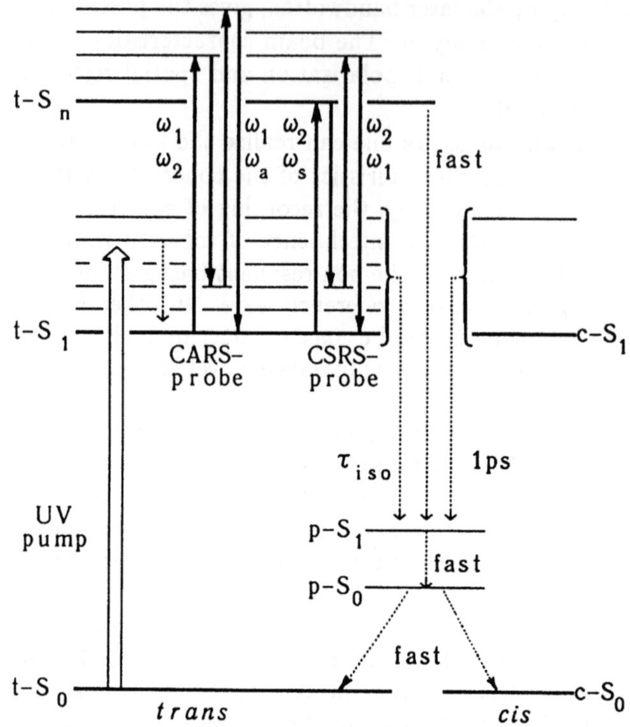

Fig. 1. The energy diagram of the stilbene molecule and the principle of the "pump-probe" experiment for transient resonant CARS/CSRS of excited electronic states.

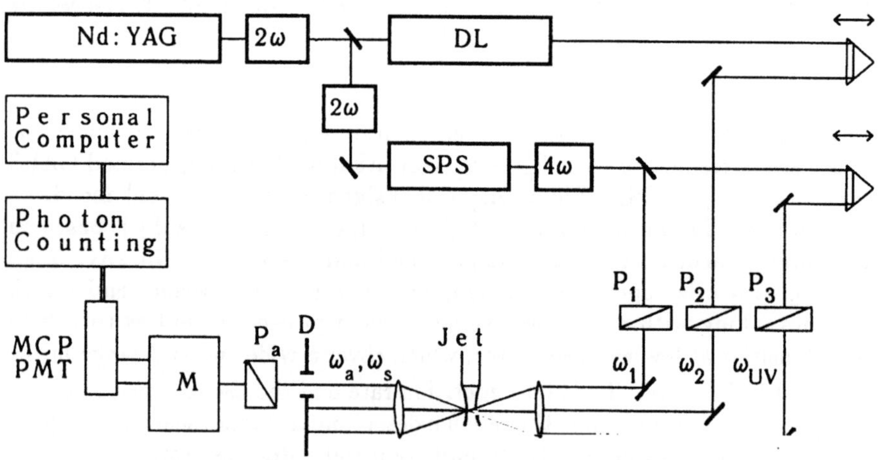

Fig. 2. Experimental set-up for picosecond excitation and transient CARS/CSRS spectrochronography of excited electronic states.
DL - dye laser; **SPS** - single pulse selector; $\mathbf{P_1 - P_3}$, $\mathbf{P_a}$ - polarizers; **D** - diaphragm; **M** - double monochromator; **MCP-PMT** - microchannel plate photomultiplier tube.

mirror and again frequency doubled in a KTP crystal with an efficiency of up to 50%. The total average power of the second harmonic after both crystals is 3.5 W, which corresponds to the peak pulse power of up to 400 kW.

The trains of second harmonic pulses from the $LiIO_3$ crystal synchronously pump a dye laser to produce the picosecond pulses with tunable frequency ω_2 for CARS/CSRS-probing. A diffraction grating (1800 lines/mm) at grazing incidence is used as an intracavity dispersive element. The wavelength is tuned in the range 550–600 nm by tilting the mirror in the first order of reflection. The output of the dye laser, which is taken off through the grating's zero order, are nearly spectral limited pulses of $\tau \approx 50$ ps duration and $\Delta\nu \approx 0.5$ cm^{-1} bandwidth ($\Delta\nu \cdot \tau \cdot 2\pi/c \leq 1$).

The single pulses of the second harmonic after the KTP crystal are selected from the trains by the Pockels cell and transformed into the fourth harmonic in a KDP crystal. The average power in the UV is up to 2.5 mW, the pulse energy is 1.0 μJ . When the pulse is focused to a waist diameter of 30 μm , the peak intensity is as high as 10^9 W/cm^2 , which exceeds the saturation intensities of many organic molecules. Thus one can obtain a population of the excited state comparable with that of the ground state. As a result, we obtain the single pulses for excitation (ω_{UV} , $\lambda = 266$ nm) and CARS probing (ω_1 , $\lambda = 532$ nm). These are separated from each other by a dichroic mirror to pass two independent optical delays.

The excitation (ω_{UV}) and probing beams (ω_1 and ω_2) are focused into a sample by a spherical mirror with 5 cm radius of curvature. The angle between the polarization vectors e_1 and e_2 of the two probing beams is adjusted using polarizers $\mathbf{P}_1$ and $\mathbf{P}_2$. To prevent undesired thermal effects and to reduce the non-resonant signal, a free flowing jet of the solution investigated is used, its thickness being about 300 μm . The flow rate of the jet is sufficient to replace completely the solution in the region of interacting beams (diameter 50 μm) between successive pulses.

The signal of coherent anti-Stokes (ω_a) or Stokes (ω_s) scattering, after passing through the diaphragm D, polarization analyzer $\mathbf{P}_a$ and double monochromator M, is detected by a fast photomultiplier tube PMT. The dispersion of the signal is recorded when tuning $\Delta\omega = \omega_1 - \omega_2$. A spectral resolution better than 0.5 cm^{-1} is determined by the laser bandwidths; a temporal resolution equal to the pulse duration (40 – 60 ps) can be achieved.

The signal detection is based on a time-correlated single photon counting system with a microchannel plate PMT [14]. The half-height width of the response function, being as low as 100 ps, allows us to strongly reject the dark noise of the PMT and to reduce the background luminescence from the excited sample, by gating the photon counts within the narrow time interval around the laser pulse.

3. Transient Resonance CARS/CSRS of S_1 trans-Stilbene

The photochemistry and photophysics of diphenylpolyenes have been the subject of many experimental and theoretical investigations, particularly directed towards understanding the mechanism of polyene *cis-trans* photoisomerization

[15, 16]. Stilbene appears to be the most interesting molecule for the study of ultrafast processes following photoexcitation. The most detailed information on the structure and dynamics of the excited electronic states of *trans*-stilbene in solutions has been obtained by means of time-resolved resonance Raman scattering [16-19].

3.1. Polarization Sensitive Resonance CARS of S_1 trans-Stilbene

Polarization measurements in coherent Raman spectroscopy are known to yield extensive information on the sample, because here we deal with the fourth-rank tensor $\chi^{(3)}_{ijkl}$ of the third-order susceptibility [5, 6, 20]. The application of polarization CARS enables the resolution, sensitivity, and accuracy to be enhanced.

The coherent background in CARS arising from the non-resonant electronic susceptibility, χ^{NR}_{solv}, of the solvent often causes a large unwanted signal which limits the sensitivity of the technique when applied to the condensed state. In the common case of low concentrations of the molecules being investigated, their resonance contribution, χ^{R}, to the net third-order susceptibility, $\chi^{(3)} = \chi^{NR}_{\text{solv}} + \chi^{R}$, of a solution may be so weak that the non-resonant background may obscure all the spectral features of interest. Polarization CARS provides the means to discriminate against coherent background and to increase the contrast of the resonances [5, 6].

The CARS signal is determined by a non-linear polarization source at the anti-Stokes frequency, ω_a:

$$\mathbf{P}^{(3)}(\omega_a) = 3(\chi^{(3)}_{1111} \cos\psi \cdot \mathbf{e}_x + \chi^{(3)}_{2112} \sin\psi \cdot \mathbf{e}_y) E_1^2 E_2^* ,$$

where E_1, E_2 are the electric fields of the probing beams ω_1 and ω_2, respectively; ψ is the angle between their polarization vectors $\mathbf{e}_1$ and $\mathbf{e}_2$ (Fig. 3).

The polarization direction of the signal due to the resonant (non-resonant) contribution is given by an angle ϕ^R (ϕ^{NR}) between $\mathbf{P}^R$ ($\mathbf{P}^{NR}$) and vector $\mathbf{e}_x$, which depends upon the ratio of $\chi^{(3)}_{ijkl}$ tensor components:

$$\phi^{R,NR} = \arctan\left(\frac{\chi^{R,NR}_{2112}}{\chi^{R,NR}_{1111}} \tan\psi\right) .$$

(The superscript $^{(3)}$ is omitted for compact representation.)

For the non-resonant susceptibility $\chi^{NR}_{2112}/\chi^{NR}_{1111} = 1/3$, and the solvent contribution $\mathbf{P}^{NR}$ is linearly polarized at an angle $\phi^{NR} = \arctan\left(\frac{1}{3}\tan\psi\right) = 45^\circ$, when $\psi = 72^\circ$. Using the difference in tensor characteristics of the two contributions we are able to suppress the coherent background by applying a polarization analyzer $\mathbf{P}_a$ in the anti-Stokes beam, crossed with the $\mathbf{P}^{NR}$ direction (Fig. 3) [5].

Figure 4 represents the wide-range transient CARS spectra of *trans*-stilbene in ethanol. Both UV-pumped (a) and unpumped (b) spectra were recorded with

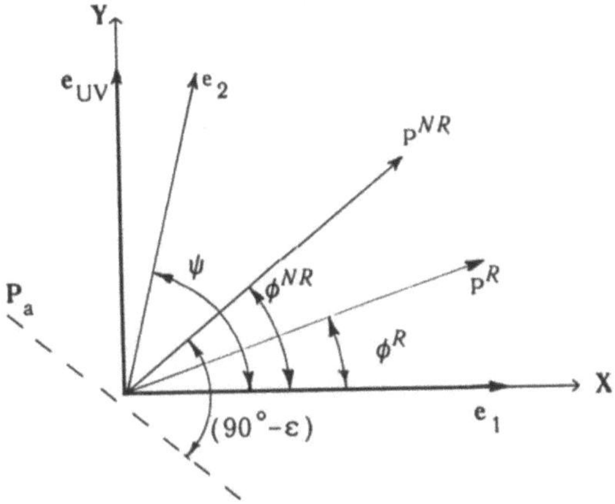

Fig. 3. The polarization vectors of exciting, e_{UV}, and CARS probing beams, e_1, e_2, and the non-linear polarization of the non-resonant signal, $\mathbf{P}^{NR}$, and of the resonance of the molecules studied, $\mathbf{P}^{R}$. $\mathbf{P}_a$ shows the direction of the polarization analyzer for non-resonant background suppression. $\psi = 72°$.

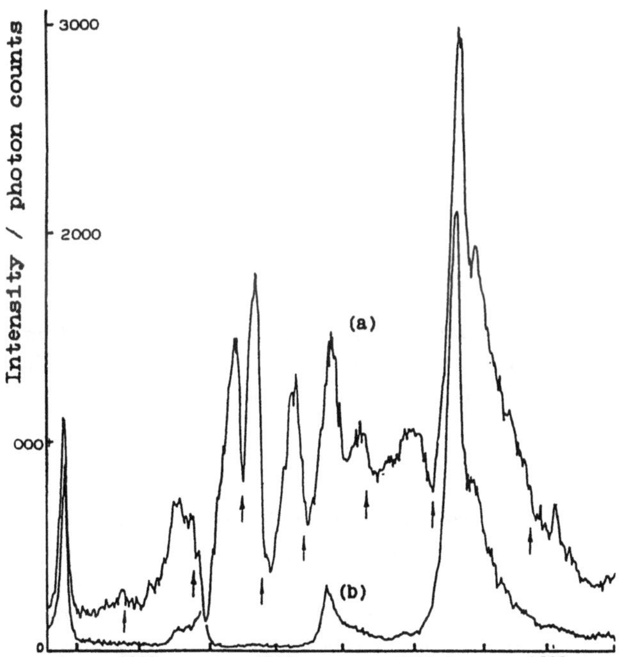

Fig. 4. Transient CARS spectra of *trans*-stilbene in ethanol with polarization suppression of non-resonant background, with UV excitation present (a) and without UV (b). Concentration 1 mM. Arrows indicate the vibrational bands of *trans*-stilbene in the S_1 state [18].

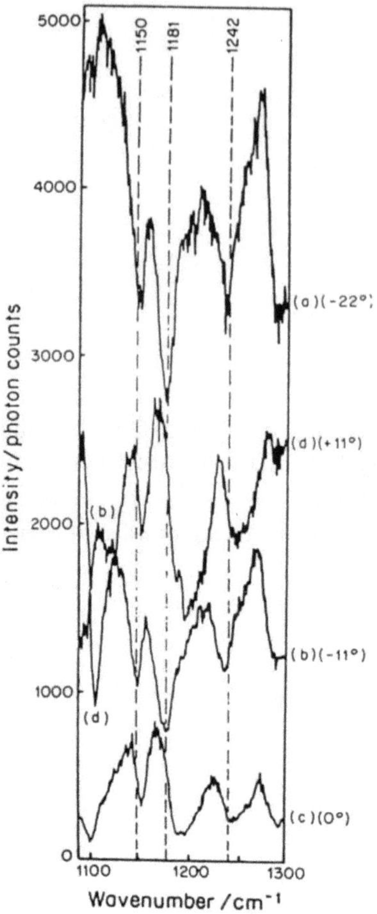

Fig. 5. Polarization-sensitive resonant CARS spectra of UV-excited *trans*-stilbene in ethanol at various settings of polarization analyzer ε: -22° (a), -11° (b), 0° (c), $+11^\circ$ (d). Vertical lines show the vibrational frequencies of *trans*-stilbene in the S_1 state.

polarization suppression of the non-resonant background of the solvent. In the spectrum (b) only the Raman bands of ethanol are present. In the spectrum (a) new signal features arise from the excited stilbene molecules. The obviously high level of CARS signal in this case is caused by the resonance enhancement, because the frequencies ω_1, ω_2 and ω_a all belong to the $S_n \leftarrow S_1$ absorption band centered at 585 nm [21, 22].

The Raman bands in the transient CARS spectrum, Fig. 4(a), look like inverted peaks (dips) at a "plateau" originating from the pure electronic resonance susceptibility χ^E_{S1} of the excited molecules [23]. The vibrational frequencies of S_1 *trans*-stilbene from the transient resonance Raman data [18] are marked by arrows in Fig. 4, showing a good correlation with the positions of the dips. The band shapes in the CARS spectrum are the result of interference between different contributions to $\chi^{(3)}$.

Figure 5 represents the set of CARS spectra obtained with different rotation angles ε of the polarization analyzer $\mathbf{P}_a$ in the anti-Stokes beam (see Fig. 3). The angle $\varepsilon = 0^\circ$ (spectrum (c)) represents the case of full suppression of the solvent non-resonant signal (same as for spectrum (a), Fig. 4). The changes in the spectra when ε is varied are caused by the variation of relative contributions of the non-resonant signal and the electronic and Raman resonances of S_1 *trans*-stilbene.

Two Raman bands of ethanol (the polarized band at 1095 cm^{-1}, $\rho^R < 1/3$, and the depolarized one at 1274 cm^{-1}, $\rho^R > 1/3$) significantly change their shapes and relative amplitudes at different angles ε. The contributions of the S_1 *trans*-stilbene Raman bands are present in all spectra (a)-(d), their band shapes changing from nearly symmetric "Lorentz-type" dips in (a) to "dispersive" in (d). The variation of their amplitudes is insignificant. This is because the polarization vectors $\mathbf{P}^R$ make large angles with the non-resonant vector $\mathbf{P}^{NR}$: $\phi^R - \phi^{NR} \geq 36^\circ$. It follows that all the S_1 Raman bands have the anomalous depolarization ratios ($\rho^R = \frac{1}{3}\tan(\phi^R) \geq 2.1$), which could be caused by the anisotropic orientation of the excited molecules created by the linearly polarized UV beam.

Thus, the method of active control of the CARS band shapes by means of polarization technique can be effectively applied to outline the spectral features of interest and to distinguish them from the unwanted (solvent) bands, when the transient excited species are studied.

3.2. Coherent Ellipsometry of Resonance Electronic Susceptibility

In the transient polarization CARS spectra a considerable contribution of the pure electronic susceptibility of S_1 *trans*-stilbene was observed [23]. Coherent ellipsometry can be used to measure both real and imaginary parts and to determine separately the different components of the $\chi^{(3)}_{ijkl}$ tensor [5]. Two linearly polarized waves, $\mathbf{P}^E$ and $\mathbf{P}^{NR}$, with the angle $\psi^E = \phi^E - \phi^{NR}$ between their polarization vectors (Fig. 6) and the phase shift θ^E, give rise to an elliptically polarized anti-Stokes signal. Measuring the dependence of the CARS intensity on the rotation angle ε of polarization analyzer, when the UV excitation is present, we obtain the parameters of the ellipse ($\mathbf{P}^E + \mathbf{P}^{NR}$). When the UV beam is blocked we measure separately the non-resonant contribution of the solvent, $\mathbf{P}^{NR}$.

The measurements were carried out at the wavenumber $(\omega_1 - \omega_2)/2\pi c =$ 1125 cm^{-1}, far enough from any Raman resonances of ethanol or stilbene (see [23] for details). Thus, we have found the relative amplitude of the electronic susceptibility $\mathbf{P}^E$ (with respect to the non-resonant signal $\mathbf{P}^{NR}$), the angle

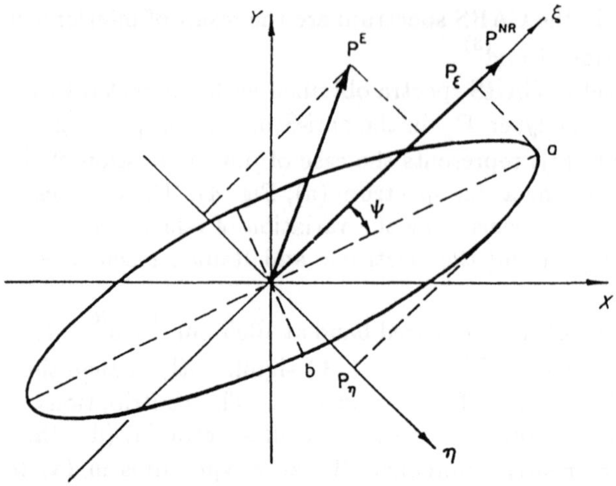

Fig. 6. Polarization vectors in coherent ellipsometry. The elliptically polarized signal is a result of interference between two linearly polarized waves: $\mathbf{P}^{NR}$ - non-resonant CARS signal due to the solvent, $\mathbf{P}^{E}$ - resonant CARS signal due to S_1 electronic susceptibility.

ϕ^E of the $\mathbf{P}^E$ vector, the depolarization ratio $\rho^E = \chi^E_{2112}/\chi^E_{1111} = \frac{1}{3}\tan\phi^E$ and the phase $\theta^E = \arctan(\mathrm{Im}\chi^E_{ijkl}/\mathrm{Re}\chi^E_{ijkl})$:

$$\mathbf{P}^E = 0.25 \cdot \mathbf{P}^{NR}\,;\ \phi^E = (86 \pm 6)^\circ\,;\ \rho^E \geq 2.7\,;\ \theta^E = -(129 \pm 8)^\circ. \tag{1}$$

The population of the S_1 state created by the UV excitation of *trans*-stilbene, $N_1 = 6.0 \cdot 10^{16}$ cm^{-3}, was measured using the optical depletion technique [24]. Taking the value of non-resonant susceptibility of ethanol from [25], $\chi^{NR}_{\mathrm{solv}} = 3.65 \cdot 10^{-14}$ esu we can calculate the value of hyperpolarizability of S_1 molecule:

$$\gamma^E = \frac{\chi^E_{S1}}{N_1 L^4} = \frac{0.25 \cdot \chi^{NR}_{\mathrm{solv}}}{N_1 L^4} = 4.0 \cdot 10^{-32}\ \mathrm{esu}. \tag{2}$$

This extremely high value of the third-order hyperpolarizability of excited *trans*-stilbene, exceeding that of the ground state by about 4 orders of magnitude, is the manifestation of the increasing degree of conjugated π-electron delocalization in the molecule under optical excitation.

3.3. Transient "Anisotropic" Resonance CSRS

For measuring the coherent Stokes Raman spectra (CSRS) of S_1 *trans*-stilbene we used the "anisotropic" scheme, in which the polarization vectors $\mathbf{e}_1$ and $\mathbf{e}_2$ of the two probing waves are perpendicular to each other, the second one being parallel to the UV polarization plane $\mathbf{e}_{\mathrm{UV}}$. Here, we deal with the "anisotropic" component $\chi^{(3)}_{2112}(\omega_s\,;\omega_2\,,\omega_2\,,-\omega_1)$ of the third-order suscepti-

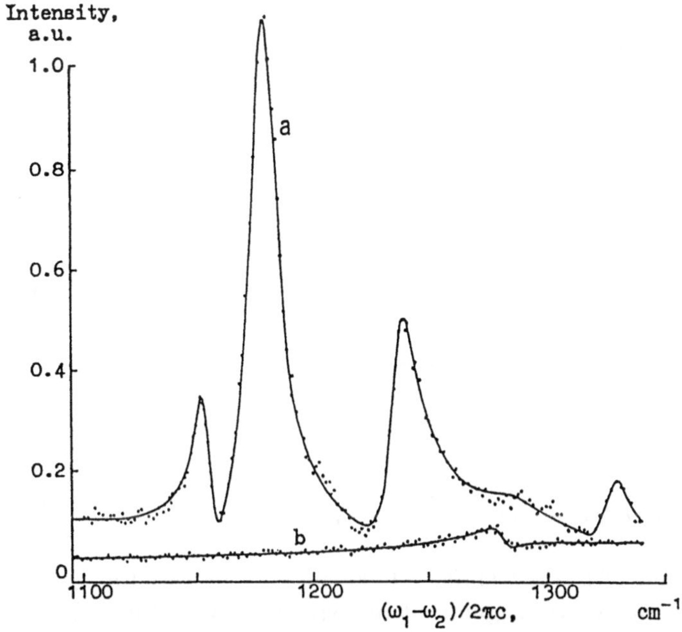

Fig. 7. Anisotropic CSRS spectra of *trans*-stilbene in ethanol with UV (a) and without UV excitation (b). Solid lines - approximation of the experimental spectra (points) according to (3)-(5). Concentration 1 mM.

bility tensor. The anisotropic orientation of the excited molecules along the electric field of the linearly polarized UV beam e_{UV} results in the anomalous depolarization ratio $\rho^E \geq 2.7$ of the S_1 *trans*-stilbene susceptibility tensor as obtained in CARS ellipsometry (1), while the depolarization ratio for the non-resonant signal is $\rho^{NR} = 1/3$. It follows that the signal-to-background ratio in the anisotropic scheme $R_\perp = (\chi^E_{2112}/\chi^{NR}_{2112})^2$ ($e_1 \perp e_2$) will be much higher than that in the isotropic configuration $R_\parallel = (\chi^E_{1111}/\chi^{NR}_{1111})^2$ ($e_1 \parallel e_2$): $R_\perp/R_\parallel = (\rho^E/\rho^{NR})^2 \geq 66$.

Another advantage of the "anisotropic" scheme is the mutually perpendicular polarizations of the generated CSRS signal ω_s and the ω_2 probe wave: $e_s \perp e_2$. This gives the opportunity for the discrimination of the CSRS signal from the strong parasitic broadband luminescence of the dye laser, which coincides in space and time with the coherent Stokes signal and covers the spectral range of interest from 550 to 630 nm.

In the coherent Stokes spectrum of the UV-excited solution, Fig. 7(a), the Raman bands of S_1 *trans*-stilbene at 1150, 1181, 1242 and 1335 cm^{-1} appear as maxima placed on the coherent background of ($\chi^{NR}_{solv} + \chi^E_{S1}$), instead of dips in a CARS spectrum, Fig. 4. This means that the imaginary parts of the complex resonance Raman contributions $\chi^{R_m}_{S1}$ in CSRS have the opposite signs in comparison with CARS [5]. In the spectrum of the unpumped solution, Fig. 7(b), only the depolarized Raman band of ethanol at 1274 cm^{-1} is present at constant level of χ_{solv}^{NR}. Both CSRS spectra, Fig. 7, measured with high resolution ($\leq 0.5\ \text{cm}^{-1}$) and good signal-to-noise ratio, were obtained under all the same experimental conditions, except for the presence (a) or absence (b) of the UV, respectively. This enables us to carry out the detailed band-shape analysis of the vibrational resonances in CSRS.

3.4. Phases of Vibrational Resonances in CSRS

The band shape of CSRS spectrum is determined by the following four different contributions to the net susceptibility $\chi^{(3)}$ of the solution: (1) non-resonant χ_{solv}^{NR} and (2) Raman $\chi_{\text{solv}}^{R_m}$ contributions of the solvent; (3) electronic χ_{S1}^{E} and (4) Raman $\chi_{\text{S1}}^{R_m}$ contributions of S_1 *trans*-stilbene, the last two being resonantly enhanced:

$$I_s^{UV} \sim \left|\chi^{(3)}\right|^2 = \left|\chi_{\text{solv}}^{NR} + \sum_m \chi_{\text{solv}}^{R_m} + \chi_{\text{S1}}^{E} + \sum_m \chi_{\text{S1}}^{R_m}\right|^2 . \tag{3}$$

In the spectral range of interest (1100 – 1300 cm^{-1}), the variation of the wavenumbers of ω_2 and ω_s is insignificant in comparison with the width of the electronic $S_n \leftarrow S_1$ transition band ($\Gamma_E \approx 800\ \text{cm}^{-1}$). Hence we can neglect the dispersion of the amplitudes of the resonant electronic and Raman susceptibilities over this spectral range and present them as follows [4-6]:

$$\chi_{\text{S1}}^{E} = A \cdot \exp(i\theta^{E}) , \tag{4}$$

$$\chi_{\text{S1}}^{R_m}(\Delta\omega) = \overline{\chi}_{\text{S1}}^{R_m} \cdot \exp(i\theta^{R_m}) \cdot (\Delta\omega - \omega^{R_m} + i\Gamma^{R_m})^{-1} , \tag{5}$$

where A and $\overline{\chi}_{\text{S1}}^{R_m}$ are real amplitudes and θ^{E} and θ^{R_m} are the phases of the electronic and Raman susceptibilities, respectively; ω^{R_m} and Γ^{R_m} are the frequency and bandwidth of the m-th vibrational mode; $\Delta\omega = \omega_1 - \omega_2$.

To determine the parameters of the vibrational resonances of S_1 *trans*-stilbene, the transient CSRS spectra were fitted by means of the least-squares procedure using (3)–(5). The values χ_{solv}^{NR} and $\chi_{\text{solv}}^{R_m}$ of the solvent contributions were determined separately by fitting the spectrum (b), the frequency dispersion of the single Raman band of ethanol at $\omega^{R} = 1274\ \text{cm}^{-1}$ being derived from (5) with $\theta^{R} = 0$ for the out-of-resonance conditions of the CSRS process in the transparent medium. The value of θ^{E} in (4) was fixed to be equal to 123° as obtained independently from the experiments on "optical depletion" [24]. The best fit parameters of the S_1 Raman bands are listed in Table 1.

To aid interpretation, we show the $\chi^{(3)}$ susceptibilities diagram on the complex plane [5, 6] in Fig. 8. Here, the non-resonant susceptibility, χ_{solv}^{NR}, is a real constant. The vector of the complex electronic susceptibility of S_1 *trans*-stilbene χ_{S1}^{E} (4) is turned to the angle θ^{E} from the real axis $\text{Re}\chi$. Complex amplitudes of the Raman resonances $\chi_{\text{S1}}^{R_m}(\omega^{R_m})$ (5) are shown by the vector,

TABLE 1. The parameters of resonant electronic and Raman susceptibilities of S_1 *trans*-stilbene, determined from CSRS spectra.

Susceptibility / Assignment*)	Electr. state	$\omega^R/2\pi c$, [cm^{-1}]	$\Gamma/2\pi c$, [cm^{-1}]	Rel. ampl.	Phase [degrees]
Non-resonant susceptibility	Ethanol	—	—	1.0	0
Electronic susceptibility	S_1	—	—	0.9	123
Vibrations:					
—	Ethanol	1274	10.0	0.2	0
C_0-Φ stretch	S_0	1193	1.5	<0.1	0
C-C-H bend	S_1	1150	3.5	0.8	183
C-H def. + C_0-Φ stretch	S_1	1181	9.0	2.5	145
C_0-Φ stretch	S_1	1242	8.0	1.5	112
C_0-C_0-H bend	S_1	1335	4.5	0.2	122

*) Refs. [18,24].

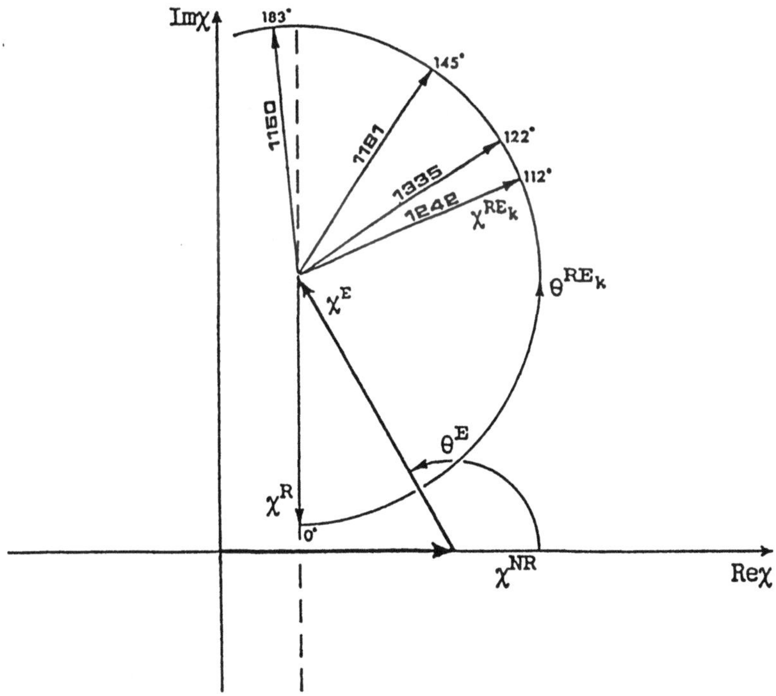

Fig. 8. The phase relations between the vibrational resonances of S_1 *trans*-stilbene, as determined in CSRS relative to the phase of resonant electronic susceptibility.

beginning at the end of $\chi_{solv}^{NR} + \chi_{S1}^{E}$ vector, and turning to the angle θ^{R_m} from the imaginary axis $-\mathrm{Im}\chi$.

The Raman resonances of ethanol which have real amplitudes ($\theta^{R} = 0^{\circ}$), are directed vertically downwards (see [5]), while those of S_1 *trans*-stilbene have different phases, θ^{R_m}. Two of them, at 1242 and 1335 cm^{-1}, have phases θ^{R_m} near θ^{E} (112° and 122°, respectively). The phases of two other bands, at 1150 and 1181 cm^{-1}, differ considerably from θ^{E} and each other (183° and 145°). These differences reflect the features of the electron-vibronic interaction in the molecule [8]. If it is weak, the phases of all vibrational resonances are the same and coincide with the phase of pure electronic resonant susceptibility. The deviation of the phase of a certain mode from θ^{E} is a manifestation of the strong vibronic coupling of this mode with the electronic transition dipole moment. The differences of these phases from each other characterize the degree of electron-vibronic interaction for different modes.

4. Picosecond Dynamics of Stilbene in the Excited S_1 State

4.1. Photoisomerization of Stilbene

The dynamics of relaxation processes induced by picosecond photoexcitation of a molecule can be revealed from the temporal evolution of the CARS signal measured at variable delays Δt with respect to the excitation pulse. We have observed no significant changes either in band shapes or in relative amplitudes of the Raman bands in time-resolved CARS spectra of *trans*-stilbene at the time scale of $|\Delta t| \leq 250$ ps [8, 23]. This means that the transient spectral features

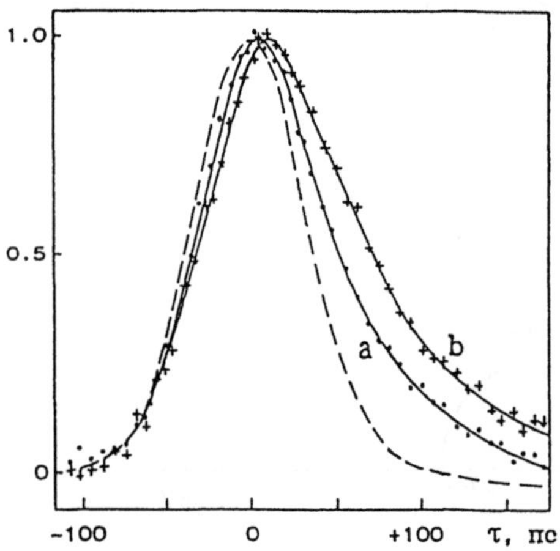

Fig. 9. Time evolution of transient CARS intensity of *trans*-stilbene in pure ethanol (a) and ethanol-glycerol mixture (b), demonstrating the viscosity dependence of the photoisomerization rate. Dashed line - CARS autocorrelation trace; solid lines - the calculated dependences for $\tau_{S1} = 60$ ps (a) and $\tau_{S1} = 130$ ps (b).

brought about by the UV pulse all belong to the one short-lived excited state of *trans*-stilbene.

The high level of the resonance CARS signal due to the excited molecules (more than 200 counts/s), with the polarization suppression of the non-resonant background enables us to record the CARS intensity dependence on Δt, Fig. 9. Curves (a) and (b) were measured at room temperature for *trans*-stilbene solutions in pure ethanol and in a mixture of ethanol with glycerol, respectively. When the viscosity increases, the S_1 lifetime should rise because of the slower rate of photoisomerization (see Fig. 1). This is clearly seen in a delayed decay of the signal in the presence of glycerol, curve (b). Deconvolution of the kinetics measured gives the values of S_1 lifetime $\tau_{S1} \approx 60$ ps and 130 ps, respectively. The observed increase of the transient CARS-signal decay time with increased viscosity is futher evidence that the relaxation of the excited state population goes through the photoisomerization of stilbene.

4.2. The Vibrational Relaxation Dynamics of S_1 trans-Stilbene

A probe pulse ω_1, resonant with the $S_n \leftarrow S_1$ transition, can effectively shorten the lifetime of molecules in the excited state. Being transferred from S_1 to the higher electronic state, the molecules (or most of them) do not return back to S_1 [24, 26], see Fig. 1. We observed the "optical depletion" of the S_1 state by measuring the attenuation of the fluorescence intensity at $\lambda = 345$ nm in the presence of a quenching green pulse. More directly, we controlled the S_1 population by measuring the dependence of CSRS signals from both UV-pumped and unpumped sample on the probe pulse intensity I_1 [24]. This way we detect only those molecules which undergo quenching within the area of the ω_1 beam waist. Thus we derived the degree of the optically induced depletion of S_1 population by factor of 5 when the intensity is varied by 3 orders of magnitude. Thus it follows that we could be able to achieve the S_1 lifetime shortening from $\tau_{S1} \approx 50$ ps [21] without depletion to a value of $\tau_{S1} \approx 10$ ps when the full pulse energy of 50 μJ is applied.

The transient CSRS spectra of UV-pumped *trans*-stilbene solution were measured at various probe intensities I_1 and constant UV excitation intensity, with simultaneous recording of the corresponding spectra with the UV beam blocked. The parameters ν^{R_m}, Γ^{R_m} and χ^{R_m} of Raman bands (frequencies, bandwidths and relative amplitudes) were obtained by fitting the experimental spectra using (3)-(5). Fig. 10 shows the results of analysis of the spectral changes in transient CSRS spectra at various intensities I_1. The band positions are presented by the frequency shift $\Delta\nu^{R_m} = \nu^{R_m}(I_1) - \nu_0^{R_m}$ with respect to their values $\nu_0^{R_m}$ at the lowest probe intensity. Clearly seen is a decrease of amplitude due to the S_1 population depletion along with a broadening and shift of the Raman bands towards the low frequencies. These spectral changes which proceed with the characteristic time of about 10 ps, reflect the kinetics of vibrational relaxation of "hot" excited molecules in the solution, demonstrating the possibility of enhancing the temporal resolution of picosecond coherent Raman methods using the optical depletion technique.

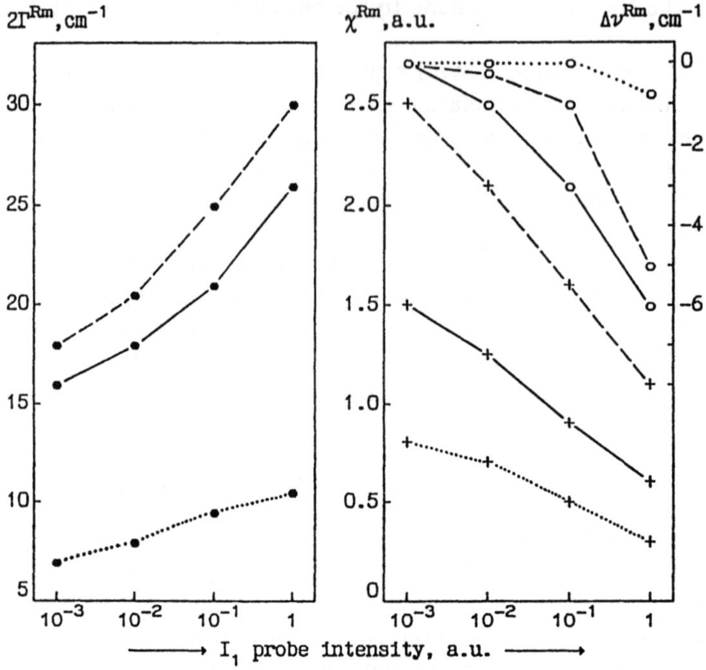

Fig. 10. The spectral changes of S_1 *trans*-stilbene Raman bands *vs* probe intensity I_1:

(• • •) bandwidths $2\Gamma^{R_m}$; (o o o) frequency shift $\Delta\nu^{R_m}$ and (+++) relative amplitudes χ^{R_m} ; solid lines - 1151 cm^{-1} ; dashed lines - 1179 cm^{-1} and dotted lines - 1242 cm^{-1} .

5. Picosecond Relaxation of Electronic Excitation in β-Carotene

The molecule of β-carotene is one of the key objects in photobiology. For nonlinear optics this molecule attracts much interest as a long one-dimensional conjugated chain with an extremely high value of non-linear susceptibility [27]. Although it has been actively studied by both linear and nonlinear resonance Raman (RR) methods, including time-resolved ones [28-30], the scheme of the relaxation of electronic and vibrational excitation as well as the structural dynamics of the molecule are still not fully defined [31-34].

The structure of the energy levels of the β-carotene molecule is presented in Fig. 11. The electronic and vibrational relaxation processes as well as the scheme of excitation and CARS probing are shown in the diagram.

Figure 12 shows the results of the pump-probe CARS experiment on β-carotene in chloroform ($C \approx 10^{-4}$ M). The probe ω_1 pulse intensity was at least 100 times less than that of the excitation pulse (532 nm). The dependence of CARS-signal intensity I_{CARS} on the time delay Δt between excitation and probing pulses was measured, the difference between two CARS-probing frequencies ω_1 and ω_2 being fixed at the vibrational resonance $(\omega_1 - \omega_2)/2\pi c =$ 1157 cm^{-1} of S_0 β-carotene. At negative delays $\Delta t < 0$, when the probing

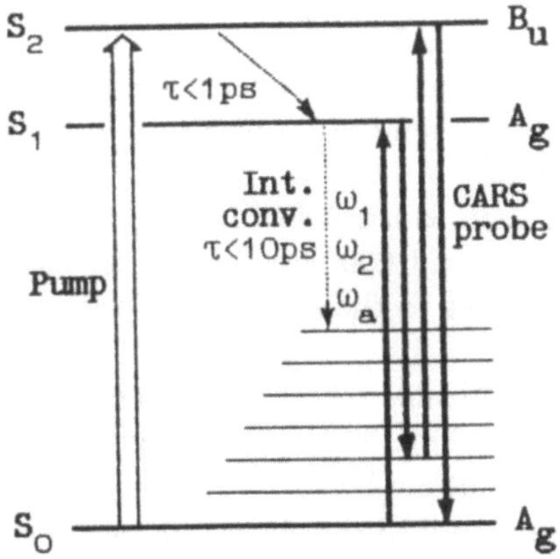

Fig. 11. The structure of energy levels, the scheme of the relaxation processes of the β-carotene molecule and the diagram of the pump-probe CARS experiment.

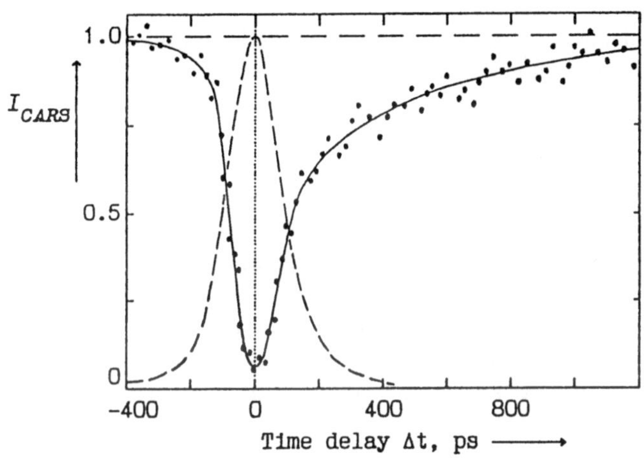

Fig. 12. Dependence of CARS intensity I_{CARS} on time delay Δt between pump and probe pulses, measured for 1157 cm^{-1} RR band of β-carotene solution in chloroform (dots), autocorrelation function of CARS signal (dashed line) and the best fit curve (solid line), calculated by convolution of the autocorrelation function with the two-exponential curve:

$$I_{CARS}(\Delta t) = \{1 - [a_1 \cdot \exp(-\Delta t/\tau_1) + a_2 \cdot \exp(-\Delta t/\tau_2)]\}^2$$

with $\tau_1 = 15$ ps, $\tau_2 = 600$ ps, $a_1 = 0.5$ and $a_2 = 0.45$.

pulses precede the excitation pulse, the high level of the signal is due to high amplitude of the ground state RR band. The minimum in the curve $I_{CARS}(\Delta t)$ at $\Delta t \approx 0$ corresponds to the maximum depletion of the S_0 ground state population, which results in a decrease of the corresponding RR signal. On increasing Δt, the CARS signal almost immediately (within the duration of the laser pulses) rises up to about the 50% level and then slowly comes up to the initial value. The fast component in the CARS signal recovery kinetics almost follows the temporal response function of the CARS setup (dashed line in Fig. 12). This correlates well with the data of transient absorption measurements [32, 33], demonstrating the lifetime of the ground state absorption recovery after picosecond excitation to be about 8 ps.

The slow component, which is clearly seen as a tail in the $I_{CARS}(\Delta t)$ dependence, has the characteristic lifetime of about 600 ps. This means that the relaxation of part of the β-carotene molecules to the initial equilibrium state proceeds over a considerably longer time.

The main difference between linear absorption spectroscopy methods and the nonlinear-optical CARS process is that the CARS signal originates from the resonant third-order molecular susceptibility $\chi^{(3)}$. The latter is known to be more sensitive to the molecular structure and to the electronic excitation of the molecule than linear molecular polarizability. In particular, the third-order hyperpolarizability of the linear organic molecules (polyenes) strongly depends on the length of the π-electron conjugated chain [27]. According to this the origin of the slow component in CARS recovery kinetics of β-carotene could be explained as follows.

The first step of the electronic excitation energy relaxation in the β-carotene molecule after internal conversion to the ground electronic state is fast intramolecular vibrational energy redistribution. Then the excess vibrational energy is dissipated to the solvent molecules, but part of it could also be transferred to the potential energy of conformational changes in the molecule. After the vibrational energy relaxation is completed the molecule remains in some non-equilibrium conformation, from which the relaxation to the equilibrium one proceeds on a subnanosecond time scale. It is this conformational relaxation process which could contribute to the slow component in the CARS recovery kinetics. It is worth noting that a similar picture of the electronic energy relaxation process was recently reported by Atkinson and coauthors [35] as being detected by picosecond excitation and time-resolved Raman measurements of the β-carotene ground state.

Acknowledgements

We gratefully thank Professor S. A. Akhmanov for support and helpful discussions, as well as Drs. V. F. Kamalov, and A. Yu. Chikishev for their contributions to the experiments and the development of the technique.

References

[1] **Time-Resolved Vibrational Spectroscopy**, Vol.4 (Eds. A. Laubereau and M. Stockburger), Springer, Berlin (1985).

[2] **Time-Resolved Spectroscopy**, Vol. 18 (Eds. R. J. H. Clark and R. E. Hester), J. Wiley, Chichester (1989).

[3] S. Matsunuma, S. Yamaguchi, C. Hirose and S. Maeda, **J. Phys. Chem.** 92, (1988) 1777.

[4] A. Lau, M. Pfeiffer, W. Werncke, *et al.*, **J. Raman. Spectrosc.** 19, (1988) 353.

[5] S. A. Akhmanov and N. I. Koroteev, **Methods of Nonlinear Optics in Light Scattering Spectroscopy**, Nauka, Moscow, (1981) (in Russian).

[6] N. I. Koroteev, **Usp. Fiz. Nauk**, 152, (1987) 493 (in Russian).

[7] S. A. Payne and R. M. Hochstrasser, **J. Chem. Phys.** 90, (1986) 2068.

[8] V. F. Kamalov, N. I. Koroteev and B. N. Toleutaev, In **Time-Resolved Spectroscopy**, Vol. 18 (Eds. R. J. H. Clark and R. E. Hester), J. Wiley, Chichester, (1989), p. 255.

[9] V. F. Kamalov and B. N. Toleutaev, In **Ultrafast Phenomena in Spectroscopy**, (Eds. Z. Rudzikas *et al.*), World Scientific, Singapore, (1988), p. 359.

[10] S. Matsunuma, N. Akamatsu, T. Kamisuki, Y. Adachi, S. Maeda and C. Hirose, **J. Chem. Phys.** 88, (1988) 2956.

[11] W. Werncke, A. Lau, M. Pfeiffer *et al.*, **Chem. Phys.** 118, (1987) 133.

[12] A. Yu. Chikishev, V. F. Kamalov, N. I. Koroteev, V. V. Kvach, A. P. Shkurinov and B. N. Toleutaev, **Chem. Phys. Lett.** 144, (1988) 90.

[13] V. F. Kamalov, B. N. Toleutaev, A. P. Shkurinov, In **Laser Scattering Spectroscopy of Biological Objects**, (Eds. J. Stepanek *et al.*), Elsevier, Amsterdam, (1987), p. 121.

[14] V. F. Kamalov, A. P. Razzhivin, B. N. Toleutaev *et al.*, **Sov. J. Quant. Electr.** 14, (1987) 1303.

[15] R. M. Hochstrasser, **Pure Appl. Chem.** 52, (1980) 2683.

[16] H. Hamaguchi, **J. Mol. Struct.** 126, (1985) 125.

[17] H. Hamaguchi, T. Urano, M. Tasumi, **Chem. Phys. Lett.** 106, (1984) 153.

[18] T. L. Gustafson, D. M. Roberts, D. A. Chernoff, **J. Chem. Phys.** 81 (1984) 3438.

[19] H. Hamaguchi, **Chem. Phys. Lett.** 126 (1986) 185.

[20] Y. R. Shen, **The Principles of Nonlinear Optics**. J. Wiley, N. Y., (1984).

[21] B. I. Greene, R. M. Hochstrasser and R. B. Weisman, **Chem. Phys. Lett.** 62, (1979) 427.

[22] K. Yoshihara, A. Namiki, M. Sumitani and N. Nakashima, **J. Chem. Phys.** 71, (1979) 2892.

[23] V. F. Kamalov, N. I. Koroteev, A. P. Shkurinov, U. Stamm, and B. N. Toleutaev, **J. Phys. Chem.** 93, (1989) 5645.

[24] V. F. Kamalov, N. I. Koroteev, A. P. Shkurinov and B. N. Toleutaev, **Chem. Phys. Lett.** 147, (1988) 335.

[25] G. R. Meredith, B. Buchgalter and C. Hanzlik, **J. Chem. Phys.** 78, (1983) 1533.

[26] M. Sumitani and K. Yoshihara, **J. Chem. Phys.** 72, (1982) 738.

[27] C. Flytzanis, In **Nonlinear Behavior of Molecules, Atoms and Ions...** (Ed. L. Neel), Amsterdam, Elsevier, (1979), p. 185.

[28] M. Asano, L. V. Haley and J. A. Koningstein, In **Time-Resolved Vibrational Spectroscopy**, (Ed. G. H. Atkinson), New York, Academic Press, (1983) p. 139.

[29] H. Hamaguchi, In **Vibrational Spectra and Structure**. Vol. 16, (Ed. J. R. Durig), Elsevier, Amsterdam, (1987), ch. 4, p. 227.
[30] L. A. Carreira, T. B. Goss and T. B. Malloy, **J. Chem. Phys.** 66 (1977) 4360.
[31] P. J. Caroll and L. E. Brus, **J. Chem. Phys.** 86 (1987) 6584.
[32] S. M. Bachilo, S. L. Bondarev and S. A. Tihomirov, **Sov. J. Appl. Spectrosc.** 50 (1989) 426.
[33] M. R. Wasielewski and L. D. Kispert, **Chem. Phys. Lett.** 128 (1986) 238.
[34] H. Hashimoto, Y. Koyama, K. Ichimura and T. Kobayashi, **Chem. Phys. Lett.** 162 (1989) 517.
[35] H. Hayashi, S. V. Kolaczkowski, T. Noguchi, D. Blanchard and G. H. Atkinson, **J. Am. Chem. Soc.** 112 (1990) 4664.

CARS Application to Monitoring the Rotational and Vibrational Temperatures of Nitrogen in a Rapidly Expanding Supersonic Flow

M. Noda and J. Hori

Mitsubishi Heavy Industries, Ltd., Nagasaki R&D Center,
1-1 Akunoura-machi, Nagasaki 850-91, Japan

Using an excimer laser based CARS apparatus, single-pulse rotational and vibrational temperatures of nitrogen molecules have been measured in a rapidly expanding supersonic flow in which nitrogen molecules were not in thermal equilibrium. The supersonic flow facility should be installed outdoors, therefore an optical fiber was used in the CARS receiving optics to protect delicate instruments, such as a monochromator and an optical multichannel analyzer, from intense vibration caused by the flow. The optical fiber also worked as a medium producing a nonresonant CARS signal which was used to normalize the resonant CARS spectrum obtained in the flow. In this experiment the validity of the newly developed excimer laser based CARS system which could be operated up to 80 Hz repetition rate was proved, because the flow continued only for 1.5 seconds and a high repetition rate of the CARS system was important.

1. Introduction

Coherent anti-Stokes Raman Spectroscopy (CARS) is now widely accepted as a versatile tool for the measurements of temperature [1-4]. But most of its applications have been to the combustion field, in which nitrogen molecules examined were in thermal equilibrium. Recently CARS has been applied to the CO_2 laser to investigate the behavior of nitrogen molecules in non-equilibrium [5]. That is a typical CARS application since there is no other way to examine the non-equilibrium situation. In this paper we report the results of a diagnosis of a supersonic flow facility based on the measurements of rotational and vibrational temperatures of nitrogen molecules in non-equilibrium using CARS. The newly developed compact CARS apparatus consisting of an excimer laser pumped two dye laser and the new easy way of obtaining a nonresonant CARS spectrum used as a reference are also described.

2. Experimental Setup

2.1 CARS Apparatus

A schematic of the excimer laser based CARS apparatus is shown in Fig. 1. An excimer laser and a dye laser were installed in a laser carrier which had two stages, a lower one for the former and an upper one for the latter, together with optics giving a collinear geometry. The dye laser, which was newly developed in order to make the original excimer laser based CARS

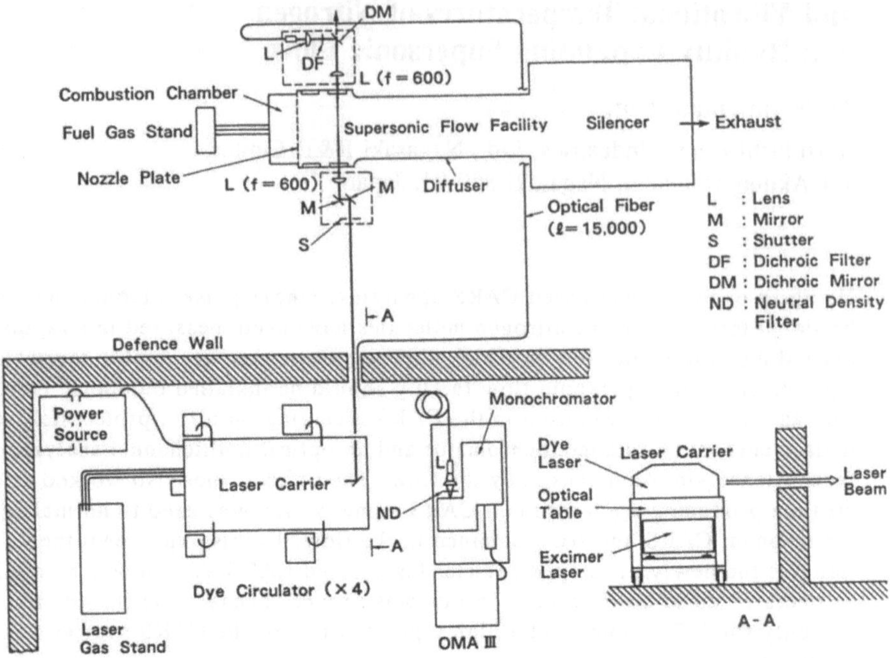

Fig. 1: Experimental setup of the excimer laser based CARS apparatus and the supersonic flow facility.

system more compact [6], consisted of a narrowband pump laser and a broadband Stokes laser. The original system had three tables, one for each laser, thus demanding a lot of space and being affected seriously by vibration, so it was almost impossible to apply it to samples in environments subject to intense vibration.

Details of the dye laser and the collinear geometry are given in Fig. 2. The excimer laser beam, delivering its energy to dye cells for the broadband and narrowband lasers by means of partially reflecting mirrors, was located at the center of the dye laser. The narrowband laser was constructed by rearranging the optical components of a Lambda Physik FL2002 dye laser. The broadband laser was newly designed to be compact and so that the two lasers could be easily combined. It had a 100% reflection rear-side coated prism as a dispersion element which could tune the center wavelength by rotating the prism. Its bandwidth could also be changed between $3cm^{-1}$ and $150cm^{-1}$ by replacing the prism by an another one with a different refractive index.

The laser energies were 250mJ at 308nm (XeCl) for the excimer laser, 15mJ for the pump laser at 480nm with $0.2cm^{-1}$ bandwidth and 8mJ for the Stokes laser at 540nm with $80cm^{-1}$ bandwidth with a BK7-prism. A photograph of the laser carrier of the CARS apparatus is shown in Fig. 3.

The pump and Stokes beams were focused at the center of the supersonic flow facility through a lens with 600mm focal length. On emerging from the flow facility, all beams were recollimated by a lens with 600mm focal length. After recollimation, the pump and Stokes components were eliminated with a

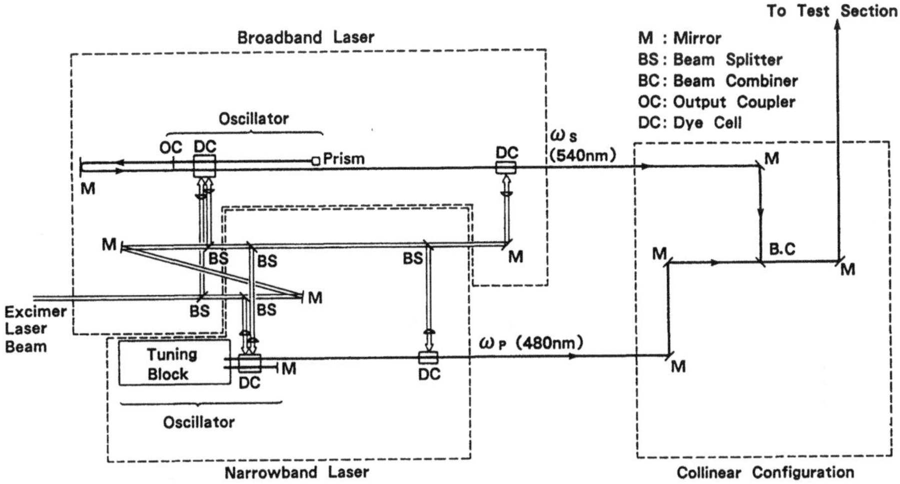

Fig. 2: Details of the excimer laser pumped dye laser which emits the narrowband and broadband radiation together with the collinear geometry optics.

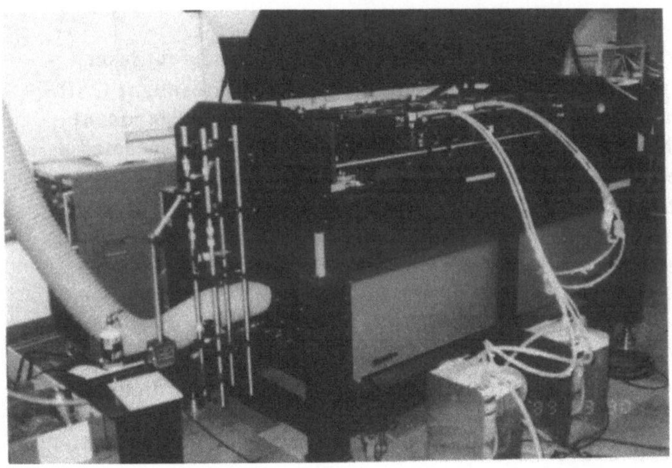

Fig. 3: A photograph of the laser carrier of the CARS apparatus.

dichroic mirror and a dichroic filter, which respectively reflected or transmitted only the CARS radiation and were set in front of a lens focusing the CARS into an optical fiber.

The optical fiber, which was 15m long and had a 100μm core diameter, was used to isolate the monochromator from hostile conditions, such as found outdoors, and intense vibration caused by a rapidly expanding supersonic flow. The optical fiber could also work as a medium generating nonresonant CARS which was used to normalize the resonant CARS spectrum obtained in the supersonic flow.

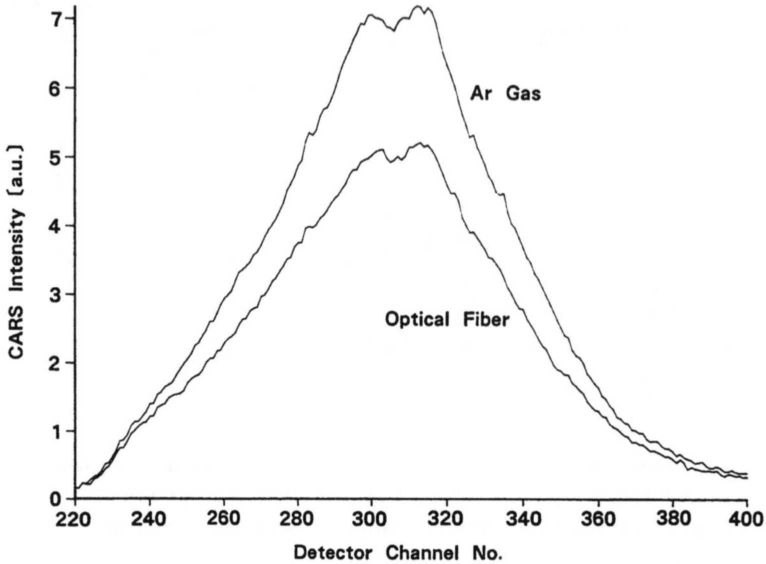

Fig.4: The nonresonant CARS spectrum obtained from the optical fiber and the argon reference cell, respectively, after 100 shots averaging.

Just a dichroic mirror was not sufficient to eliminate the incident laser components, so that the remaining components generated nonresonant CARS at the edge of the optical fiber. This was a nuisance, because the resonant CARS generated in the flow mixed with the nonresonant CARS generated at the fiber edge. Therefore the dichroic filter which transmitted the CARS but reflected the incident laser beams was set additionally behind the dichroic mirror to eliminate the incident laser beams completely. When the nonresonant CARS had to be measured just before or after the resonant CARS measurement, the dichroic filter was replaced by an another one which transmitted the incident laser beams but reflected the CARS. This was a very convenient method compared with using a reference cell, in terms of necessary components and optical alignment. It only requires the additional dichroic filter and the replacement.

The nonresonant CARS spectra obtained from the optical fiber and from the argon reference cell respectively after averaging over 100 shots are shown in Fig.4. These are almost the same, which showed that the nonresonant CARS spectrum obtained from the optical fiber could be used for normalizing the CARS spectrum. The small differences between the two were due to the time difference of the measurements, because two CARS spectra could not be obtained simultaneously in this measurement system.

2.2 Supersonic flow facility

The supersonic flow facility consisted of a fuel gas stand, a combustion chamber, a supersonic nozzle plate, a diffuser and a silencer as shown in Fig.1. Two pairs of windows for the CARS measurements were fixed to window plates which were isolated vibrationally from the body of the flow facility by bellows. Their positions were 10cm and 24cm downstream from the nozzle plate. The distance between the windows facing each other was about 800mm and it required a long focal length focusing lens such as 600mm lens mentioned before for the laser beams.

As fuel gas, N_2O, CO and H_2 were employed and mixed in the injector and ignited by electric discharge. This combustion mode was referred to as premixed combustion. The combustion gas, of which the components were about 60% N_2, 30% O_2, a few % of CO_2 and a small amount of H_2O, was expanded rapidly through the nozzle, which made the flow supersonic.

In a " downstream-mixing-after -combustion " mode, a few % of CO_2 gas was mixed additionally right after the nozzle and the fuel gas component ratio was also changed a little from the ratio of the premixed combustion mode.

3. Temperature Measurement

CARS measurements were performed about 10cm or 24cm downstream from the supersonic nozzle under conditions of both premixed combustion and downstream-mixing-after-combustion. In each case static pressure at the measuring point was about 60 torr and nitrogen partial pressure was about 40 torr. In these tests attention was focused primarily on the feasibility and quality of CARS generation rather than on measurement accuracy. But it was very important to confirm that there was no pump-laser-induced population change [7] with the laser power, as mentioned above, in these measurements, since that would lead to a serious error in vibrational temperature.

The N_2 CARS spectrum obtained 10cm downstream from the nozzle in the supersonic flow is shown as a solid line in Fig.5 (the premixed combustion) and in Fig.6 (the downstream-mixing-after-combustion) together with the theoretical spectrum (dashed lines). Since nitrogen molecules in the supersonic flow were not in thermal equilibrium, the fitted theoretical spectrum had two different rotational and vibrational temperatures. These were 300K and 1000K respectively in the case of the premixed combustion and 250K and 1100K in the case of the downstream-mixing-after-combustion. As shown in Fig.5 and Fig.6, measurements and calculations were in good agreement, however, the accuracy of the vibrational temperature was not good at about 1000K. This is because in the thermally non-equilibrium case the feature of the vibrational temperature appears only in the very weak hot band of the CARS spectrum, besides, its derivative with temperature is very small compared with the main band at about less than 1000K. Nevertheless, it was found from Fig.5 and Fig.6 that the vibrational temperature was far less than 1400K, which was the design operating point of the supersonic flow facility. This indicated that there was some heat loss at the supersonic nozzle or differences of relaxation processes between experiment and analysis, because the combustion temperature was deduced to be about 2000K from the combustion analysis for the given pressure of the combustion chamber and the flow rate. On the other hand, the tendency of the rotational temperature in the case of the premixed combustion to be about 50K higher than that in the case of the downstream-mixing-after-combustion agreed with the analysis.

Considering the results of the CARS measurements, some improvements have been made to the supersonic nozzle and in the selection of the combus-

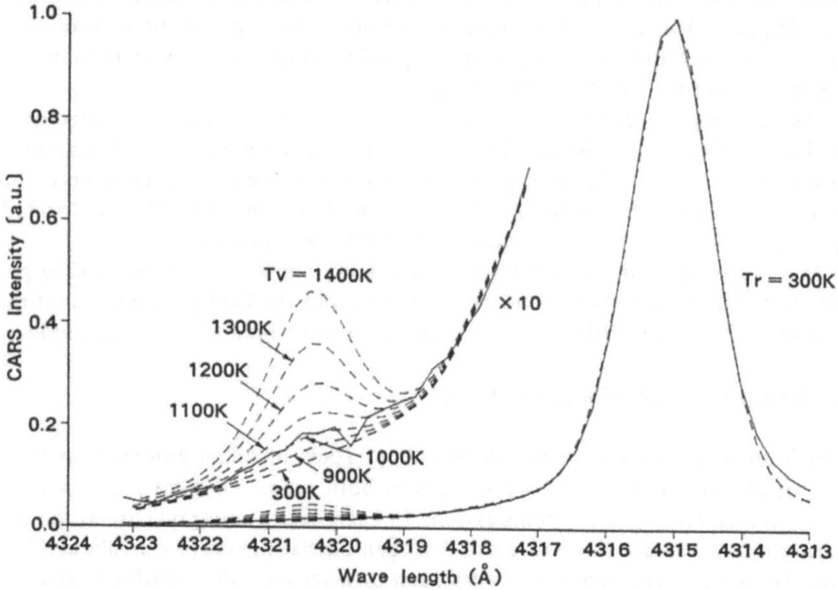

Fig. 5: N_2 CARS spectrum obtained 10cm downstream from the nozzle in the supersonic flow of the premixed combustion (solid line) together with the theoretical spectrum (dashed line).

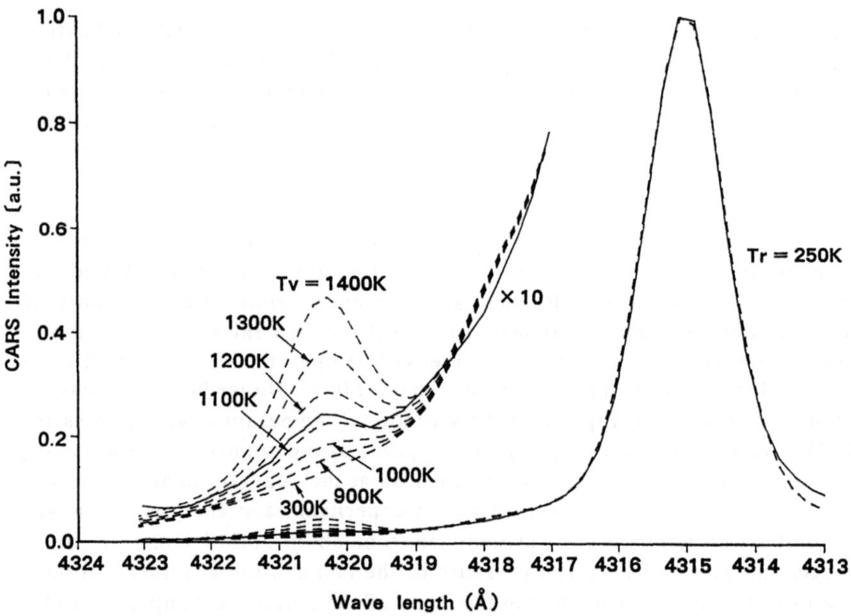

Fig. 6: N_2 CARS spectrum obtained 10cm downstream from the nozzle in the supersonic flow of the downstream-mixing-after-combustion (solid line) together with the theoretical spectrum (dashed line).

tion condition. As can be seen from the above discussion, the important diagnosis of the supersonic flow facility could not be done without the CARS measurements.

4. Summary

Coherent anti-stokes Raman Spectroscopy (CARS) has been successfully demonstrated in the supersonic flow generated from the combustion gas through the supersonic nozzle. Rotational and vibrational temperatures deduced from the N_2 CARS spectrum obtained from the supersonic flow showed the thermally non-equilibrium state of nitrogen molecules. The accuracy of those data was not too good, however, they did allow the diagnosis of the flow facility and ways it could be improved were found. It was the first time that an excimer laser based CARS apparatus had been used for a facility installed outdoors. Therefore, a rigid and compact laser carrier in which the heavy excimer laser and the dye laser could be fixed was developed and an optical fiber for receiving CARS radiation was adopted to ensure that the CARS apparatus could be used in a hostile environment. The dye laser which could emit two beams, broadband and narrowband radiation, was also developed to fit in the laser carrier. The new CARS apparatus could be operated up to 80Hz repetition rate. This high repetition rate operation allowed an analysis of vibrational temperature fluctuation or a transient process even in the short period of only 1.5s for which the supersonic flow examined could continue.

With improvements of the details of the CARS apparatus based on this experiment, the next CARS measurements in a supersonic flow will be performed in the near future. Then, more accurate rotational and vibrational temperatures in many combustion conditions will be measured and more quantitative discussion presented. Recently, CARS application to transient phenomena have been increasing and an excimer laser based CARS system will play a more important role in the future.

Acknowledgements. The authors thank J.Kisaki and K.Nanba for giving us the opportunity of applying CARS to the facility reported here.

References

1. A.C.Eckbreth, Combust. Flame **39**, 133 (1980)
2. M.Aldén and S.Wallin, Appl. Opt. **24**, 3434 (1985)
3. M.Péalat, P.Bouchardy, M.Lefebvre and J.P.Taran, Appl. Opt. **24**, 1012 (1985)
4. M.Noda, A.Gierulski and G.Marowsky, Proc. of Joint Conference of Western States and Japanese Sections of the Combustion Institute., Honolulu, 355 (1987)
5. N.Wenzel, K.Kishimoto, H.Große-Wilde and G.Marowsky, 9th European CARS Workshop, Dijion, March 19-20 1990
6. B.Lange, M.Noda and G.Marowsky, Appl. Phys. **B49**,33 (1989)
7. A.Gierulski, M.Noda, T.Yamamoto, G.Marowsky and A.Slenczka, Opt. Lett. **12**, 608 (1987)

Part IV

Selected Applications of Coherent Raman Techniques for Diagnostics of Gaseous and Liquid Media

CARS Diagnostics of High-Voltage Atmospheric Pressure Discharge in Nitrogen

I.V. Adamovich[1], *P.A. Apanasevich*[2], *V.I. Borodin*[1], *S.A. Zhdanok*[1], *V.V. Kvach*[2], *S.G. Kruglik*[2], *M.N. Rolin*[1], *A.V. Savel'ev*[1], *A.P. Chernukho*[1], *and N.L. Yadrevskaya*[1]

[1]A.V. Luikov Heat and Mass Transfer Institute, Byelorussian Academy of Science, SU-220728 Minsk, USSR

[2]B.I. Stepanov Institute of Physics, Byelorussian Academy of Science, SU-220602 Minsk, USSR

1. INTRODUCTION

Chemical reactions that occur in the nonequilibrium plasma of a gas discharge evoke great interest both from the applied and from the fundamental viewpoints. Usually, nonequilibrium conditions are realized in a low-pressure plasma. Experimental and theoretical work demonstrates the possibility of producing a nonequilibrium plasma in a high-voltage atmospheric pressure discharge (HVAPD), which is promising in such plasmochemical processes as nitrogen oxide synthesis [1], metal nitriding [2], and oxidation of high temperature superconducting ceramics [3]. This evokes interest in different diagnostic methods when applied to HVAPD plasma. Unfortunately, the high nonuniformity and large temperature gradients in HVAPD make it impossible to use the traditional methods of diagnostics.

Despite its complexity, the nonlinear optical method of coherent anti-Stokes Raman scattering (CARS) has proven to be a perfectly suitable technique for nonperturbing local measurements in gas systems and plasma.

This contribution presents the preliminary results of investigations of molecular nitrogen in HVAPD plasma. The CARS technique was applied to measure the electronic ground state vibrational and rotational temperatures of N_2 molecules. The excited electronic state $C^3\Pi_u$ was studied by emission spectroscopy.

2. EXPERIMENTAL

A HVAPD was burnt in the gas flow between movable cathode and fixed anode. The interelectrode gap length was equal to 4 cm. From the cathode side, the positive column of the discharge was stabilized by a quartz tube, of 8 mm ID. The outer tube surface was under free heat transfer conditions. A test gas (high purity nitrogen) entered the discharge region from the cathode side and was then exhausted into the atmosphere.

Eds.: G. Marowsky and V.V. Smirnov

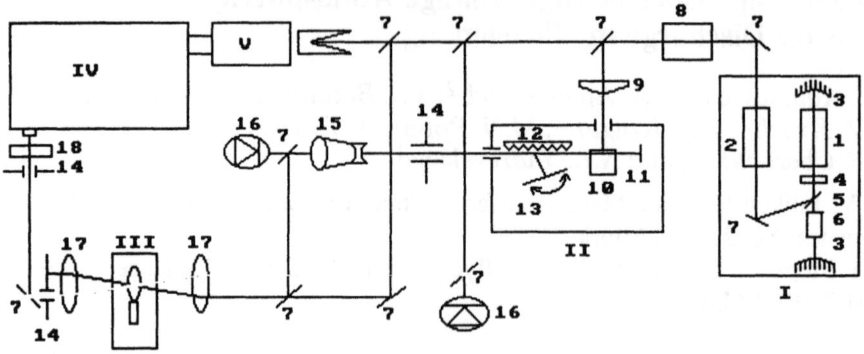

Fig.1. Schematic of a CARS spectrometer: I,YAG:Nd laser; II,dye laser; III,discharge; IV,monochromator; V, photomultiplier; 1,oscillator head; 2,amplifier head; 3,resonator mirrors; 4,λ/4 plate; 5,polarizer; 6,Pockels cell; 7, dielectric mirrors; 8, second harmonic generator; 9, cylindrical lens; 10, flowing dye cell; 11,dye resonator mirror; 12,grating; 13,tuning mirror; 14,diaphragms, 15, telescope; 16,photodiodes; 17,focusing lenses; 18,colour filters.

The region of free burning (~1 cm) has been tested by the CARS and emission techniques.

CARS spectra were obtained with a single-channel laser spectrometer (Fig.1) based on repetitive pulsed lasers: Q-switched YAG:Nd laser I and tunable narrow band dye laser II. The lasers have been designed to meet the high standards usually imposed on temporal, spatial and power characteristics of laser equipment for quantitative measurements with nonlinear scattering techniques. The YAG:Nd laser consists of oscillator 1 and single-stage amplifier 2. The concept of unstable resonators is the background of the oscillator optical layout [4]. The unstable telescopic resonator is formed by fully reflecting concave and convex mirrors 3 although energy extraction is produced with dielectric polarizer 5 and phase quarterwave plate 4. The resonator optical arrangement permits diffraction limited beams to be delivered with an output pulse energy up to 250 mJ, the pulse duration being 15 ns. To narrow the laser spectral linewidth an intracavity Fabry-Perot etalon is installed between mirror 3 and polarizer 5. Depending on the etalon thickness the halfwidth was lowered from the original 0.8 cm^{-1} to 0.1 cm^{-1} at 532 nm without significant drop (less than 20%) in output energy.

When passing through amplifier stage 2, YAG:Nd radiation at the fundamental frequency is doubled with a type II KDP crystal (the energy conversion efficiency reaches 40%). Partially the second harmonic power was used to realize the CARS diagnostics arrangement; the rest of the radiation was used for transverse optical pumping of the dye laser II. The dye laser was arranged according to the grazing incidence scheme [5]. The zero-diffraction output with a fully reflecting mirror 11 was used, otherwise Fresnel reflection coupling from a glass wedge was used when the superluminescence background may deteriorate the results. In the second case, to increase the pulse peak power the longitudinally pumped dye amplifier was used. The spectral linewidth can be varied from 0.1 cm^{-1} to 0.5 cm^{-1} by tuning the grating incidence angle. The spectral line position was reproduced with accuracy better than the linewidth. The step motor driver scanned the dye laser frequency by rotating mirror 13.

Second harmonic radiation of the YAG:Nd laser was split by mirrors 7 into two equal fluxes and, together with the dye laser output, was focused into the discharge by an achromatic lens 17 (f=70 mm). The divergence of the dye laser beam, its waist dimension and position were monitored by a Galilean telescope 15 having a magnification 3x. The coplanar BOXCARS arrangement was used: $\mathbf{K}_{cars}-\mathbf{K}_{pr}=\mathbf{K}_{p}-\mathbf{K}_{s}$, where $\mathbf{K}_{pr}$, $\mathbf{K}_{p}$ and $\mathbf{K}_{s}$ are the wave vectors for the probe, pump at 532 nm and the Stokes laser waves, respectively.

To avoid CARS spectra distortions due to saturation effects, the pulse energies were lowered to 10 mJ at 532 nm and to 1 mJ for the dye laser output.

When passing the discharge, the laser beams were recollimated and spatially rejected by the screen and diaphragms 14. The spectral filtering of the scattered light was made by dichroic mirrors 7 and 18. The final spectral discrimination was accomplished by monochromator IV. The photomultiplier tube V was used to detect the CARS signal while photodiodes 16 controlled the energy of the exciting laser pulses. The photomultiplier I_{cars} and photodiode I_s, I_p responses were digitized, and the scattered light intensity was rationed according to the law $I_{cars}/I_p^2 {\bullet} I_s$ for each laser shot. The CARS signal was averaged over each scanning step.

The photon counting technique was implemented to measure the emission spectra of the 2^+ system over a frequency range 20000-25000 cm^{-1}; the spectral resolution was 1 cm^{-1}; the storage time was 2 sec/point. The spectrometer control, data acquisition and storage were performed by a CAMAC assembly and by a personal computer.

3. MATHEMATICAL MODEL FOR THE NITROGEN HVAPD PLASMA

A mathematical model for HVAPD has been constructed taking into account the processes of vibrational-translational relaxation of N_2 molecules on N_2 and N, heterogeneous relaxation and recombination on the wall, dissociation of vibrationally excited N_2 molecules and vibrational excitation of nitrogen by electron impact. As was shown in [1], the values of the reduced electric field typical of HVAPD are over the range E/N=10-50 Td where, according to [6], more than 90% of the discharge energy is used for N_2 vibration excitation. So, the model does not take into account the electronic excitation of N_2 molecules and electron impact dissociation. The vibrational nitrogen kinetics was analysed in the harmonic oscillator approximation.

Rate constants for V-T relaxation of N_2 molecules were taken from [7]. A rate constant of N_2 dissociation from all the vibrational levels was borrowed from [8] and was fitted to the thermal rate constant from [9]. The probabilities of heterogeneous N recombination and vibrational N_2 relaxation on the discharge tube wall were taken, respectively, from [10,11].

The mathematical model represents the system of equations of a laminar boundary layer for a gas velocity, translational and vibrational temperatures, nitrogen atom concentration, and ionization degree y_e. In the present work, the electron kinetics processes have not been taken into account. Necessary data on electron energy balance in a discharge, as well as a drift velocity W_{dr} and ambipolar diffusion coefficient D_a were obtained using the methods of [6] as functions of E/N. The values of a longitudinal pressure gradient and electric field strength (it was assumed that $\partial P/\partial r = \partial E/\partial r = 0$; the cathode region was not considered) were found from the mass flow G and current I conservation conditions.

The ionization degree was determined from the kinetic equation

$$\frac{\rho}{\mu} U \frac{\partial y_e}{\partial z} + \frac{\rho}{\mu} V \frac{\partial y_e}{\partial r} = \frac{1}{r}\frac{\partial}{\partial r}\left(r\frac{\rho}{\mu} D_a \frac{\partial y_e}{\partial r}\right) + \frac{\rho^2}{\mu^2}\left[k_i y_e - \left(k_r^d + N k_r^t \right) \cdot y_e^2 \right] \quad (1)$$

assuming the exponential dependence for the ionization rate constant

$$k_i = A \cdot \exp\left(- \frac{B}{E/N} \right) , \quad (2)$$

where A is the preexponential coefficient for the direct ionization rate constant [12]; B the parameter determined by fitting a design value of the discharge voltage to an experimental one. In equation (1), μ is the molecular mass of a mixture; e the electron charge; k_r^d and k_r^t are the constants of dissociative and three-body recombination [12]; N, U, V the number density, longitudinal and transverse velocities, respectively.

4. RESULTS AND DISCUSSION

Figure 2 shows a CARS spectrum of the molecular nitrogen Q-band from the axial part of HVAPD. The spectrum

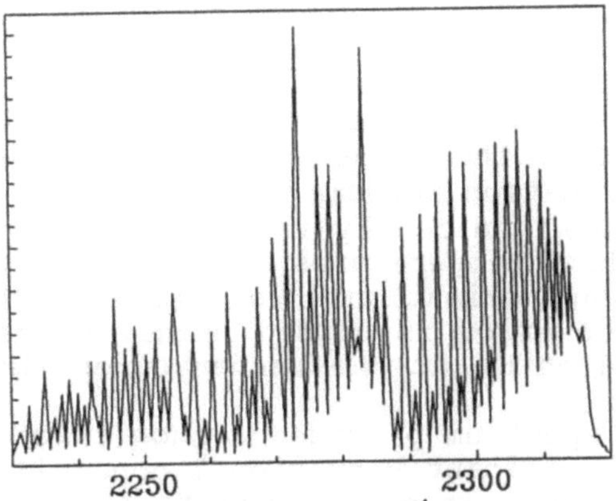

Fig.2. CARS spectrum (Q-band) of nitrogen molecules on the discharge axis: G = 2200 g/h, I = 40 mA.

was obtained using the parallel linear polarizations of the laser beams. The spatial resolution was 70x70x500 μm, the spectral resolution was 0.2 cm^{-1}. In this spectrum, the vibrational-rotational lines are displayed for three vibrational transitions (1-0, 2-1, 3-2) in the ground electron state $X^1\Sigma_g$. It is clearly seen that in HVAPD, contrary to a glow discharge [13], high values of both the vibrational and rotational temperatures in the axial part of the discharge lead to complicated coherent interference between rotational components of the different vibrational transitions. In this case, reliable data on the vibrational-rotational population distribution cannot be obtained without careful numerical fitting of the experimental spectrum.

Relative vibrational $T_{v,v'}$ and rotational T_r temperatures have been used as parameters to construct Q-band N_2 spectrum. The magnitudes $T_{v,v'}$ and T_r were varied to attain a minimum of the functional

$$F = \sum_i (C\,\chi_i \chi_i^* - I_i)^2 / I_i^2, \tag{3}$$

where I_i is the experimental CARS intensity profile, χ_i is the nonlinear susceptibility, and C is the scaling factor.

The χ_i magnitude was calculated as a sum over different rotational transitions:

$$\chi_i = \sum_{m,k} \frac{(N_m - N_k)(d\sigma/d\Omega)_{mk}}{\omega_{mk} - \omega_i - i\Gamma_{mk}} + \chi_N, \tag{4}$$

where N_m , N_k are the relative populations of the energy levels m and k; $(d\sigma/d\Omega)_{mk}$ is the Raman cross-section for an appropriate transition; Γ_{mk} is the Raman halfwidth of a line; χ_N is the nonresonance component of susceptibility. The Raman cross-sections were calculated from the Teller-Placzek theory [14]; the values of Γ_{mk} were determined with the approximation formula from [15].

Functional (3) was minimized by the gradient descent method, thus yielding temperatures T_r and $T_{v,v'}$, as well as χ_N. The average approximation error was 15%. The authors consider that the error is caused mainly by the absence of reliable data on Γ_{mk} at the temperatures above 2000 K.

The results of the spectrum processing show that at the discharge axis a significant vibrational nonequilibrium T_v=3000 K was observed, when referred to a comparatively high translational temperature T_r= 2600 K (vibrational temperatures of the first three levels coincided within experimental error). That points to the high rate of vibrational N_2 excitation due to an electron impact in the axial HVAPD region. In this situation, the above processes compete effectively with rapid V-T relaxation at high temperatures. The latter is caused by high radial nonuniformity of an electron density and by a considerable energy contribution to vibrational degrees of freedom of molecular nitrogen [6].

Vibrational populations and rotational temperatures of the state $C^3\Pi_u$ were obtained by fitting of the emission spectra. The spectrum for 15 vibrational bands was constructed, which revealed the rotational structure of each band. The populations of the vibrational levels of $C^3\Pi_u$ (v=0-4) and the rotational temperature T_r were found independently by fitting the model and experimental spectra (Fig.3).

Values of the rotational temperatures obtained by CARS and emission spectroscopy agree well with the results of numerical simulations (Fig.4). Figures 4,5 show that the investigated discharge is characterized by high radial nonuniformity. The gas temperature on the discharge axis several times exceeds its bulk temperature (curves 2 in Fig.4). The increasing temperature across the discharge radius with a gradient of 5000 K/cm (Fig.5a) must sharply augment the ionization frequency and, hence, must yield a strongly nonuniform electron density profile. Calculated and experimental values of a translational temperature along the discharge axis are in both qualitative and quantitative agreement. The theoretical curve lies 10%-30% above the experimental points. Apparently, this difference may be due to the fact that the model used does not take into account the influence of the associative nitrogen ionization with the following mechanism [16]:

$$2N_2(X^1\Sigma_g^+, v\geq 16) \longrightarrow N_2(a'^1\Sigma_u^-) + N_2(X^1\Sigma_g^+, v-\Delta v) \qquad (5)$$

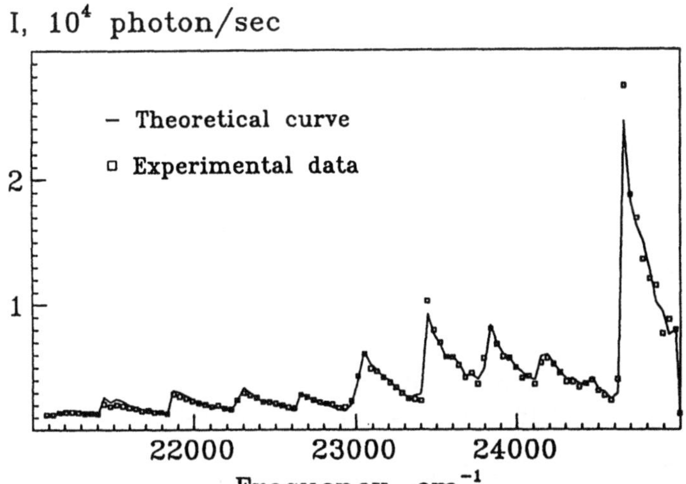

Fig.3. Best fit (solid line) of the experimental emission spectrum: T_v=3480 K, T_r=2450 K.

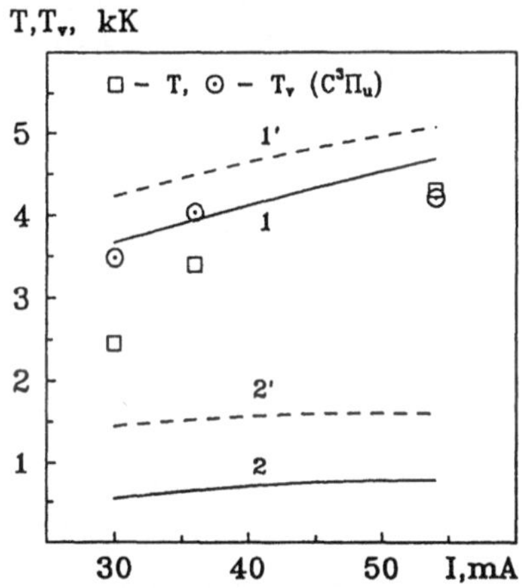

Fig.4. Comparison of experimental and theoretical data on discharge temperatures: G=850 g/h; points, experimental emission data; solid lines, theoretical data for translational temperatures; dashed lines, theoretical data for vibrational temperatures; 1, discharge axis; 2, bulk temperatures.

$$N_2(a'^1\Sigma_u^-) + N_2(a'^1\Sigma_u^-) \longrightarrow N_4^+ + e \ , \qquad (6)$$

as well as the dependence of the ionization rate coefficient upon the N_2 vibrational temperature. It was shown [17] that vibrational excitation gives rise to the ionization rate. From Fig.5b it is seen that the drop of the vibrational temperature is slower than translational

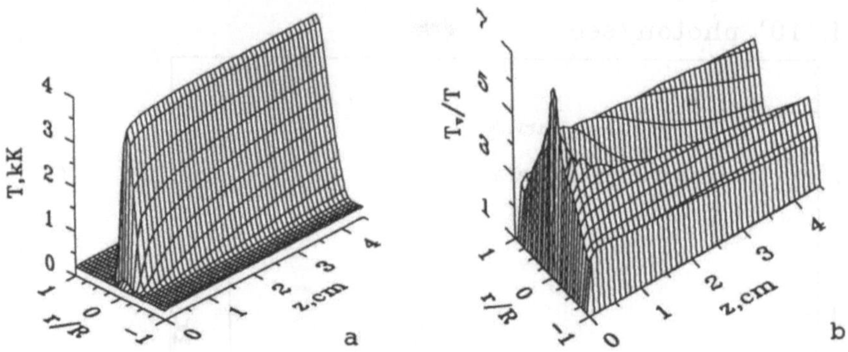

Fig.5. Theoretical space distribution of plasma parameters for discharge conditions investigated by the CARS method:
a, translational temperature;
b, ratio of vibrational and translational temperatures;

one. That is caused by slow V-T nitrogen relaxation in cold regions near the wall. Taking account of the associative ionization and the dependence between the ionization rate and vibrational excitation will lead to an expansion of the current channel and to a decrease of the axial translational gas temperature. Another reason for a difference between the experimental and simulated data may be associated with the effects of finite spatial resolution of measurements under high discharge nonuniformity.

The results of the numerical simulations show that the measurements were performed in the region of intense vibrational relaxation, where the difference between translational and vibrational temperatures is not pronounced. The study of the radial vibrational temperature distribution of N_2 molecules at distances of r/R~0.5-0.7 from the discharge axis is of great interest. According to the numerical calculations there is a substantial vibrational nonequilibrium in this region. This is, apparently, the reason for the successful use of this discharge in various technological applications [1-3].

5. CONCLUSIONS

Preliminary results are presented on CARS measurements of vibrational and rotational temperatures of the N_2 electronic ground state in high voltage atmospheric pressure discharge plasmas.

The data on vibrational and rotational N_2 temperatures in the excited electronic state $C^3\Pi_u$ were obtained by emission spectroscopy. A good agreement was observed between the rotational temperatures in the ground and excited $C^3\Pi_u$ states .

Numerical simulation of two-dimensional HVAPD nitrogen plasma was made in the laminar boundary layer approach. The experimental and numerical data are in a fair agreement.

The results of our investigations prove the possibility of obtaining a vibrationally nonequilibrium plasma in the investigated discharge.

REFERENCES

1. V.I.Borodin, S.A.Zhdanok and A.P.Chernukho, In Proc. Int. School-Seminar: Mod. Probl. of Heat and Mass Transfer in Chem.Technology,Vol.3,ITMO Press, Minsk, (1987) p.87
2. S.A.Zhdanok, E.M.Vasilieva and L.A.Sergeeva, Sov. J. Eng.Phys. 58, (1990) 101.
3. V.I.Borodin, J.A.Bumai, V.A.Zhdanok and S.A.Zhdanok, In Heat and Mass Transfer at Phase and Chem.Changes, ITMO Press, Minsk, (1990) p.3
4. P.A.Apanasevich, V.V.Kvach, V.G.Koptev,V.A.Orlovich, A.A. Stavrov and A.P. Shkadarevich, Sov. J. Quant. Electron. 14, (1987) 265.
5. M.Littman and H.Metcalf, Appl.Opt. 17, (1978) 2224.
6. N.L.Aleksandrov, F.I.Islamov, I.V.Kochetov, A.P.Napartovich and V.G.Pevgov, Sov.Fiz.Vyssok.Temp. 19, (1981) 22.
7. G.D.Billing and E.R.Fisher, Chem.Phys.43 (1979) 395.
8. P.Hammerling, G.D.Tear and B.Kivel, Phys.Fluids 2, (1959) 422.
9. O.E.Krivonosova, S.A.Losev, V.P.Nalivaiko et al., Plasma Chemistry, Vol.14 (Ed. B.M.Smirnov), Energoizdat, Moscow, (1987) p.3.
10. V.D.Berkut, N.N.Kudryavtsev and S.S.Novikov, Sov. Khim.Fiz. 7,(1988) 979.
11. V.V.Breev, S.V.Dvurechensky, A.T.Kukharenko and S.V. Pashkin, Sov.Zh.Prikl.Mekh. i Tekhn. 1,(1988) 27.
12. A.Kh.Mnatsakanyan and G.V.Naidis, Plasma Chemistry, Vol.14 (Ed.B.M.Smirnov), Energoizdat, Moscow, (1987) p.227.
13. A.A.Vakhterov, A.A.Ilyukhin, V.Yu.Yurov, N.I.Lipatov, V.V.Smirnov, Pisma v Zh.Tekh.Fiz., 11, (1985), p.3
14. G.Placzek and E.Teller, Z. Phys. 83, (1933) 143.
15. R.Hall, J.Appl.Spectrosc. 34,(1980) 700
16. A.V.Berdyshev, I.V.Kochetov and A.P.Napartovich, Sov.Plasma Physics 14,(1988) 741.
17. N.L.Aleksandrov, A.M.Konchakov and E.E.Son, Sov.Teplofiz.Vyssok.Temp. 17,(1979) 210.

CARS in Aerospace Research

B. Attal-Trétout, P. Bouchary, N. Herlin, M. Lefebvre, P. Magre, M. Péalat, and J.P. Taran

Office National d'Etudes et de Recherches Aérospatiales,
BP 72, F-92322 Châtillon Cedex, France

1 - INTRODUCTION

Coherent anti-Stokes Raman scattering (CARS) has become one of the indispensable tools of aerospace research. Both in low density gaseous reactive flows and in high pressure combustion, it can return high quality non-intrusive measurements of species densities and temperatures. Certainly, the most common application of CARS in analytical chemistry has been combustion diagnostics. Flames and combustors of all kinds have been investigated. The work was started on small-scale devices [1-3], but soon large scale industrial furnaces [4] , piston engines [5, 6] , jet engine combustors [7-9] and supersonic combustors [10] were studied. Variants of the technique, like simultaneous multiple species detection, have been proposed and tested [11-13]. At the same time, attention was drawn to several weaknesses of CARS, like saturation [14-17] and lack of spatial resolution [18], which may cause substantial errors and are frequently not given enough attention. Roughly speaking, however, application to combustion has matured, and a substantial fraction of the technical research on CARS now centers on refining lineshape reduction at high pressures [19-23] and on the application of resonance-enhanced CARS to detection of radicals at high pressures [24].

A very interesting application of CARS is the study of low pressure gases. In particular, discharges [25,26], chemical vapor deposition (CVD) reactors [27-29], and supersonic flows [30] provide interesting challenges. In these media, many elementary physical and chemical processes can be observed; their analysis and modelling are often easier and more complete than in combustion. The low pressure gas field has actually become one of the prime areas of CARS research at ONERA.

This report covers work performed recently in the area of CVD. In addition, instrumental research has been undertaken to perform single-shot measurements in short duration rarefied hypersonic flows; a two-line CARS method has been selected and the results are also given. Some progress also has been made in the area of instrumentation for combustion diagnostics, with the simultaneous measurement of temperature and concentrations of N_2 and of one other species with excellent sensitivity. Finally, the use of Resonance-Enhanced CARS (RECARS) in the detection of OH at high pressure has come closer to practical implementation in real combustors as shown by several recent developments.

2 - TECHNICAL CONSIDERATIONS AND EXPERIMENTAL SETUP

The CARS setup in use at ONERA has been designed for simultaneously measuring the temperature and the relative concentrations of two species in turbulent flames. Its schematic diagram is presented in Fig.1. It combines a classical multiplex arrangement for N_2 bandshape analysis, which can give the temperature in a single laser shot, and a narrowband detection capability based on a tunable dye laser which can be used either separately

Springer Proceedings in Physics, Volume 63 **Coherent Raman Spectroscopy**
Eds.: G. Marowsky and V.V. Smirnov

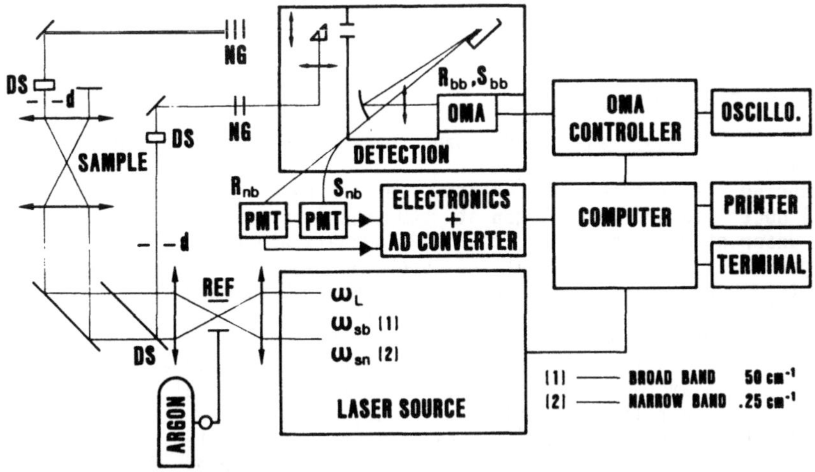

FIGURE 1 : Measurement system (the BOXCARS arrangement is shown and the reference and the signal channels are set up in "series"); NG: neutral glass filter; DS: dichroic filter; d: diaphragm; PMT: photomultiplier tube; OMA: Optical multichannel analyser.

for spectroscopic studies or rarefied gas investigations or in conjunction with the broadband arrangement for the detection of a chemical species different from N_2 with excellent sensitivity (e.g. 0.1 % at 2000 K in a single shot).

Note that this combination of lasers is similar to that proposed by Goss and Switzer [11] also for turbulent flame studies. Note also that several groups have recently proposed methods to detect two species or more using two broadband dye lasers [12, 13], but all of these methods work well only for the detection of majority species.

Sources

The core of the system is a single-mode Q-switched Nd:YAG laser chain which delivers 800 mJ of IR in 15 ns pulses. This output is frequency-doubled in a KDP crystal, giving a total of 220 mJ at 532 nm. Some 40 mJ are used to pump the broadband (ω_{2b} frequency) and the narrowband (ω_{2n}) frequency oscillators. About 50 mJ are used to pump each of the dye amplifiers, whereas the remainder (80 mJ) is used as the ω_1 CARS beam. This is amply sufficient for most studies, especially in plasmas at low pressure; note that the laser beams have to be focused using long focal length lenses at such energies to limit high field effects. However, in practical experiments done in atmospheric pressure flames which require tight focusing for better spatial resolution, the ω_1 beam energy is reduced to ≈ 40 mJ at the probe volume to avoid dielectric breakdown. Note that the transverse and longitudinal single-mode character of the ω_1 laser is considered to be essential to optimize the spatial resolution, to reduce the magnitude of saturation and Stark shifts, and to facilitate the analysis of spectral profiles.

The resonator of the narrowband dye laser comprises a 2400 lines per millimeter holographic grating at grazing incidence and a flat rotating back mirror. The wavelength is selected by tilting this back mirror and is adjustable from 560 nm to 700 nm. The linewidth (full width at half maximum, FWHM) can be adjusted between 0.07 and 1 cm^{-1} . The broadband dye laser central frequency is tuned to 607 nm in order to excite N_2. A 4-μm-thick air-spaced Fabry-Pérot etalon facilitates tuning of this laser and reduces its spectral width to 50 cm^{-1} (FWHM). Both dye lasers emit 4 mJ per pulse at the probe volume.

CARS Optical Arrangement

Referencing, BOXCARS, and nonresonant background cancellation are used systematically.

- Referencing is arranged "in series", i.e. the beams are first focused in a small duct flushed with argon at atmospheric pressure and then in the medium under study. In flames, four CARS signals are created: R_{bb}, R_{nb} in the argon cell and S_{bb}, S_{nb} at the probe volume, where R, S and their subscripts bb and nb stand for reference, sample, broadband and narrowband respectively. In most other cases, only the narrowband signals are used.
- Planar BOXCARS is used. With 250 mm focal length achromats and 8 mm beam separation, the spatial resolution at the sample probe volume is 3 mm long and 20 μm in diameter. Note that our definition of spatial resolution is quite conservative following earlier work [18]. In fact, we take the length of the probe volume over which 99 % of the signal is created. We shall see below that even this strict definition is not always adopted. Following the same definition, the probe volume obtained with 500 mm focal length achromats is 12 mm long and 40 μm in diameter if the same beam separation is used. In rarefied gases, which present weak gradients, collinear CARS or BOXCARS with 1 m lenses are used frequently.
- Nonresonant background cancellation is performed at the sample probe volume for both the narrowband and the broadband CARS signals by using the conventional 60° polarization orientations for the two Stokes beams and for the analysis of the two anti-Stokes beams. Cancellation of background is rarely necessary in rarefied gases.

Signal Detection

In turbulent flame studies, reference and signal CARS beams are dispersed in two separate 0.8 m spectrometers each equipped with a concave, 50 mm diameter, 2100 g/mm holographic grating. The nitrogen spectrum S_{bb} and the reference spectrum R_{bb} formed in the output planes are imaged with a magnification of 4 on the targets of two photodiode arrays (EGG). The resolution and the dispersion of each of the spectrometer-detector systems are respectively 0.8 cm^{-1} (FWHM) and 0.167 cm^{-1}/channel. A mirror is positioned in each of the spectrometers in order to intercept the narrowband signal (S_{nb}) and reference (R_{nb}) and to reflect them towards liquid fiber light guides (Bodson CLL 4-180, core diameter:4 mm). The light guides are used to direct the radiation onto two 10-dynode photomultipliers (PMT:RTC XP 2012). A mini computer (COMPAQ 386/25) reads the narrowband data, controls the scanning procedure of the photodiode arrays, fires the laser synchronously with the detector scanning and processes the data at a repetition rate of 3 Hz. The processing is done online by a software which subtracts background noise levels and divides each S_{bb} spectrum by the R_{bb} reference spectrum and each S_{nb} reading by its R_{nb} reference; the single-shot narrowband and broadband ratios are also individually processed online. The temperature T is first inferred from the shape of the N_2 spectrum according to a procedure which has been described in a previous paper [31]. Then, the nitrogen number density is deduced from the N_2 spectrum and that of the other species from the narrowband signal. In the rarefied gas studies, only the narrowband channels are used and the lasers can be fired at 5 Hz. Double monochromators mounted in the subtractive modes can be installed in replacement for the spectrographs to gain in compactness.

In spite of all the precautions taken, it is fair to admit that the instrument suffers, like many other CARS setups, from several deficiencies which affect the measurements:

• high field effects, namely vibrational saturation and Stark shifting, preclude usage of high laser energies; for instance energies in excess of 10-50 mJ at ω_1 and 1-5 mJ at ω_2 cannot be used in flames with diffraction-limited beams; otherwise, the signal is reduced and the data analysis becomes very complex;
• the spatial resolution, although much improved by the BOXCARS technique, is inadequate to resolve turbulence microscales in most aerodynamic flows and flames; obtaining adequate spatial resolution, e.g.by crossing the beams at a larger angle, results in a signal loss which severely reduces the detection sensitivity and measurement accuracy;
• density gradients in the flow or flame under study degrade the pump beam quality and preclude concentration measurements if they are too strong; however, this problem is not too grave in rarefied gases.

3 - APPLICATIONS TO RAREFIED GASES

3.1 - CVD

CVD constitutes a nice field of application to CARS. Recently work was done at ONERA on the analysis of a model reactor for the deposition of SIC on a graphite susceptor from the precursor tetramethylsilane (TMS). The reactor was a 55 cm long, vertical stainless steel water-cooled duct. The susceptor was a rectangular graphite plate, 2.5 cm wide by 5 cm high, mounted flush with one of the reactor walls to minimize aerodynamic disturbance, and Joule-heated to 1100-1500 K. The TMS was injected at the bottom of the reactor at pressures in the range 10 to 150 Pa and velocities close to 1 m/s.

Scanning CARS was used and the data processing was performed according to the procedure described in ref.[25]. The probe volume, which was about 2 cm long and 200 μm in diameter, was aligned parallel to the susceptor so as to permit exploration of the hot boundary layer, down to 0.3 mm from the surface. The study revealed the disappearance of the TMS and the formation of a large and hot H_2 fraction. Also, a phenomenom of thermal slip was observed. This section reports on the latter results, which will be covered fully in a separate publication [29]. A special study was made using pure H_2 and three types of susceptors, i.e. graphite 5890, SiC-coated graphite, and graphite covered with a 0.3 mm-thick tungsten plate mechanically attached to it.

Fig. 2 gives the Boltzmann diagrams of the rotational populations for H_2 at 0.5 mm from the three types of surfaces. In these three cases, rotational temperatures can be deduced from the Boltzmann plots because the data are fitted by a straight line. Whatever the distance from the surface, similar diagrams are observed for pressures between 26 Pa and 130 Pa. It is then possible to plot temperature profiles for H_2/C, H_2/SiC and H_2/W (Fig.3 + 4) . One also observes the existence of a temperature gap close to the surface. This gap ranges from 200 K to 600 K, figures which are somewhat larger than the temperature uncertainty (50 K). Also, the temperature gap depends on the nature of the surface (Fig.3) and on the pressure (Fig.5). However, the temperature profiles may still be modelled by solving the Navier-Stokes equations if the boundary conditions to the hot surface are modified to take into account the temperature jump [32].

The behavior of the N_2/W system is quite different. At 53 Pa, when the distance x from the surface is larger than 10 mm, the Boltzmann diagram also yields a rotational temperature (Fig.6). But, for $x < 1$ mm, i.e. at a distance less than 2 mean free paths, the populations do not follow a Boltzmann distribution. It then does not make sense to assign a temperature to the points close to the surface. The limit which separates the equilibrium and non-equilibrium zones comes closer to the W surface when the N_2 pressure is increased.

The quite different behavior of the rotational distribution of H_2 and N_2 close to the W surface can be related to the difference existing between

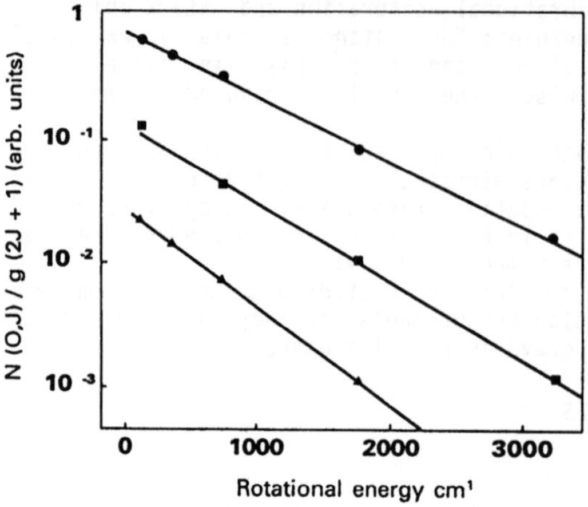

FIGURE 2 : Boltzmann diagrams for H_2/C(•), H_2/SiC (▪) and H_2/W (Δ) systems; N(0,J): population of the Jth rotational state (vibrational state v = 0); g: nuclear spin degeneracy factor; surface temperature: 1520 K; H_2 pressure: 130 Pa for H_2/W: surface temperature: 1170 K; H_2 pressure: 53 Pa.

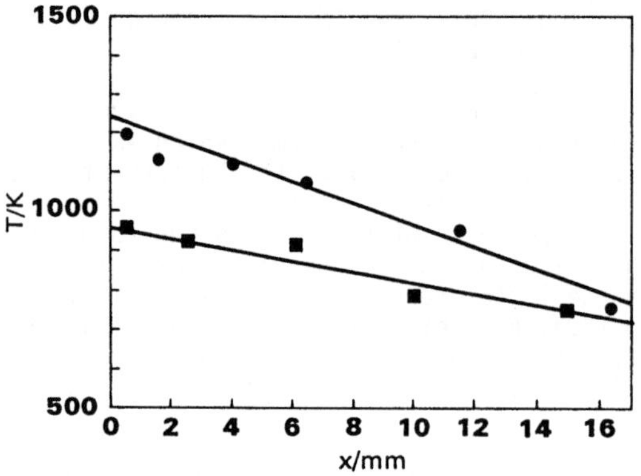

FIGURE 3 : Rotational temperature profiles of H_2 in front of a SiC surface (▪) or a C surface (•); surface temperature is 1520 K and H_2 pressure is 130 Pa. The solid lines are calculated profiles [32].

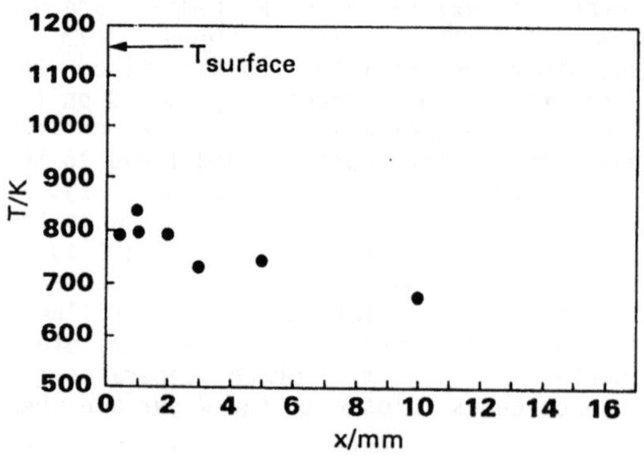

FIGURE 4 : Rotational temperature profile of H_2 in front of a W surface; surface temperature is 1170 K and H_2 pressure is 53 Pa.

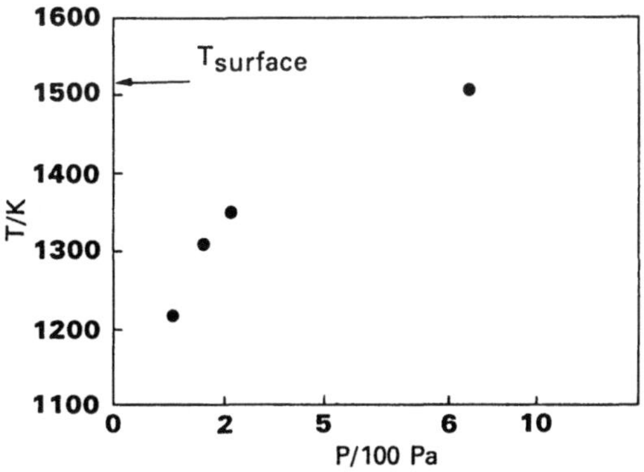

FIGURE 5 : H_2 Rotational temperature at 0.5 mm from a C surface versus pressure. The surface temperature is 1520 K.

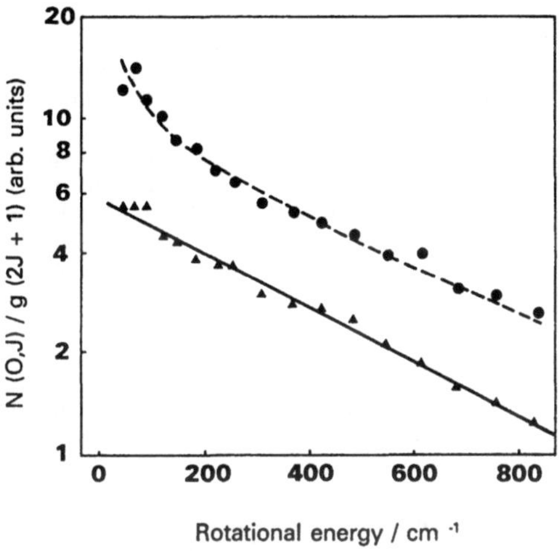

FIGURE 6 : Boltzmann diagrams for N_2/W systems. N(0,J) is the population of the Jth rotational state (vibrational state v = 0). Surface temperature is 1170 K, N_2 pressure 53 Pa and the distance from the surface is 0.5 mm (•) or 10 mm (Δ). Solid line: calculation at 778 K; dashed line: calculation with mixture of 10 % of the molecules at 100 K and 90 % at 1000 K.

the interaction potentials of H_2 and N_2 on W. From earlier kinetic studies on W(100) using thermal (diffusive) molecular beams, it was concluded that the systems N_2 on W(100) and H_2 on W(100) represented two extremes in the relative importance of trapping into weakly bounded precursor states prior to non-activated dissociative adsorption [33]. Assuming that hydrogen molecules desorb with a rotational temperature equal to that of the surface, accommodation being complete due to the dissociative adsorption, two rotational populations are expected: one at the temperature of the impinging molecules and one at the temperature of the surface. When pressure is increased, intermolecular collisions thermalise both distributions. However, within one mean free path λ from the surface (λ = 0.8 mm for H_2 at 1000 K) , the initial distributions are not yet perturbed. This situation is simply described by the Maxwell model. In this model, one considers that one half of the molecules is going toward the surface, and the other half is coming back. Of the second part, a fraction

ρ (where ρ is the sticking coefficient) is at the surface temperature, and a fraction $(1-\rho)$ at the temperature of the impinging molecules.

Therefore, one should see two rotational distributions, one at the surface temperature and one at a temperature somewhat lower than the surface temperature. Experimentally, the rotational distribution yields a unique rotational temperature, lower than the surface temperature. The deviation of the rotational distribution from the Boltzmann one is in fact not significant enough to be seen by CARS given the CARS accuracy. Also, the accommodation coefficient can be derived from the formula

$$\Delta T = \lambda \frac{2 - \alpha}{\alpha} \left.\frac{\partial T}{\partial x}\right|_{x = 0} ,$$

where $\left.\frac{\partial T}{\partial x}\right|_{x = 0}$ is the slope of the temperature profile perpendicular to the surface at the surface. In this simple model, α is equivalent to ρ. For the H_2/W system, it is found to equal 0.1.

For the H_2/C and H_2/SiC systems, the same analysis can be done, the rotational distribution revealing in both cases only one temperature. The accommodation coefficients α are respectively found to be 0.15 and 0.05. Moreover, for the H_2/C system, we have verified that α is independent of the pressure, as expected. To our knowledge, this is the first report of accommodation coefficients for these systems.

For N_2 on W, at incident gas temperature up to 500 K and for a crystal surface temperature as high as 1300 K, dissociative adsorption only proceeds appreciably from that fraction of the incident molecules which are trapped into a precursor state. Theories of the kinetics of formation of the chemisorbed layer assume that chemisorption is preceded by adsorption into the physisorbed layer, usually called the "precursor state", in which diffusion may occur across the surface to sites where adsorption takes place [34].

The rotational distribution shown in Fig.6 can be modelled by assuming the superposition of two different rotational distributions in Boltzmann equilibrium. However it is not possible to fit the experimental distribution with a group of molecules at the surface temperature (1200 K) according to the Maxwell model. This demonstrates that the accommodation of the molecule on the surface is not complete. The best fit is obtained for 90 % of molecules at 1000 K and 10 % molecules at 100 K. The molecules at 1000 K can be considered as a mixture of incident and scattered molecules, whereas the molecules at 100 K could be desorbed from the precursor state. A more detailed analysis of the phenomena will be found in [29].

3.2 - Diagnostics in Hypersonic Facilities

The detection sensitivity of CARS is not always good enough, especially in rarefied gas flows. Hypersonic facilities represent a difficult challenge, since, in addition, temperatures can be very high and flow durations are frequently quite short, prohibiting the use of the more sensitive scanning spectroscopy.

In some cases, multiplex CARS remains possible. A simulation was attempted in a 200 W microwave discharge in nitrogen. The pressure was 6×10^3 Pa. The spectrum of Fig.7 was obtained using a planar BOXCARS beam configuration, without nonresonant background suppression and averaging a small number of shots ($\simeq 10$) to facilitate the visual comparison with a theoretical simulation computed assuming Boltzmann equilibrium at a temperature of 400 K. The first hot band is easily observed; from its intensity, the vibrational temperature is deduced (~ 2400 K). In this pressure and static temperature range, the signal to noise ratio is quite satisfactory for single shot measurements of rotational and vibrational

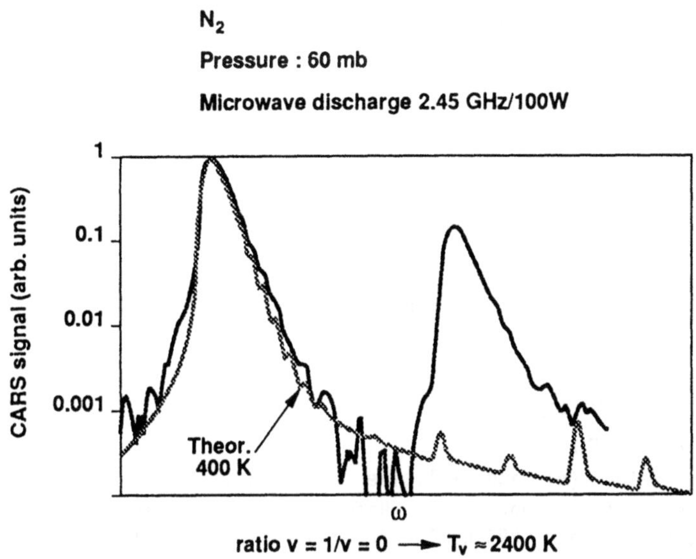

FIGURE 7 : Multiplex CARS spectrum of N_2 in microwave discharge at 6×10^3 Pa.

temperatures, as well as density. If the static temperature were raised to 2000 K, then the limiting N_2 pressure would be about 2 10^4 Pa at the same signal to noise ratio. Sensitivities are typically 10^2 to 10^3 times better using scanning CARS. Hence the proposal to use instead of one broadband Stokes laser, two narrowband lasers tuned to two separate Q-branch lines of the same species [35]. In this scheme, one obtains both the temperature (from the relative amplitudes of the two lines) and the density of the species. To reduce the problems associated with random signal fluctuations caused by fluctuations in spectral content of the dye lasers, the referencing is done in a cell containing the same species at a comparable temperature and pressure to have very similar lineshapes. This strategy offers another advantage with regard to saturation. In fact, most of the low density CARS experiments are conducted under conditions of moderate or high saturation to extract as much signal as possible from the medium [15, 25]. Using the same gas for the reference and matching carefully the power densities in the flow and the reference, one can cancel out the influence of laser power fluctuations on the signal. The advantage of referencing in the same gas was investigated. The sample (S) and reference (R) signals were generated in cells filled with 10^3 Pa of pure N_2 at room temperature. The ratio Q_1 of the two signals obtained at the peak of the Q(8) line was measured with an S of 10^4 photoelectrons and R of 5 10^3 photoelectrons on the average, Q_1 was fluctuating by 2.6 %, against a fluctuation of 1.7 % expected assuming that shot noise was the only source of fluctuations; meanwhile, S and R were undergoing fluctuations of ± 40 %.

Figure 8 gives the average value of $Q_1^{1/2}$ and the rms of Q_1 vs N_2 pressure in the signal cell, when the latter is varied from 10^2 to 10^3 Pa; notice the linear dependence of $Q_1^{1/2}$ and the fact that the fluctuations are nearly constant, except at lower pressures, where shot noise increases because of the substantial signal reduction. Finally, using the second dye laser, another ratio Q_2 can be generated on another line.

From the ratio $Q = Q_1/Q_2$, it is possible to determine the gas temperature in the sample cell using the Boltzmann population coefficients. The demonstration is shown in Fig.9, filling both the reference cell and

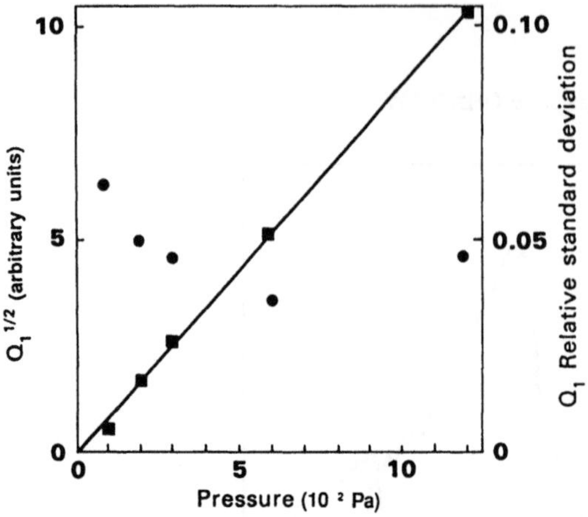

FIGURE 8 : Average value (▪) and relative standard deviation (•) of Q_1 versus N_2 pressure in the signal cell. Reference cell pressure: 10^3 Pa Temperature in both channels: 293 K. The frequency difference $(\omega_1 - \omega_2)$ is tuned on the Q(8) line.

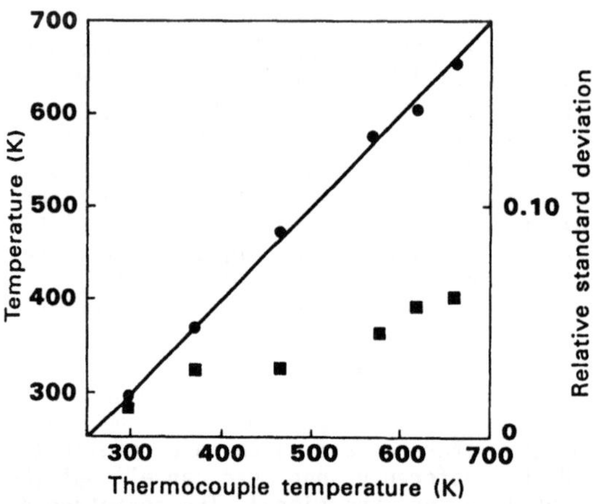

FIGURE 9 : Average (•) and relative standard deviation (▪) of single shot dual line CARS temperature versus temperature measured by a thermocouple. Reference and signal cell pressure: 12.5 x 10^2 Pa . Reference cell temperature: 293 K.

the sample cell with 1.25 10^3 Pa of N_2 at 293 K, and then varying the temperature of the sample cell. The Q(16) line was used in conjunction with Q(8). Figures 8 and 9 demonstrate that fairly accurate density and temperature measurements are possible in rarefied gas flows to rather low pressures.

Note, however, that it was necessary to adapt the power densities at the foci in the reference and sample cells, and that the temperatures in the two cells must be reasonably close to each other (i.e. within a factor of two) so that the Doppler-broadened lineshapes have similar widths. Finally, the results presented here were obtained with short focal length achromats, viz 500 mm for the reference and 250 for the sample, in order to have a reasonably short spatial resolution of 2-3 mm in the latter. The consequence was that power densities reached 10^{11} W/cm^2 at ω_1 and 2.6 x 10^8 W/cm^2 at ω_2 in the focal regions, causing substantial saturation and signal loss. Use of larger focal lengths without changing beam diameters would permit better signal generation efficiency, hence detection sensitivity, but at a cost in spatial resolution. It is hoped that N_2 densities as low as 10^{16} cm^{-3} can be monitored and temperatures measured in the static temperature range 300-600 K, with spatial resolution of 1-3 cm. This technique has not been tested in wind tunnels so far, but its anticipated performance raises hopes that it will be useful in a large number of situations.

4 - APPLICATIONS TO COMBUSTION

4.1 - Multiple Species Detection by CARS

Application of CARS to combustion requires much caution, especially when it comes to measuring reactant concentration. The technique of in situ referencing using N_2 seems promising [36]. A validation experiment has been performed in the burnt gases of a laminar, stable, premixed C_2H_4-air flame. When necessary, for the sake of simulating the effect of turbulence on the measurements, a turbulent jet of hot air is blown on the laser beams. The flame is stabilized at the exit of a bundle of 1000 tubes of 0.8 mm internal diameter and 1 mm external diameter. The set of elementary tubes is contained in an externally water-cooled pipe. The overall diameter of the bundle is 36 mm. Detectivity tests were performed on O_2. The influences of saturation and turbulence were studied.

Detectivity

The detectivity measurement is performed without turbulence simulation using 250 mm focal length lenses in order to achieve a good spatial resolution. The probe volume is positioned on the burner axis and 5 mm above it. Figure 10 presents the averaged temperature, and the O_2 mole fraction in the burnt gases versus equivalence ratio in the range $\Phi = 0.6 - 1.5$. Solid lines give the theoretical values assuming chemical equilibrium. The agreement between experimental and equilibrium temperature curves is satisfactory. The difference between the two temperatures has a maximum value of 150 K at $\Phi = 1.1$. This reflects the heat losses to the head of the burner. For $\Phi < 1$ the experimental O_2 mole fraction curve is slighty below the equilibrium one. For rich flames, i.e. for $\Phi > 1.1$, a nearly constant concentration of 0.4 % is found experimentally for O_2. This value is far above the calculated equilibrium value of 60 ppm and causes one to question the nature of the CARS signal. To understand this origin, a partial scan of nearby Q-branch lines was done; this scan revealed that the rotational temperature of the O_2 detected is of the order of 700 K, indicating that one is seeing O_2 from the room air, perhaps partially entrained by the flame edge. It was checked that the anomalous signal is not leakage from the collinear CARS beam, which can be quite strong, and may be scattered into the direction of the BOXCARS beam. The signal is actually created 15-20 mm away from the focus, although the spatial resolution is only about 3 mm, because the radial intensity profile of the pump beams does not decrease abruptly enough. Thus there remains sufficient overlap of the BOXCARS pump beams at such distances to create a detectable signal with narrowband CARS (this signal is, here, 10^{-5} - 10^{-6} times smaller than that which would be created if the O_2 were at the focus). This finding illustrates the fact that the concept of spatial resolution in CARS is a subtle one, because of the "fuzzy" nature of the contours of the probe volume, and that the spatial resolution must be measured for each

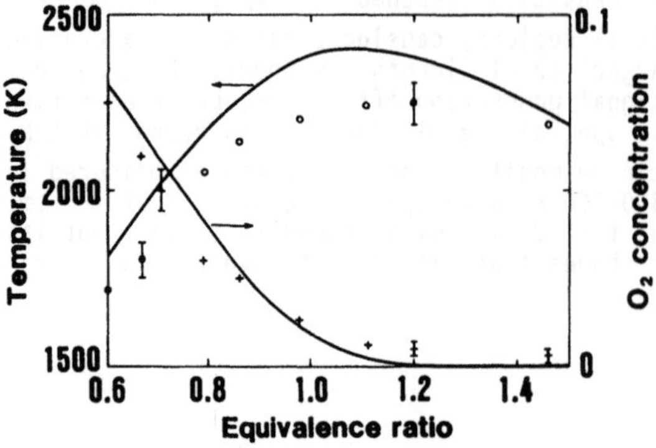

FIGURE 10 : Average temperature (•) and O_2 concentration, (+) versus equivalence ratio in a laminar flame (solid lines are the equilibrium curves)

experimental situation. Finally, while the current demonstrated detectivity limit is about 0.4 %, the actual limit in a homogeneous burner of larger scale is probably in the range of 0.1 - 0.2 % at 2000 K.

Saturation

These experiments also show saturation interference to be present. By making adjustments in both the laser energies and the focal length of the beam focusing lens, it is possible to achieve variations from 180 to 700 GW/cm^2 for laser power density I_1 and from 10 to 80 GW/cm^2 for I_2 at the probe volume.

No measurable effect was seen on the temperature within 40 K, i.e. within measurement accuracy. This is not entirely surprising since saturation is practically independent of rotational quantum number within a Q-branch [15] , so that the Q-branch associated with each vibrational quantum number v preserves its shape. Furthermore, the change in vibrational state populations, which alters the relative amplitudes of the Q-branches within a spectrum, and which has been reported by some in multiplex CARS [14], can be minimized. As a matter of fact, this change can be made the same for the first two or three states, i.e. those states which are normally detected in multiplex CARS. For this, one notes that the rate of loss of population out of vibrational state v into v + 1 due to saturation is proportional to v + 1. This is done by taking advantage of the bell shape of the spectrum emitted by the broadband dye laser. One then tunes the laser's frequency and adapts its width so that the product of v + 1 by the spectral density is approximately constant for the v = 0,1 and, to a lesser extent, v = 2 bands.

The effect of high fields on the concentration measurements is more obvious. This effect varies with temperature. For instance, the value found for the nitrogen mole fraction in the burnt gases is 0.66 ± 0.03, instead of 0.76 expected for equilibrium, at the lowest power densities that could be used to ensure acceptable detection sensitivity, i.e. I_1 = 180 GW/cm^2 and I_2 = 10 GW/cm^2. The mole fraction appears to decrease even further, down to 50 %, when the power densities are raised to their maximum values. The study also reveals that the signal reduction is mainly a function of I_1, and not of the product $I_1 I_2$, suggesting the Stark effect as the primary cause of disturbance. Finally, the study also shows that the ratio of O_2 and N_2 mole fractions is independent of the power densities over the range studied, within experimental error.

Turbulence Studies

The studies are undertaken to explore the potential of CARS, and, more precisely, of our approach of simultaneous two-species measurement, in a turbulent flame. To this end, we simulate the beam propagation conditions encountered, for instance, in an industrial burner. This is done in the same flame by monitoring the same parameters as before at the same point. For that purpose we create artificial turbulence on the optical path of the laser beams by blowing a turbulent hot air flow of 3 cm diameter and with a temperature of 800 K between the focusing lens and the probe volume. Ideally, we should get the same results as before if turbulence had no effect. However, we observe that the optical perturbations cause considerable signal fluctuations and loss of more than 50 % of the laser shots because of insufficient signal level. Note that this is quite severe and that, if more shots were lost in a real turbulent combustor, the results of the CARS measurements would become questionable; we thus simulate here a quite extreme situation on the borderline of the applicability of CARS. The incidence of the fluctuations on the oxygen and nitrogen concentrations probed in the post-flame region is quite severe (Fig.11a). We see that the mole fractions of N_2 and O_2, which are both affected by the optical perturbations, fluctuate between 25 and 70 % and 3 and 13 % respectively; this is far more than when the perturbations are absent, a case also shown in the figure. However, we also note that beam steering effects alter similarly the oxygen and nitrogen signal intensities. This is demonstrated in Fig.11b, where the ratio of O_2 and N_2 single-shot mole fractions is plotted versus the temperature found for each measurement.

Figure 11b presents a view of the current potential of CARS regarding measurement accuracy in laminar and in turbulent combustion. Ideally, if measurements were perfect, one would get a single point since we are probing a stable flame. The scatter of the data points essentially reflects CARS measurement uncertainties. One sees that in the two cases, i.e. with and without perturbation of the beam propagation, the temperature scatter remains approximately the same. We have $\bar{T}$ = 1624 K with $\sigma(T)$ = 48 K without perturbation and $\bar{T}$ = 1630 K with $\sigma(T)$ = 60 K with perturbation; in the latter case, only 30 % of the shots can be processed, the signal strength being too small for the others to be kept. Similarly, the standard deviation of the concentration ratio increases only from 17 to 20 % although it reaches 30 % for both O_2 and N_2 separately.

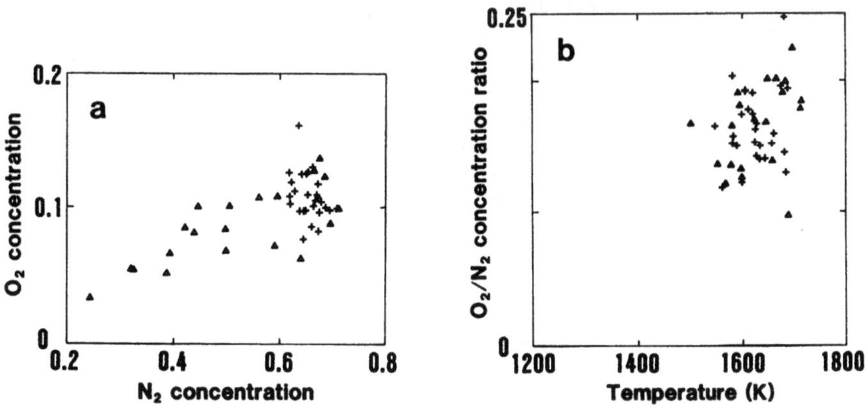

FIGURE 11a,b : Single-shot mole fraction measurements, with and without the effect of turbulence: a- O_2 mole fraction vs N_2 mole fractions; b- O_2 mole fraction relative to N_2 mole fractions vs temperature; (+) no perturbation of laser beams; (Δ) with perturbation of laser beams

Thus, it now appears feasible to perform instantaneous spatially resolved measurements of temperature and of the concentrations of two species simultaneously in turbulent combustion; these measurements should be possible with an accuracy of 3-5 % for the temperature using a majority species like N_2 as the first species and 20 % for a reactant like O_2 as the second species, and with detection sensitivity approaching 1000 ppm at 2000 K. However, the measurement of the concentration of the second species is referenced to that of the majority species, which we assume to be either constant or easily calculable; this restricts the application of the method to premixed flames or to flames where N_2 is mixed with the fuel. On the other hand, the technique is fairly free from both saturation and beam disruption by density gradients over a reasonable range of experimental parameters.

4.2 - Detection of Trace Species by RECARS

RECARS is a variant of conventional CARS where the lasers are tuned to selected absorption lines of the species to detect. Gains in detection sensitivity can reach 1 - 1000 depending on the chemical species under study. The theory of the effect, which is beyond the scope of this paper, has been treated elsewhere [37]. Recently, the study of OH has been performed in a flat flame burner capable of operating up to 20 atm and fuelled with a methane/air mixture [24] . To demonstrate the full sensitivity of the technique, the excitation of the CARS spectra is done using three lasers [38]. The three beams are tuned into the UV in order to excite the 0-0, 0-1, and 1-1 bands of the A-X system of OH, in the range of 310 to 340 nm. The beams are applied to the sample in a BOXCARS arrangement, achieving a spatial resolution of about 10 mm. Each laser pulse has an energy of 1-3 mJ. Spectra are recorded by scanning CARS and reduced numerically (Fig.12). They show the feasibility of sensitive OH detection. Profiles of OH above the flame front have been plotted versus distance from the burner. By detuning slightly one of the pump lasers, it has been possible to isolate a line associated with the upper vibrational state and to obtain a temperature (typically 2200 K) from 1 to 15 atmospheres (Fig.13). Figure 14 shows OH densities measured 6 mm above the burner, normalized to that obtained at 1 atm, versus pressure and for three values of the equivalence ratio. The experimental values compare well with those obtained from equilibrium flame calculations. The scatter of the data is larger at the higher pressures, because the signal decreases and because the burner is less stable as the flame approaches the extinction limit. At $\phi = 1$ and at 1 bar, the experimental OH mole fraction is about 3.5×10^{-3} while the detection sensitivity is 7 ppm. However, because of the fast pressure dependence of the signal, this sensitivity degrades quickly and is only 500 ppm at 20 bars, for our experimental conditions; some improvements are to be gained by increasing laser powers. A study shows that LIF may be able to operate to 100 bars and that RECARS and LIF can be used up to 30 bars at least [39]. However, note that, at high pressure, quenching is a difficult problem in LIF, since the quenching parameters are not well known yet in flames at high pressure; this makes quantitative concentration measurements delicate over a wide range of temperature, composition and pressure [40].

Multiplex RECARS is also possible. Figure 15 shows an experimental spectrum obtained under conditions similar to those of Fig.13, but with a Stokes radiation of width $\sim 3\ cm^{-1}$ in the UV range , dispersing the anti-Stokes spectrum and detecting with a multichannel detector. No reference channel was used in that experiment. Signal fluctuations induced by one-photon saturation were easily observed from shot to shot on the CARS profile.

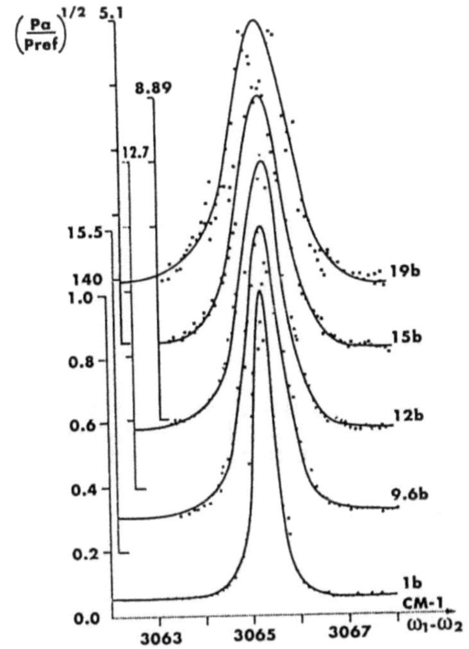

FIGURE 12 : RECARS experimental spectra of the O_1 (7.5) line of OH at triple resonance in flat $CH_{4/air}$ flame from 1 to 19 atm; solid line: theory.

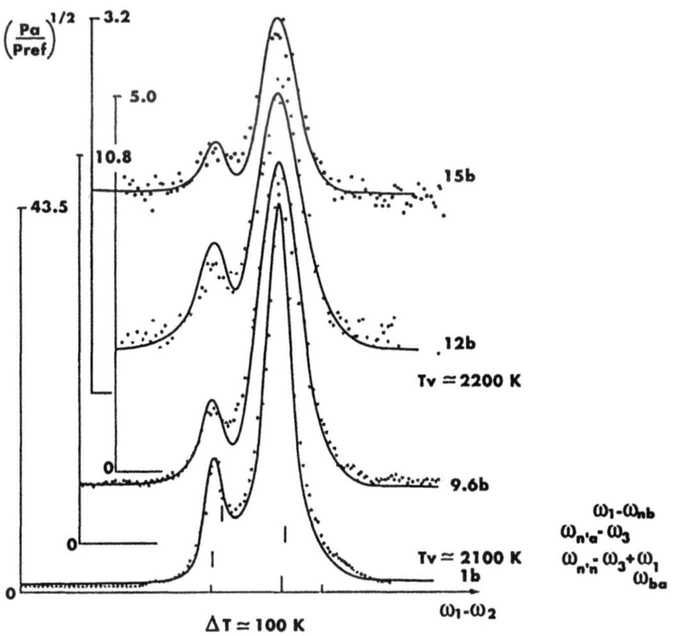

FIGURE 13 : RECARS spectra of same OH line as in Fig.12 with slight detuning of one pump laser, from 1 to 15 atm.

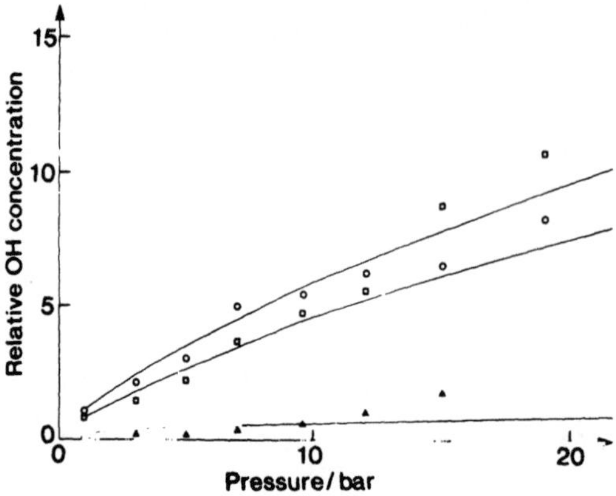

FIGURE 14 : OH concentration in flame at 6 mm above the burner vs pressure : (o) ∅ = 1; (□) ∅ = 0.83; (Δ) ∅ = 1.21; solid line: theory. Data points are presented normalized to the concentration at 1 bar and ∅ = 1.

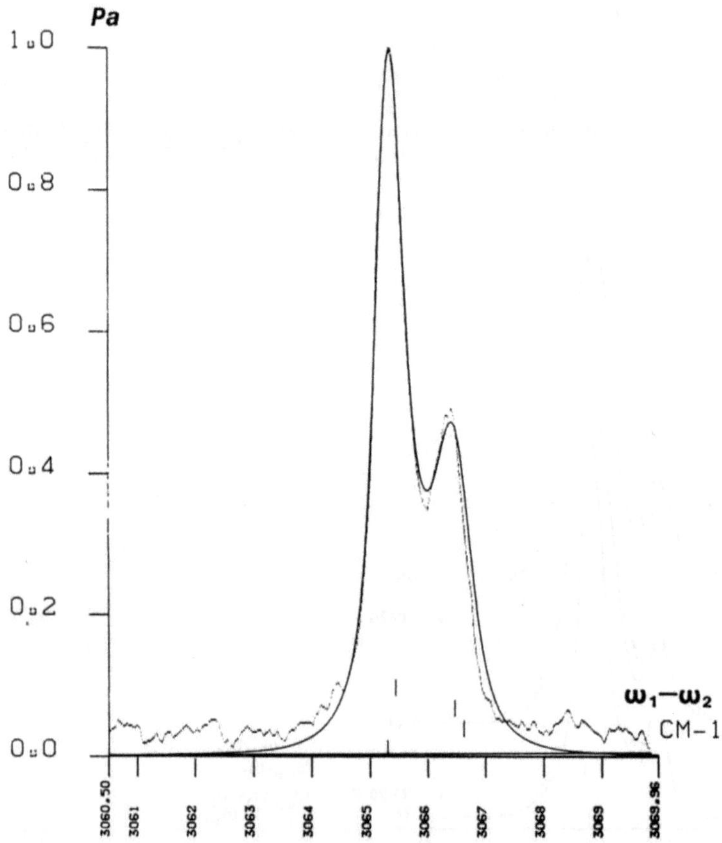

FIGURE 15 : Multiplex RECARS spectrum under conditions similar to those of Fig.13, at 1 atm and for ~ 1 cm^{-1} detuning of one pump laser; 30 shots are averaged; spectral resolution ~ 0.4 cm^{-1}.

5 - CONCLUSION

The role of CARS in aerospace research is slowly growing. CARS can perform useful tasks in rarefied gases as well as in moderate or high pressure combustion. The study of hypersonic flows is certainly one of the most promising for the next few years, since real gas effects and the high enthalpy chemistry should rapidly become tractable. It is also worthwhile pointing out that the role of CARS had been primarily limited to temperature measurements in combustion. Our last results show that, provided certain precautions are taken, sensitive and fairly precise concentration measurements should also be possible, simultaneously with the temperature measurements even under conditions where turbulence causes the loss of an appreciable number of laser shots. At atmospheric pressure, detection sensitivities may reach 1000 ppm in a single shot; even better results will be achieved with RECARS, in the range of 1 to 100 ppm depending on the radical to be detected. However, implementing the latter technique is delicate and requires further development before it can become applicable to industrial environments.

REFERENCES

[1] - P. Régnier and J.P. Taran, Appl. Phys. Lett. 23, 240 (1973).

[2] - F. Moya, S.A.J. Druet, and J.P. Taran, Opt. Commun. 13, 169 (1975).

[3] - K. Müller-Dethlefs, M. Péalat, and J.P. Taran, Ber. Bunsenges.Phys. Chem. 85, 803 (1981).

[4] - M. Aldén, Sov. J. Quantum Electron. 18, 746 (1988).

[5] - D.A. Greenhalgh, D.R. Williams, and C.A. Baker,"CARS Thermometry in a Firing Production Petrol Engine", Proc. Autotech. Conference, I. Mech E (1985).

[6] - M.J. Cottereau, F. Grich, and J.J. Marie, Appl. Phys. B 51, 63 (1990).

[7] - G.L. Switzer, W.M. Roquemore, R.B. Bradley, P.W. Schreiber, and W.B. Roh, Appl. Optics 18, 2343 (1979).

[8] - A.C. Eckbreth, G.M. Dobbs, J.H. Stufflebeam, and P.A. Tellex, Appl. Opt. 23, 1328 (1984).

[9] - R. Bédué, P. Gastebois, R. Bailly, M. Péalat, and J.P. Taran, Comb. Flame, 57, 141 (1984).

[10] - T.J. Anderson, G.M. Dobbs, and A.C. Eckbreth, Appl. Opt. 25, 4076 (1986).

[11] - L.P. Goss, and G. Switzer, "Laser Optics/Combustion Diagnostics", Final Report AFWAL-TR-86-2023 (1986).

[12] - A.C. Eckbreth, T.J. Anderson, and G.M. Dobbs, Appl. Phys. B 45, 215 (1988).

[13] - R.R. Antcliff, and O. Jarrett, Jr., Rev. Sci. Instrum. 58, 2075 (1987).

[14] - A. Gierulski, M. Noda, T. Yamamoto, G. Marowsky, and A. Slenczka, Opt. Lett. 12, 608 (1987).

[15] - M. Péalat, M. Lefebvre, J.P. Taran, and P.L. Kelley, Phys. Rev. A 38, 1948 (1988).

[16] - V.N. Zadkov and N.I. Koroteev, Chem. Phys. Letters 105, 108 (1984).

[17] - R.P. Lucht and R.L. Farrow, J. Opt. Soc. Am. B 6, 2313 (1989).

[18] - J.P. Boquillon, M. Péalat, P. Bouchardy, G. Collin, P. Magre, and J.P. Taran, Opt. Lett. 13, 722 (1988).

[19] - R.J. Hall, J.F. Verdieck, and A.C. Eckbreth, Opt. Commun. 35, 69 (1980).

[20] - M.L. Koszykowski, R.L. Farrow, and R.E. Palmer, Opt. Lett. 10, 478 (1985).

[21] - L. Bonamy, J. Bonamy, D. Robert, B. Lavorel, R. Saint-Loup, R. Chaux, J. Santos, and H. Berger, J. Chem. Phys. 89, 5568 (1988).

[22] - F.M. Porter, D.A. Greenhalgh, P.J. Stopford, D.R. Williams, and C.A. Baker, Appl. Phys. B 51, 31 (1990).

[23] - See also papers by Millot et al. and Temkin et al. in this volume.

[24] - B. Attal-Trétout, S.C. Schmidt, E. Crété, P. Dumas, and J.P. Taran, J. Quant. Spectr. Radiat. Transfer 43, 351 (1990).

[25] - B. Massabieaux, G. Gousset, M. Lefebvre, and M. Péalat, J. Physique 48, 1939 (1987).

[26] - See also papers by Devyatov et al. and Kruglik et al. in this volume.

[27] - R. Lückerath, P. Balk, M. Fischer, D. Grundmann, A. Hertling, and W. Richter, Chemtronics 2, 199 (1987).

[28] - See paper by Smirnov, Volkov, Yazan and Marowsky in this volume.

[29] - N. Herlin, M. Péalat, M. Lefebvre, P. Alnot and J. Perrin, "Rotational energy transfer on a hot surface in a low pressure flow studied by CARS", to be published.

[30] - See papers by Barth and Huisken and by Ilyukhin et al. in this volume.

[31] - M. Péalat, P. Bouchardy, M. Lefebvre, and J.P. Taran, Appl.Opt. 24, 1012 (1985).

[32] - N. Herlin, M. Péalat, M. Lefebvre, and M. Parlier, J. Physique, Colloque C5, 50 (1989), C5 - 843.

[33] - D.A. King, CRC Reviews in Solid St. and Material Sc. 7, 167 (1979).

[34] - D.A. King and M.G. Wells, Proc. Roy. Soc. London, A 339 245 (1974).

[35] - M. Péalat and M. Lefebvre "Temperature measurement by single-shot dual-line CARS in low-pressure flows", to be published.

[36] - M. Péalat, P. Magre, P. Bouchardy, and G. Collin "Simultaneous temperature and sensitive two-species concentration measurements by single-shot CARS", Appl. Opt., to be published.

[37] - J.P. Taran, and S.A.J. Druet, Prog. Quantum Electron. 7, 1 (1981).

[38] - B. Attal-Trétout, P. Berlemontand, and J.P. Taran, Molec. Phys. 70, 1 (1990).

[39] - K. Kohse-Höinghaus, U. Meier, and B. Attal-Trétout, Appl. Opt. 29, 1560 (1990).

[40] - C.D. Carter, J.T. Salmon, G.B. King, and N.M. Laurendeau, Appl. Opt. 26, 4551 (1987).

Coherent Rotational and Vibrational Raman Spectroscopy of CO_2 Clusters

H.-D. Barth and F. Huisken

Max-Planck-Institut für Strömungsforschung,
Bunsenstr. 10, W-3400 Göttingen, Fed. Rep. of Germany

Abstract: Coherent Stokes Raman scattering (CSRS) has been employed to study the rotational and vibrational spectroscopy of carbon dioxide clusters in the expansion of a supersonic jet. In the low-frequency range, around 72 and 88 cm^{-1}, the cluster spectrum features two peaks which can be assigned to librational modes of symmetry E_g and F_{g-}. This observation indicates that the gas-phase CO_2 clusters adopt the structure of crystalline CO_2. Comparison with the data obtained in solid CO_2 shows that the cluster librations are in thermal near-equilibrium with the monomer rotations. The vibrational spectroscopy of CO_2 clusters has been studied in the $(\nu_1, 2\nu_2)$ Fermi dyad. In the ν_1 component a strong cluster peak, red-shifted by 9.7 cm^{-1} from the monomer line, is observed which can be assigned to large CO_2 clusters with crystalline structure. Weaker and less red-shifted spectral features, significant at milder expansion conditions, are attributed to small CO_2 clusters, i.e. dimers, trimers, and tetramers.

1. Introduction

During the last few years there has been increased interest in the investigation of loosely bound atomic and molecular clusters which are conveniently generated by adiabatic expansion of the desired gas into a vacuum. Generally the motivation for their investigation is the desire to understand how physical and chemical properties vary if one goes from molecular to condensed phase systems.

Much work in cluster research has been devoted to the interaction of clusters with photons of various wavelengths, i.e. to their spectroscopy. Such investigations may provide extremely valuable information on the structure and energetics of these species. Several techniques have been developed and some of them have already been applied to CO_2 clusters. Very detailed information on the structure of the CO_2 dimer has been obtained by infrared predissociation [1,2] and absorption [3] spectroscopy. Fourier transform spectroscopy [4] has been applied to large CO_2 clusters, whereas coherent anti-Stokes Raman spectroscopy (CARS) has been employed to study dimers and larger CO_2 clusters [5]. The latter technique has been shown to be particularly powerful for the investigation of clusters because of its universality [6].

Springer Proceedings in Physics, Volume 63 **Coherent Raman Spectroscopy**
Eds.: G. Marowsky and V.V. Smirnov

Recently we have built a supersonic beam coherent Raman spectrometer to study relaxation processes and cluster formation in the expansion zone of supersonic jets. Detailed information has already been obtained on the formation and spectroscopy of large NH_3 and C_2H_4 clusters [7,8]. Whereas in the past the spectrometer has been operated exclusively in the spectral region around 3000 cm^{-1}, the apparatus has been reconstructed and enables now also the study of low-frequency Raman transitions below 100 cm^{-1}. This is accomplished using the three-dimensional, so-called BOXCARS [9] geometry. First results on N_2 and CO_2 clusters obtained with this spectrometer have been published very recently [10,11]. In the study on CO_2 clusters [11] we have been able to observe two librational modes of different symmetry.

In this contribution we present our recent results on gas-phase CO_2 clusters obtained in both the rotational spectral region below 100 cm^{-1} and the vibrational region around 1300 cm^{-1}. In the rotational spectrum the two strongest librational modes are investigated. Their observation indicates that the structure of the large CO_2 clusters in the jet is the same as in crystalline CO_2. Vibrational spectra have been measured in the (ν_1 , $2\nu_2$) Fermi dyad at 1285 cm^{-1} and 1388 cm^{-1}. In both components strong spectral features arise when the stagnation pressure is sufficiently high. They are red-shifted from the monomer lines and can be assigned to intramolecular vibrations in large CO_2 clusters. Weaker and less perturbed lines are attributed to smaller clusters and, in particular, to the CO_2 dimer.

2. Experimental

The supersonic beam coherent Raman spectrometer has already been described in detail elsewhere [7,11,12]. Therefore, we will focus only on the most relevant features important for this experiment. The laser system incorporated a frequency-doubled and -tripled Nd:YAG laser. Whereas the 532 nm radiation was split into two beams providing the pump for the coherent Raman process, the 355 nm radiation was used to pump a tunable dye laser which was operated with Coumarin 500. With a tuning range of 495-532 nm, Raman shifts between 0 and 1400 cm^{-1} were obtained. Since the dye laser frequency was on the anti-Stokes side of the pump beam, coherent Stokes Raman scattering (CSRS) was employed to perform rotational and vibrational Raman spectroscopy of CO_2 monomers and clusters. Typical pulse energies were up to 70 mJ at 532 nm and 5 mJ for the tunable anti-Stokes beam. With bandwidths of 0.11 and 0.12 cm^{-1} an overall resolution of 0.2 cm^{-1} was obtained. For rotational Raman spectroscopy, which is dominated by anisotropic scattering, we used perpendicularly polarized pump and Stokes beams, in order to reduce the stray light. For the investigation of the intramolecular cluster vibrations, the polarizations have been chosen to be parallel.

The pulsed supersonic jet was formed from a slightly modified Bosch automobile fuel injection valve operated at room temperature. In order to enhance the cluster formation, a 0.6 mm diameter diverging nozzle with an opening angle of 30° (MBB ERNO, 0.5 N) has been attached to the tip of the valve. These nozzles are usually employed as attitude control thruster nozzles for satellite positioning. The original length of the nozzle throat was 7 mm. However, during the course of this experiment, it has been shortened several times up to 2.1 mm. This has been done to shift the cluster distribution towards smaller mean sizes. Through this nozzle both neat CO_2 and various mixtures of CO_2 in He (1-10 %) have been expanded at stagnation pressures of up to 20 bar.

Using the three-dimensional beam geometry [9], the three laser beams were focused onto the axis of the supersonic jet by a common f = 200 mm quartz lens. The interaction region was approximately a cylinder of 0.05 mm diameter and 1 mm length. The Stokes photons which were generated in the case of resonance were separated from the generating laser beams by means of spatial filters and a monochromator. CSRS spectra were recorded using a fully computerized gated integrator.

3. Results and Discussion

Rotational CSRS

A pure rotational CSRS spectrum of CO_2 monomers is shown in the upper panel of Fig. 1. It has been obtained 1 mm downstream from the nozzle exit with neat CO_2 expanded at a stagnation pressure of 1 bar. At such mild expansion conditions, only the rotational

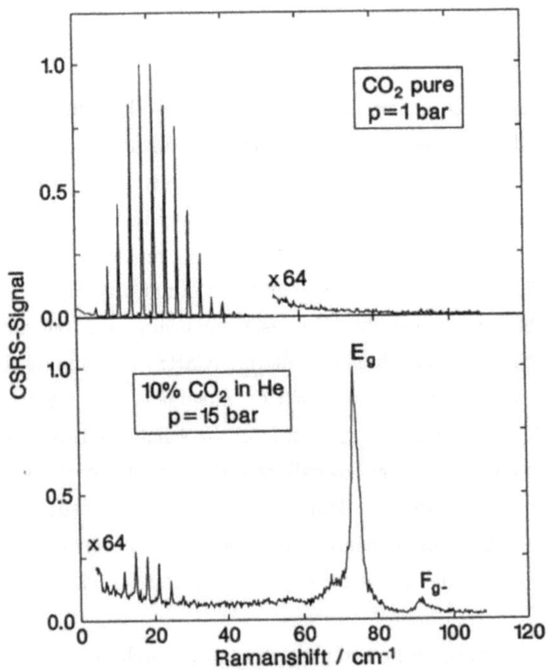

Fig. 1. Low-frequency CSRS spectra measured 1 mm downstream from the nozzle exit with pure CO_2 and a mixture of 10 % CO_2 in He. The peaks at 74.2 and 91.5 cm^{-1} are assigned to librational motions in large CO_2 clusters.

transitions of the CO_2 monomer are observed. Rotations from J = 0 to J = 26 are nicely resolved. These lines have been used to calibrate the wave number axis by matching the measured line positions with the theoretical positions calculated with B_0 = 0.3903 cm^{-1} [13]. Comparison with calculations further shows that the monomer rotations are in thermal equilibrium and can be described by a temperature of T_{rot} = 100 K. The spectral range above 50 cm^{-1} has been recorded with increased sensitivity. No features which can be attributed to clusters are observed.

In order to produce conditions which favour the formation of clusters, we employed a mixture of 10 % CO_2 in He and expanded this mixture at a considerably higher stagnation pressure of p = 15 bar. The CSRS spectrum measured under these conditions is shown in the lower panel of Fig. 1. Compared to the upper spectrum, the monomer rotational lines are significantly smaller, showing that the density of uncomplexed CO_2 molecules has drastically decreased. On the other hand, two new features which are absent in the upper spectrum are now observed at $\tilde{\nu}$ = 74.2 and 91.5 cm^{-1}. Since these features arise only at sufficiently high stagnation pressure and become stronger if the pressure is further increased, it seems clear that they must be assigned to the excitation of some specific mode in larger CO_2 clusters.

Anderson and Sun [14] have carried out Raman studies of polycrystalline samples of CO_2 under equilibrium conditions. In the frequency range below 150 cm^{-1} they observed three lines at 75.5, 94, and 134 cm^{-1} with intensity ratios of 100 : 25 : 12.5 which they assigned to the three Raman active librational modes E_g , F_{g-}, and F_{g+}. Since our peaks coincide almost exactly with the positions observed by Anderson and Sun, we adopt their assignment and ascribe the two peaks to the E_g and F_{g-} librations. Further support for the correctness of this assignment is provided by the intensity ratio of our peaks. Considering the quadratic dependence of the CSRS signal on the number density and taking the square root we obtain a corrected amplitude ratio of 100 : 21.7 which is very near to the value published by Anderson and Sun.

This is the first time that the weaker F_{g-} librational mode has been observed in gas-phase CO_2 clusters. Compared to the spectra of Nibler and co-workers [6,15], who observed only the stronger E_g libration, our spectra are characterized by an improved signal-to-noise ratio. Our observation shows that the structure of the CO_2 clusters generated in the expansion is the same as in crystalline CO_2. This is in contrast to the observations of Barnes and Gough [4] who concluded from their data that their CO_2 clusters were liquid-like. However, these workers had different experimental conditions. They employed infrared absorption spectroscopy and used windows which they placed very near to the nozzle. Thus, there were interactions between the expansion and the windows, probably causing a considerably higher cluster temperature.

We have also investigated the dependence of the librational peaks on the separation from the nozzle exit, i.e. on the internal cluster temperature. In this experiment we used the short, 2.1 mm long conical nozzle. The results obtained with a 10 % mixture of CO_2 in He, expanded at 20 bar, are shown in Fig. 2. From the uppermost to the bottom spectrum the

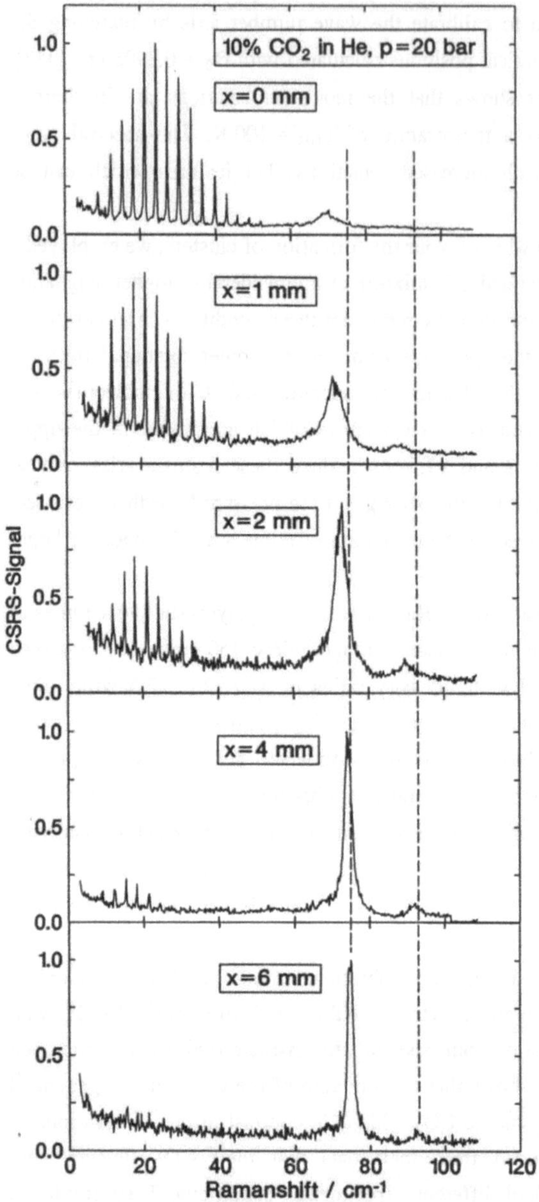

Fig. 2. CSRS spectra measured at various positions in the jet. With increasing separation from the nozzle and decreasing internal temperature the librational peaks move to larger wave numbers.

distance has been varied from x = 0 to x = 6 mm. With increasing distance the following observations are made: The monomer rotational lines drastically decrease in intensity and their envelope shifts to smaller wave numbers. In contrast, the librational lines gain in intensity and their positions move to larger wave numbers. In order to demonstrate this progressive shift, the dashed vertical lines have been drawn into the figure. They mark the peak positions which correspond to the lowest temperature achieved in this experiment. In addition to the frequency shifts, a narrowing of the librational transitions is clearly noted.

These experimental results can be interpreted as follows: With progressive expansion the number density of CO_2 monomers continually decreases. At the same time the relative density of CO_2 clusters drastically increases as evidenced by the strong growth of the main librational peak. Whereas at x = 0 only a few clusters are observed, at x = 6 mm almost all CO_2 molecules are found in the complexed state. As the distance from the nozzle is enlarged, the rotational temperature of the CO_2 monomers decreases. We have calculated theoretical CSRS spectra assuming a thermal distribution for the monomer rotations. Comparison with these calculations yields rotational temperatures between T_{rot} = 135 K at x = 0 and T_{rot} = 40 K at x = 6 mm. With decreasing temperature the librational lines move to larger frequencies and became narrower. This is in agreement with Raman studies carried out in crystalline CO_2 under equilibrium conditions [14,16]. Here the same behaviour is observed, suggesting that the physical explanation is the same. With decreasing temperature the intermolecular spacings between the CO_2 molecules in the cluster become smaller, thus leading to a stronger interaction, a larger force constant and, therefore, a larger excitation frequency. The narrowing of the librational lines can be explained by lifetime considerations. The excited librations couple to the phonon bath. At higher temperature the density of states is larger, the coupling more efficient and, hence, the lifetime shorter. With decreasing temperature the lifetime becomes larger, resulting in a reduced linewidth. That our lines are somewhat broader than observed in crystalline CO_2 [14,16] can be explained by the fact that, in our experiment, the signal is generated by a distribution of clusters with different sizes and perhaps different temperatures. This causes an additional broadening of the librational transitions.

From the measurements in crystalline CO_2 at equilibrium [14,16], a one-to-one relation between peak position and temperature can be established. Thus, with this knowledge and the measured librational line positions it is possible to assign a cluster temperature to each of our spectra. In Fig. 3 the librational peak positions are plotted as a function of temperature. The open circles are the results of various Raman measurements in crystalline CO_2 at equilibrium conditions [16]. These data can be used to derive temperatures for our cluster spectra. In Fig. 3, however, we chose a different presentation. The solid points, which are the result of this work, are obtained by combining the measured librational peak positions with the monomer rotational temperatures determined for the same spectrum, i.e. we assumed that monomer and cluster temperatures were the same. The good agreement with the crystalline data points justifies this assumption and shows that in our jet the monomer rotational and cluster librational temperatures are very much the same. However,

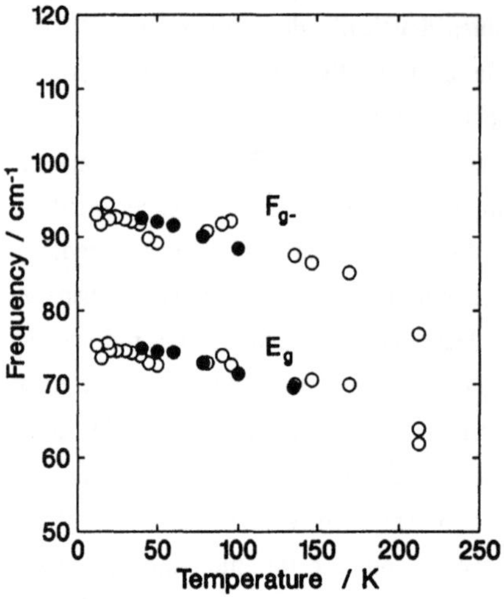

Fig. 3. Librational peak positions as a function of temperature in crystalline CO_2 at equilibrium conditions (open circles, Ref. 16) and in the jet (solid circles).

since the temperature dependence of the librational peak positions is not very strong, we admit that a similar good agreement would be obtained if the clusters were assumed to be 10 - 20 K warmer than the monomers. Such a difference between monomer and cluster temperature has been observed for N_2 clusters [17].

Vibrational CSRS

Using the same experimental setup, i.e. without changing the laser dye, we could also study the vibrational spectroscopy of CO_2 monomers and clusters in the (ν_1 , $2\nu_2$) Fermi dyad. In the gas phase the ν_1 component has its origin at 1285.4 cm^{-1}, while the rotationless $2\nu_2$ component is observed at 1388.2 cm^{-1} [18]. Originally the higher frequency peak was assigned to ν_1 and the lower to $2\nu_2$ [18,19]. However, a more consistent correlation with vibrational constants is achieved with the reversed assignment also used here [14].

First we investigated the ν_1 component, employing various dilutions of CO_2 in He. The result obtained with a 5 % mixture of CO_2 in He is shown in Fig. 4. The spectra were recorded as a function of the stagnation pressure directly at the exit of the 2.1 mm long conical nozzle. In each spectrum the most intense peak has been normalized to one. The frequency scale has been calibrated with the aid of a spectrum measured with neat CO_2 at p = 10 bar. The individual rotations of the ν_1 O-branch which could be observed in this spectrum have been matched with the line positions calculated from the spectroscopic constants ν_1 = 1285.41 cm^{-1} [18], B_0 = 0.3903 cm^{-1} [13], and B_1-B_0 = 2.66 x 10^{-4} cm^{-1} [18].

At a stagnation pressure of p = 2 bar only the ν_1 Q-branch of the CO_2 monomer is observed. Individual rotations could not be resolved. From the narrow peak, however, it can be deduced that only a few rotational states are populated and that, therefore, the temperature

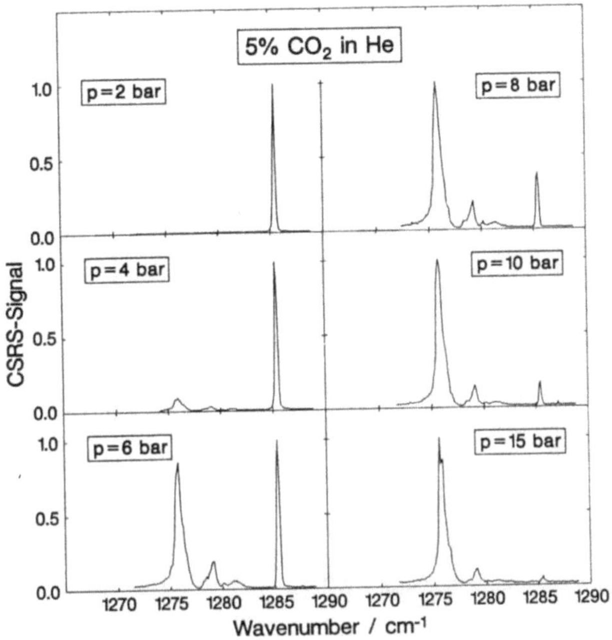

Fig. 4. Investigation of the ν_1 component as a function of the stagnation pressure for a 5 % mixture of CO_2 in He. The narrow peak at 1285.4 cm^{-1} is the monomer ν_1 Q-branch; all other features can be assigned to clusters.

must be quite low. When the stagnation pressure is raised to p = 4 bar, three new structures arise at 1275.7, 1278.8, and 1281.0 cm^{-1}. With increasing pressure these new peaks further gain in intensity, whereas the monomer peak at 1285.4 cm^{-1} continually decreases. This pressure behaviour suggests that the new structures must be assigned to vibrational transitions in CO_2 clusters. At the maximum pressure of p = 15 bar the cluster peak at 1275.7 cm^{-1} has developed to the dominant spectral feature, whereas the monomer peak is barely visible. This indicates that, at this pressure, condensation is so efficient that almost all CO_2 molecules are clustered.

Since the three cluster peaks have different pressure dependences, we assign them to different cluster sizes. Relative to the other peaks, the 1275.7 cm^{-1} peak grows continually with increasing pressure. It is the most red-shifted peak and its position coincides exactly with the transition observed in crystalline CO_2 [20]. Therefore, it seems obvious to assign this peak to the excitation of the ν_1 vibration in large CO_2 clusters having adopted the structure of crystalline CO_2. That CO_2 clusters with crystalline structure are actually in the jet has been shown before by rotational spectroscopy (see preceding section). The least red-shifted peak at 1281.0 cm^{-1} is most pronounced at p = 4 bar. At this pressure its amplitude is 25 % of the large cluster peak. With increasing pressure its significance gradually decreases, and at p = 15 bar it is less than 3 % of the large cluster peak. This

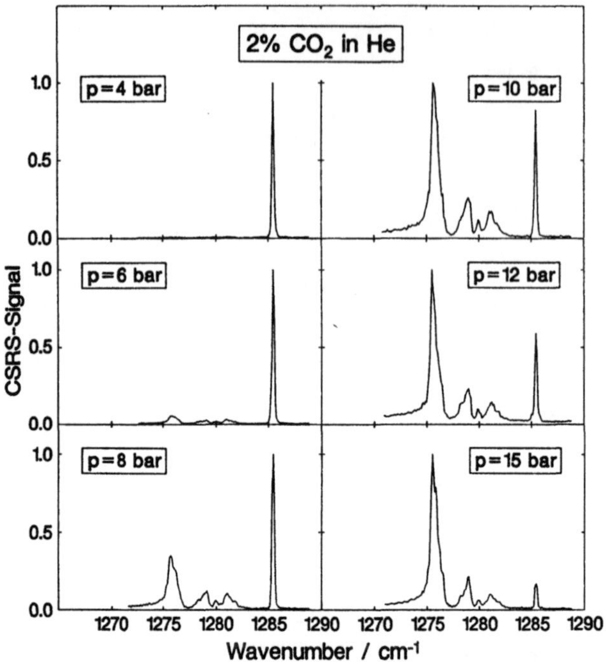

Fig. 5. Investigation of the ν_1 component as a function of the stagnation pressure for a 2 % mixture of CO_2 in He. Note that the less red-shifted spectral features are enhanced compared to Fig. 4.

pressure dependence suggests that the peak at 1281.0 cm^{-1} must be assigned to small CO_2 clusters, perhaps CO_2 dimers. A similar behaviour is observed for the intermediate structure at 1278.8 cm^{-1}. Compared to the large cluster peak, its amplitude varies from 33 % at p = 4 bar to 11 % at p = 15 bar. Correspondingly, this peak must be also due to smaller CO_2 clusters. On careful inspection of the spectra in Fig. 4 two other reproducible structures may be discerned: a small peak at 1279.9 cm^{-1} and a shoulder at the low-frequency side of the 1278.8 cm^{-1} peak.

In order to obtain more distinctive information on the assignment of the smaller cluster peaks, we have also investigated a mixture of 2 % CO_2 in He. This more dilute mixture favours the formation of smaller clusters under otherwise identical conditions. The measured spectra are shown in Fig. 5. Compared to the spectra of Fig. 4, higher stagnation pressures must be employed to observe the cluster features with the same intensity. In addition, the cluster peaks at 1278.8 and 1281.0 cm^{-1} are more pronounced now. This supports strongly the assumption that they must be assigned to small CO_2 clusters. At intermediate pressures the two peaks at 1278.8 and 1281.0 cm^{-1} are roughly equal in height, whereas at high pressures the 1278.8 cm^{-1} peak surpasses the other. This pressure dependence suggests that the 1278.8 cm^{-1} peak is composed of two contributions or has to be attributed to larger CO_2 clusters than the 1281.0 cm^{-1} peak.

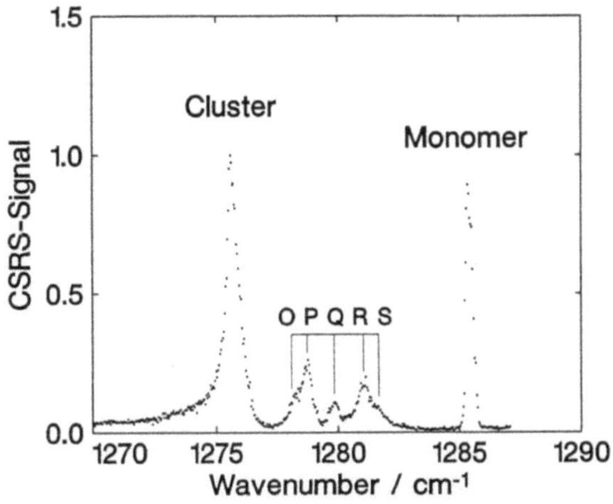

Fig. 6. Best spectrum obtained with 2 % CO_2 in He at p = 10 bar. The intermediate symmetric structure, which resembles the unresolved (O, P, Q, R, S) band contour of a prolate symmetric top for transitions between totally symmetric vibrational states, is assigned to small CO_2 clusters of different size.

The spectra shown in Fig. 5 clearly reveal the fine structures mentioned earlier: the small central peak at 1279.9 cm^{-1} (see the 10 bar spectrum) and the shoulder on the red side of the 1278.8 cm^{-1} peak. In addition, another shoulder on the blue side of the 1281.0 cm^{-1} peak becomes clearly discernable. It should be noted that all these finer structures appeared to be very reproducible.

Another series of spectra has been recorded with a mixture of 1 % CO_2 in He. Here even higher stagnation pressures must be supplied to observe the same cluster features. In agreement with their assignment to small clusters, the structures between 1277 and 1283 cm^{-1} were also more pronounced compared to the large cluster peak.

Careful inspection of the spectra of Fig. 5 measured at p = 8, 10, and 12 bar reveals that the structures between 1277 and 1283 cm^{-1} are remarkably symmetric. This symmetry is particularly suggestive in the spectrum of Fig. 6 which has also been obtained with a mixture of 2 % CO_2 in He expanded at p = 10 bar, but which is different from the spectrum shown in Fig. 5.

The symmetric structure of the three peaks between 1277 and 1283 cm^{-1} with shoulders on both the red and blue wings resembles very much the non-resolved band contour of a prolate symmetric top for Raman transitions between two totally symmetric vibrational states. Placzek and Teller [21] have calculated band shapes for such transitions as a function of the ratio of the rotational constants C/A. In their simulation they used the same rotational constants for the lower and upper vibrational states, and therefore the individual Q-branch lines fell together and could not be considered on the right scale. Our spectrum shows good qualitative agreement with the two simulations carried out for C/A = 0.25 and C/A = 0.1 (see Fig. 1 d and e of Ref. 21), but best agreement would probably be obtained

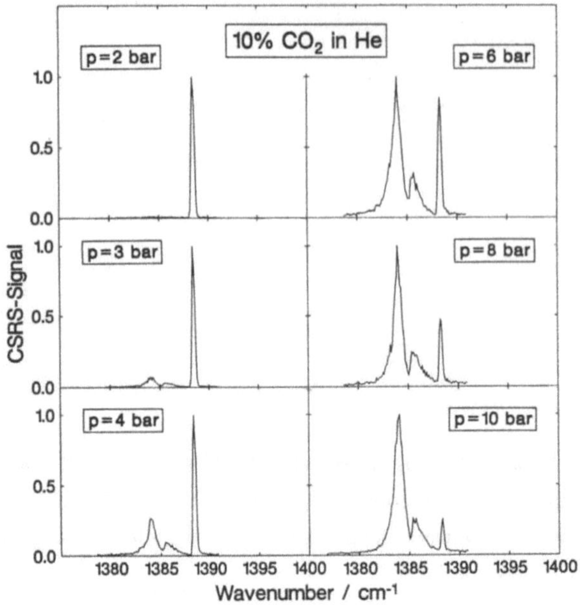

Fig. 7. The $2\nu_2$ component of the Fermi dyad as a function of stagnation pressure. The narrow peak at 1388.2 cm^{-1} is the monomer $2\nu_2$ Q-branch. The red-shifted structure arising at higher pressure is assigned to large CO_2 clusters.

with C/A between 0.1 and 0.25. Because of this striking resemblance, we were tempted to assign the symmetric structure observed between 1277 and 1283 cm^{-1} to the unresolved O-, P-, Q-, R-, and S-branches of the totally symmetric stretch vibration of the CO_2 dimer as indicated in Fig. 6.

However, very recent studies carried out in our and Nibler's laboratory [22] show that the various peaks designated in Fig. 6 with O, P, Q, R, and S have distinctively different pressure dependence. Therefore, it seems clear that they cannot be assigned to a single cluster size, i.e., the appealing interpretation given above is not correct. Instead it is more appropriate to assign the least red-shifted peak at $\tilde{\nu}$ = 1281.0 cm^{-1} to the CO_2 dimer and the further red-shifted features to CO_2 trimers and tetramers. It should be noted, however, that additional measurements are required to corroborate this interpretation.

Our results are in perfect agreement with the previously published measurements of Nibler et al. [5] but show an improved signal-to-noise ratio. Nibler and co-workers observed four cluster features at 1275.3, 1278,7, 1279.8, and 1281.3 cm^{-1} which must be compared to our positions at 1275.7, 1278.8, 1279.9, and 1281.0 cm^{-1}. In agreement with our assignment, they assigned the peak at 1275.3 cm^{-1} to large CO_2 clusters and the peak at 1281.3 cm^{-1} to CO_2 dimers. The intermediate peak at 1278.7 cm^{-1} was tentatively attributed to CO_2 trimers.

Finally we have investigated the $2\nu_2$ component of the Fermi dyad at 1388.2 cm^{-1}. We used a 10 % mixture of CO_2 in He and measured CSRS spectra as a function of the stagnation pressure directly at the nozzle exit. The result is shown in Fig. 7. The narrow peak at 1388.2 corresponds to the excitation of the $2\nu_2$ vibration in the CO_2 monomer (Q-branch). This is the dominant spectral feature at low stagnation pressure. When the pressure is increased, a new double-structure evolves at 1384.0 and 1385.8 cm^{-1}. This structure gains in intensity with increasing stagnation pressure and is the dominant feature at $p = 10$ bar. At this high pressure only a few monomers are observed. The most red-shifted peak at 1384.0 cm^{-1} coincides exactly with the position observed in crystalline CO_2 [20]. With the same arguments as used before, we assign it to the $2\nu_2$ excitation in large CO_2 clusters. The smaller, less disturbed peak at 1385.8 cm^{-1} has in all spectra approximately the same amplitude ratio to the larger cluster peak. Therefore, we believe that it must also be ascribed to larger clusters. However, the nature of this peak is not yet clear, perhaps it is due to some combination band. No features assignable to small CO_2 clusters could be observed. However, we have not yet tried more dilute mixtures, which favour the formation of smaller clusters as we saw in our studies of the ν_1 component.

4. Conclusion

In the present experiment, coherent Stokes Raman scattering has been employed to study the rotational and vibrational spectroscopy of carbon dioxide clusters. In the rotational region two spectral features are observed which arise at sufficiently strong stagnation conditions. They are assigned to the E_g and F_{g-} librational modes probed in large CO_2 clusters. Their observation implies that the CO_2 clusters formed in our experiment have adopted the same structure as encountered in crystalline CO_2. By comparing the positions of the librational peaks observed in our experiment with those determined under equilibrium conditions, cluster temperatures can be derived. On the other hand, the monomer rotational lines permit the determination of a monomer rotational temperature. Comparing these temperatures, we obtain the somewhat surprising result that monomer rotations and cluster librations are in near-equilibrium and cool at approximately the same rate.

In order to gain information about the vibrational spectroscopy of CO_2 clusters, we have studied the $(\nu_1, 2\nu_2)$ Fermi dyad around 1285.4 cm^{-1} (ν_1) and 1388.2 cm^{-1} ($2\nu_2$). In the ν_1 component we observed several spectral features which could be assigned to CO_2 clusters. The strongest peak, red-shifted by 9.7 cm^{-1} from the monomer line, coincides exactly with the position measured for ν_1 in crystalline CO_2. This finding is in agreement with the earlier observations in the librational region and corroborates the result that large CO_2 clusters adopt the structure of crystalline CO_2. If very dilute mixtures are employed, a weaker symmetric structure consisting of five peaks is observed between the monomer and cluster peaks. Although it resembles very much the vibrational band contour of a prolate symmetric top we have indication that these weak features must be assigned to small CO_2 clusters of different sizes (probably dimers, trimers, and tetramers). Further studies with improved resolution aimed to provide a consistent assignment are under way. In the $2\nu_2$ component only features attributable to large CO_2 clusters could be observed. It should be interesting to investigate this spectral region with more dilute mixtures and to search for dimer features.

In summary, we have shown that coherent Raman spectroscopy constitutes a powerful technique to monitor the formation of CO_2 clusters in a supersonic jet and to characterize the conditions under which strong clustering occurs. Furthermore, detailed information about the spectroscopy of small and large CO_2 clusters is obtained, yielding valuable insight into the structure of these species.

Acknowledgement

The authors wish to thank Professor H. Pauly for his continuous support. The computer calculations have been carried out at the Gesellschaft für wissenschaftliche Datenverarbeitung, Göttingen.

References

[1] K.W. Jucks, Z.S. Huang, D. Dayton, R.E. Miller, and W.J. Lafferty, J. Chem. Phys. 86, 4341 (1987)

[2] K.W. Jucks, Z.S. Huang, R.E. Miller, G.T. Fraser, A.S. Pine, and W.J. Lafferty, J. Chem. Phys. 88, 2185 (1988)

[3] M.A. Walsh, T.H. England, T.R. Dyke, and B.J. Howard, Chem. Phys. Lett. 142, 265 (1987)

[4] J.A. Barnes and T.E. Gough, J. Chem. Phys. 86, 6012 (1987)

[5] G.A. Pubanz, M. Maroncelli, and J.W. Nibler, Chem. Phys. Lett. 120, 313 (1985)

[6] J.W. Nibler and G.A. Pubanz, in: Advances in Non-linear Spectroscopy, eds. R.J.H. Clark and R.E. Hester (Wiley, New York, 1988) p. 1

[7] H.-D. Barth and F. Huisken, J. Chem. Phys. 87, 2549 (1987)

[8] H.-D. Barth, C. Jackschath, T. Pertsch, and F. Huisken, Appl. Phys. B 45, 205 (1988)

[9] J.A. Shirley, R.J. Hall, and A.C. Eckbreth, Opt. Lett. 5, 380 (1980)

[10] H.-D. Barth, F. Huisken, and A.A. Ilyukhin, Appl. Phys. B 52, 84 (1991)

[11] H.-D. Barth and F. Huisken, Chem. Phys. Lett. 169, 198 (1990)

[12] F. Huisken and T. Pertsch, Appl. Phys. B 41, 173 (1986)

[13] A. Weber, in: Raman Spectroscopy of Gases and Liquids, ed. A. Weber (Springer, Berlin, 1979), p. 71

[14] A. Anderson and T.S. Sun, Chem. Phys. Lett. 8, 537 (1971)

[15] J.W. Nibler, in: Applied Laser Spectroscopy, eds. M. Inguscio and W. Demtröder, NATO ASI Series (Plenum, New York, 1989)

[16] H. Olijnyk, H. Däufer, H.-J. Jodl, and H.D. Hochheimer, J. Chem. Phys. 88, 4204 (1988)

[17] R.D. Beck, M.F. Hineman, and J.W. Nibler, J. Chem. Phys. 92, 7068 (1990)

[18] I. Suzuki, J. Molec. Spectrosc. 25, 479 (1968)

[19] G. Herzberg, Molecular Spectra and Molecular Structure, Vol. 2 (Van Nostrand, New York, 1945)

[20] R. Ouillon, P. Ranson, and S. Califano, J. Chem. Phys. 83, 2162 (1985)

[21] G. Placzek and E. Teller, Z. Physik 81, 209 (1933)

[22] J.W. Nibler, private communication (1990)

Degenerate Four-Wave Mixing in Combustion Diagnostics

T. Dreier[1], *D.J. Rakestraw*[2], *and R.L. Farrow*[2]

[1]Physikalisch Chemisches Institut, Universität Heidelberg, Im Neuenheimer Feld 253, W-6900 Heidelberg, Fed. Rep. of Germany

[2] Combustion Research Facility, Sandia National Laboratories, Livermore, CA 94550, USA

1. INTRODUCTION

Since the availability of high power tunable laser sources, optical spectroscopy has gained widespread application as a diagnostic tool to detect stable and reactive species in combustion processes under a variety of environmental conditions [1]. Sensitive laser spectroscopic techniques like laser induced fluorescence (LIF) or resonance enhanced multiphoton ionization (REMPI) in the past predominantly have been applied to low pressure systems: burners that operate under several mbar of total pressure show - due to lower reaction rates - flame zones of increased spatial extent that make concentration profiling of intermediate species more accurate.

At atmospheric pressure spectroscopic methods that rely on excited state emission of laser prepared molecules suffer from electronic quenching of these states by collisions with bath gas molecules. The accompanying lowering in detection sensitivity at higher pressures is partly offset by higher concentrations of the radicals under study so that other techniques like laser absorption can be used. Even spontaneous Raman scattering is being employed for the quantitative detection of major species concentration or temperature measurements in laminar and turbulent flames. In hostile environments, incoherent methods like laser induced fluorescence or spontaneous Raman scattering become more and more difficult to use effectively: In order to collect enough signal photons and still do measurements at high spatial resolution one requires optics of large apertures used in close proximity to the probing laser beams. This makes these methods useful only for easily accessible combustion processes. One interesting possibility to circumvent these drawbacks has been the successful introduction of fiber optic probes to couple in the laser beams and simultaneously collect the signal photons through the same optics [2]. Another problem of incoherent detection methods is that they are susceptible to broadband scattered light within the wavelength region and solid angle of the light detectors employed. This means that sooting flames that contain large amounts of strong blackbody radiators generated by particles and soot precursors heated up in the gas flow are not very suitable candidates for these methods, even though electronic gating of viewing time can discriminate strongly against background radiation. This is especially difficult if the laser used for the generation of signal photons is itself scattered by solid particles in the beam path. Generating the signal in a coherent beam that can be detected remotely is one means by which these problems can be circumvented. This way, discrimination against isotropically scattered background light, if the wavelength of the signal is close to or in coincidence with that of the exciting laser light, can be done by spatial isolation of the signal beam before additional spectral separation of the signal photons is performed in a monochromator or other filtering stage. Therefore, in recent years nonlinear optical methods have proven to be very attractive in combustion diagnostics. In this case the beam characteristics of the signal wave result from multiple wave mixing interactions of the incoming beams in the nonlinear medium and require phase matching so that the signal wave can build up within the interaction length of the overlapping beams.

Nonlinear optical techniques generally suffer from not being very sensitive as compared to methods that use excitations within absorption bands of atoms and molecules. Resonance enhancements of higher order wave mixing processes have been used successfully to increase the detection sensitivity by several orders of magnitude [3]. By using three indi-

Eds.: G. Marowsky and V.V. Smirnov

vidually tunable frequency doubled dye lasers, the OH radical, an important intermediate in combustion processes, has been detected in atmospheric [4] and higher pressure [5] flames. Using conventional CARS [6] the OH radical, probed as a product of HNO_3 laser photolysis at 193 nm, has been detected at room temperature. The concentration of OH in these experiments has been estimated as approximately $1.7 \cdot 10^{16}$ cm^{-3}. This limit has been lowered by the resonance CARS experiment to $1 \cdot 10^{13}$ cm^{-3} or 2-4 ppm at 2200 K. Resonance CARS has been demonstrated for iodine vapor [7] and C_2 [3] as well. The experimental complexity together with a sophisticated spectral interpretation still keeps this technique from being used routinely.

This article mainly deals with another closely related nonlinear optical process involving the third order susceptibility $\chi^{(3)}(\omega)$: Degenerate Four Wave Mixing (DFWM). This parametric process is widely used in measuring ultrafast relaxation and diffusion mechanisms in liquids [8] and solids [9]. Additionally, its phase conjugation property makes it attractive for experiments in adaptive optics [10]. It also has, because of the Doppler-free nature of the interaction process, potential applications in high resolution spectroscopy [11]. The basic experimental realization of this technique is by tuning the output radiation of a single laser within the absorption transitions of atoms and molecules and receiving a coherent signal beam at the same frequency as the laser beams entering the sample. This technique has been demonstrated to constitute a sensitive detection method for atoms in flames [12]. Small concentrations of excited species formed by collision assisted transitions have been detected by Ewart and O'Leary [13]. The same group first demonstrated the use of DFWM in monitoring selected transitions of the OH radical in an atmospheric pressure flame [14].

This article presents a more extended investigation of using DFWM spectroscopy for combustion diagnostics applications. Results on monitoring OH, NH and NO, important intermediate species in combustion, are reported. Useful information about flame parameters like temperature and species distributions are derived. Additionally preliminary results are given about the basic physical properties important in DFWM saturation and collisional deactivation.

2. THEORETICAL DESCRIPTION

2.1 Basic principles of DFWM

The purpose of this section is to give a short overview and an intuitive physical picture of the physics that leads to the generation of the signal wave in a DFWM process. The general derivation of the relevant equations is much too complex and beyond the scope of this article and can be read in standard textbooks on electrodynamics or special issues on phase conjugation. An excellent review about these topics and references to more detailed work is given by Fisher [15].

In degenerate four wave mixing, the interaction of three input beams of identical frequency, ω, with a nonlinear medium, produces a fourth coherent signal beam with the same frequency ω. A schematic diagram of the wave mixing geometry used in our and most other experiments on phase conjugation is shown in Fig. 1. In this geometry two pump waves, with electric field vectors $\mathbf{E}_f$ ("forward") and $\mathbf{E}_b$ ("backward"), are counterpropagating to each other and cross a third wave, $\mathbf{E}_p$ ("probe") at a small angle θ.

The nonlinear polarization responsible for the radiated signal wave and given by

$$\mathbf{P}^{(3)}(\omega) = \varepsilon_0 \chi^{(3)} \left[\mathbf{E}_f(\omega) \mathbf{E}_p^*(\omega) \mathbf{E}_b(\omega) + \mathbf{E}_b(\omega) \mathbf{E}_p^*(\omega) \mathbf{E}_f(\omega) \right] \qquad (1)$$

couples these three waves to a fourth wave $\mathbf{E}_c$ ("signal") which counterpropagates to $\mathbf{E}_p$. This geometry meets the phase matching requirements for efficient generation of a signal wave during the interaction process for all angles θ and has the unique property that $\mathbf{E}_c$ is proportional to the complex conjugate of $\mathbf{E}_p$. In the above expression (1) $\chi^{(3)}$ is the third order nonlinear susceptibility of the medium. A strong resonance enhancement of the signal wave therefore occurs whenever the frequency ω of the laser radiation matches an allowed

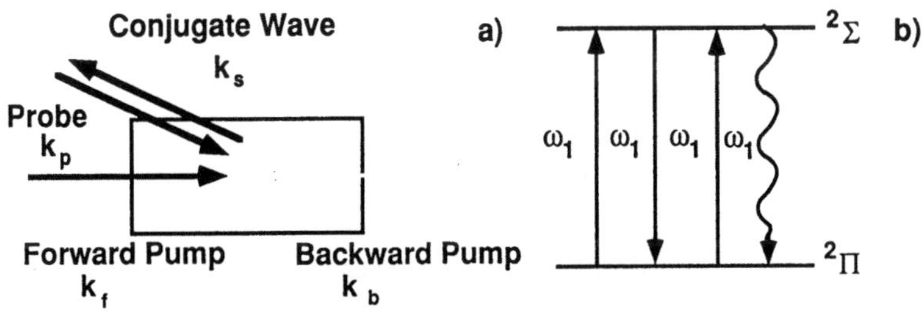

Figure 1. Schematics of the degenerate four wave mixing process: a) Beam geometry; $\mathbf{k}_i$ denote the wave vectors of the interacting beams. b) Energy level diagram with resonance at ω.

molecular transition, because due to the degeneracy of the wave mixing process, as shown in fig. 1b, a triple resonance is always maintained.

The response of the system is determined by the nature of the susceptibility tensor components. In addition it is helpful to give a qualitative description of the four wave mixing interaction in understanding the generation of $\mathbf{E}_c$: An interference of $\mathbf{E}_p$ with $\mathbf{E}_f$ (first term in eq. (1)) or $\mathbf{E}_b$ (second term in eq. (1)) gives rise to a sinusoidal modulation of the medium response in the beam overlap region. In an absorbing medium this response is governed by the variations of the nonlinear index of refraction in the high and low intensity regions of the volume diffraction gratings from which the backward or forward pump beam is reflected off in a Bragg type scattering, producing the signal wave $\mathbf{E}_c$. The orientation of the Bragg grating written in the medium by $\mathbf{E}_p$ and $\mathbf{E}_f$ or $\mathbf{E}_b$, respectively, is determined by the wavevector difference $\mathbf{q} = \mathbf{k}_f - \mathbf{k}_p$ ($\mathbf{k}_b - \mathbf{k}_p$) so that the signal wave propagates exactly in the reverse direction of $\mathbf{E}_p$ by Bragg reflection.

If the polarization vectors of all three input waves are parallel, the signal wave - due to conservation of photon angular momentum - is polarized parallel to the incoming waves. In this case not only does the first term in equation (1) make a contribution to the four wave mixing signal but also the second term that describes the Bragg reflection of the forward pump off a grating formed by the interference of the backward pump and probe beams. However, the fringe spacing Λ ($\Lambda = \lambda/(2\sin(\theta/2))$) of this grating for most practical angles is close to λ/2 (half the wavelength of the laser radiation) so that washout of this grating by molecular motion, especially in high temperature gases, with the accompanying reduction in scattering efficiency, is significant [8]. The contribution from the second term can be eliminated if $\mathbf{E}_f$ and $\mathbf{E}_p$ have parallel polarizations that are oriented at 90° with respect to $\mathbf{E}_b$. Then no interference pattern will be generated between the crossed polarized waves $\mathbf{E}_b$ and $\mathbf{E}_p$, and $\mathbf{E}_c$ is polarized parallel to $\mathbf{E}_b$ [16]. This geometry furthermore has the advantage of efficient discrimination of stray light generated by $\mathbf{E}_f$, $\mathbf{E}_p$ by using a polarizer in the signal beam path.

2.2 Simple two level atom

An early detailed theoretical description of the nonlinear polarization generated by the four wave mixing process was given by Abrams and Lind [17]. Investigations of more complicated systems followed and included Doppler broadening [18,19], saturation phenomena [19], collision effects [20], multiple level schemes [19] and effects of broadband laser sources [21]. More or less all these complications arise in the experiments described below. A full treatment of DFWM under these conditions, therefore, is still lacking.

Assuming negligible absorption of the pump beams (with equal intensity I), but allowing for absorption of the probe (I_p) and phase conjugate signal (I_c) beams the line center signal intensity is given by [22]

$$I_c = RI_p = \left| \frac{\beta \sin \gamma L}{\gamma \cos \gamma L + \alpha \sin \gamma L} \right|^2 I_p \qquad (2)$$

where R is the phase-conjugate reflectivity, α is the attenuation coefficient:

$$\alpha = \alpha_0 \frac{1+2I/I_s}{(1+4I/I_s)^{3/2}} , \tag{3}$$

β is the nonlinear coupling coefficient:

$$\beta = i\alpha_0 \frac{2I/I_s}{(1+4I/I_s)^{3/2}} \tag{4}$$

and $\gamma^2 = \beta^2 - \alpha^2$. α_0 is the line center field absorption coefficient:

$$\alpha_0 = (\omega/2c)\Delta N_0 \mu^2 T_2/\hbar\varepsilon_0 , \tag{5}$$

where ΔN_0 is the population difference in the absence of applied fields, μ^2 is the square of the transition dipole moment (proportional to line strength), and ε_0 is the permittivity of vacuum. I_s is the line center saturation intensity:

$$I_s = \varepsilon_0 c\hbar^2 / 2T_1T_2\mu^2 \tag{6}$$

where T_1 and T_2 are the population and coherence decay times for the transition. Equation (2) can be simplified in the limit of low absorption, $\alpha_0 L$. In this case, $\gamma L << 1$ and

$$I_c = \frac{|\beta\gamma L|^2}{\gamma^2} I_p = |\beta L|^2 I_p = \alpha_0^2 L^2 \frac{4(I/I_s)^2}{(1+4I/I_s)^3} I_p \tag{7}$$

at line center. We see that the signal is dependent on the square of the line center absorption coefficient (and thus on the square of the unperturbed population difference). The limits of Eq. (7) for weak and strong pump powers, respectively, are:

$$I_c = \alpha_0^2 L^2 4(I/I_s)^2 I_p \qquad (I<<I_s), \tag{8}$$

$$I_c = \alpha_0^2 L^2 (I_s/16I) I_p \qquad (I>>I_s) . \tag{9}$$

3. EXPERIMENTAL

The experimental setups used for the measurements reported in this review differ mostly in the laser sources employed for the specific needs of the measured quantity. The important features of a DFWM experiment therefore can be seen in the general setup depicted in Fig. 2. Tunable laser radiation is generated by the frequency doubled output of a dye laser (LUMONICS Hyperdye 3000) pumped by the second harmonic output of a Nd:YAG laser (Spectra Physics DCR 2A). The frequency doubling units are either computer controlled (LUMONICS Hypertrack 1000) or self-tracking (INRAD) devices to keep the uv-output radiation at a maximum during tuning of the dye laser. The uv-output beam subsequently is split by a beamsplitter into collinear, counterpropagating forward and backward pump beams. A second beamsplitter and turning prism direct a low intensity probe beam under a small crossing angle (typically 4°) into the overlap region of the two pump beams to define a measurement volume of about 35 mm^3 if 1.5 mm diameter beams are used. To increase the spatial resolution of the setup one of the beams can be collimated to a smaller diameter. Polarization rotation devices (Fresnel rhombs, half wave plates) and polarizers in the beam paths enable the intensity of the individual beam legs to be varied and their polarization directions to be rotated with respect to each other arbitrarily.

The phase conjugated DFWM signal generated in the probe volume is split off with a 30% reflectivity beamsplitter and directed to a photomultiplier. Stray and background light is effectively prevented from reaching the detector by directing the signal beam several meters through apertures, a polarizer and focussing the beam onto a 50 μm pinhole. In many

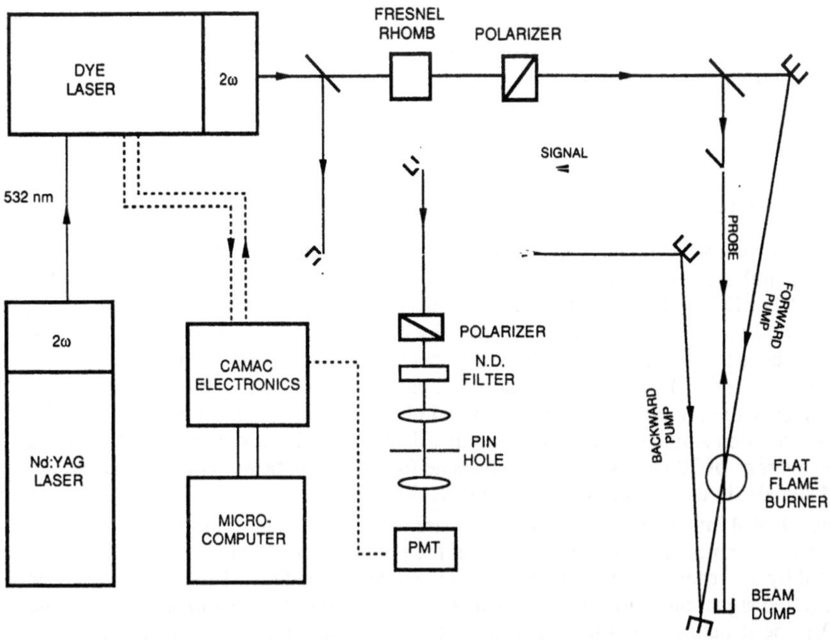

Fig. 2 Experimetal setup of beam arrangement and detection system as used in DFWM spectroscopy of flame radicals.

TABLE 1. Flame conditions and temperature measurement results using CARS and DFWM diagnostics.

Flame	Volume flows [l/min]	T_{CARS} [K]	T_{DFWM} [K]
C_3H_8/air	(C_3H_8): 5	1703 ± 20	1730 ± 60
	(air) : 48		
CH_4/air	(CH_4) : 9	2100 *	2170 ± 110
	(air) : 74		
$NH_3/O_2/N_2$	(NH_3) : 2.1	2100 ± 150	2042 ± 106
	(O_2) : 1.5		
	(N_2) : 1.0		

* : W. Stricker: private commun.

cases it is useful to observe the laser induced fluorescence signals with a second photomultiplier at right angles to the overlap region of the intersecting beams. All electrical signals are preprocessed using standard CAMAC electronics and subsequently stored in a laboratory computer (DEC Microvax II). The computer also controls the scanning of the dye laser by tuning slowly across the relevant lines while scanning fast in between lines. A photodiode continuously monitors the laser intensity of the probe beam. The energies of the individual beams are measured with a fast thermopile detector (Molectron J3-05).

For most experiments stoichiometric premixed air fed flames at atmospheric pressure were burned using propane, methane or ammonia as fuel gases (see table 1). For combustion under reduced pressures the McKenna-type burner was enclosed in a rectangular stainless steel vacuum chamber equipped with a variable exhaust valve to regulate the pressure under flow conditions of 2.3 slpm NH_3 and 6.1 slpm of O_2.

4. RESULTS

4.1 OH detection and thermometry

The OH-radical is known to be an important intermediate in combustion processes because of its high reactivity with other species. The OH takes part in the oxidation step that converts CO to CO_2 in fuel-rich combustion zones. The easily accessible electronic absorption transitions in the near ultraviolet additionally make this radical an attractive species for optical diagnostics. An assigned [23] DFWM spectrum of the OH in the spectral region of the A $(^2\Sigma^+)$ - X $(^2\Pi)$ (0,0) electronic absorption band taken 9 mm above the burner surface of a 1 atm pressure propane/air flame is shown in Fig. 3a (bottom solid trace) [24]. A LIF spectrum recorded simultaneously is shown as a dashed line. Each spectral point in the figure is an average of 20 laser shots to compensate for the strong power dependence of the DFWM signals (in the absence of saturation, $E_c \sim E_p^* E_f E_b$).

The DFWM intensities depend on the square of the third order nonlinear susceptibility and therefore are proportional to the square of the population difference of the two levels connected by the resonant electronic transitions. We also found (see Table 2) - by comparing integrated signal intensities from two different spectroscopic transitions that probe the same OH fine structure state - that their ratio depends on the square of the respective rotational line strength. With this assumption rotational populations within the R_1, R_2-branches of the OH radical were determined via the relation

$$I_c \propto (B_{ij} N_{OH}(v'',J''))^2 \tag{10}$$

where $N_{OH}(v'',J'')$ is the OH rotational state number density and B_{ij} the one photon line strength for the probed molecular transition. Assuming further a Boltzmann population distribution within the rotational levels of OH (a valid assumption at atmospheric pressure flames) temperatures were derived from the slope of a straight line fitted to a halflogarithmic plot of population versus energy ΔE_N of the rotational levels using the relation

$$N_N(v''=0) \propto (2J+1) exp(-\Delta E_N hc / kT) \tag{11}$$

with $J = N \pm 1/2$. This result is shown in Fig. 3b.

The temperatures measured using DFWM have been compared with coherent anti-Stokes Raman scattering (CARS) thermometry. In numerous investigations CARS has proven to give accurate temperature readings when this information is derived from nitro-

TABLE 2. Comparison of measured ratios of main (I_{R1}) and satellite (I_{R21}) DFWM line intensities with calculated ratios of the corresponding line strengths (LS). Pump beam energy: 844 µJ, probe beam energy: 32 µJ.

N	$LS(R_1)/LS(R_{12})$	$(I_{R1(N)}/I_{R21(N)})^{1/2}$
1	0.667	0.68
2	1.19	1.3
3	1.84	1.8
4	2.63	2.7
5	3.55	3.5

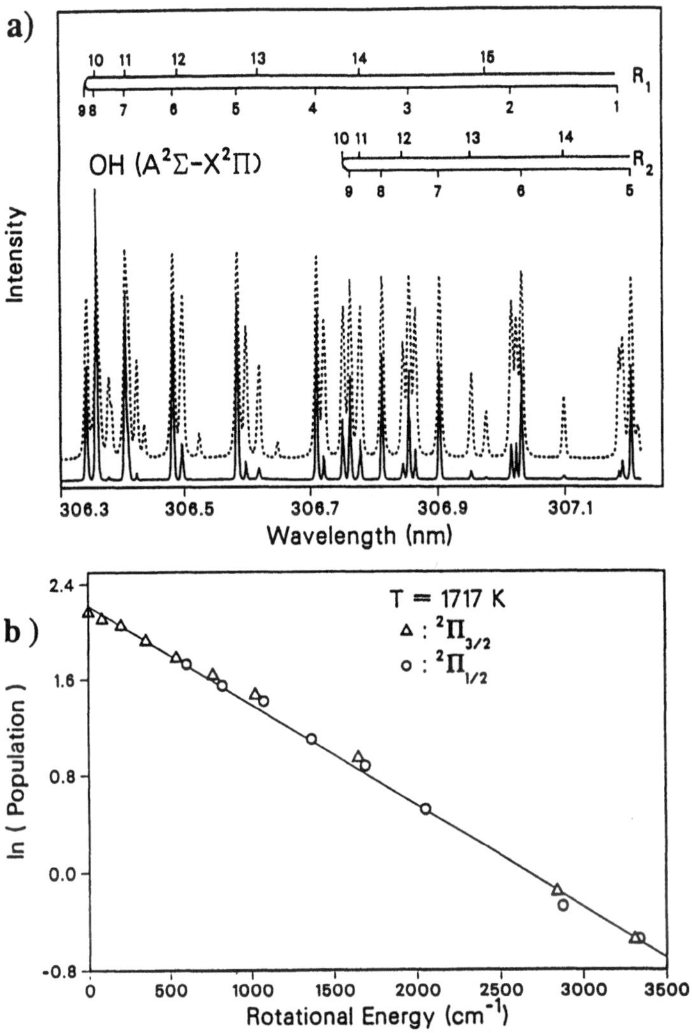

Figure 3. a) Simultaneous recording of the DFWM (solid line) and LIF (dashed line) spectra for the OH $A(^2\Sigma^+)$ - $X(^2\Pi)$ transition. Pump beam energy: 780 µJ; probe beam energy: 74 µJ. For clarity the LIF spectrum is offset with respect to the DFWM spectrum baseline by an arbitrary amount. The vertical scale is in arbitrary intensity units. b) Boltzmann plot of OH population distribution derived from DFWM spectrum as shown in a). In the analysis R_1 (triangles) and R_2 (circles) transitions are used. Solid line: least squares linear regression through the data points.

gen Q-branch spectra [25]. In most flames nitrogen is present in sufficient concentrations that high quality CARS spectra can be acquired which are then compared to computer generated spectra in a least squares fitting routine [26].

Details of implementing CARS spectrometers experimentally and the procedures taken for thermometry can be found in the literature [25, 27]. Extensive literature also exists about the physics underlying the CARS signal generation process as well as phenomena that influence its spectral shape in different media [25, 28]. All these effects have to be considered in the modeling codes for CARS spectral shapes.

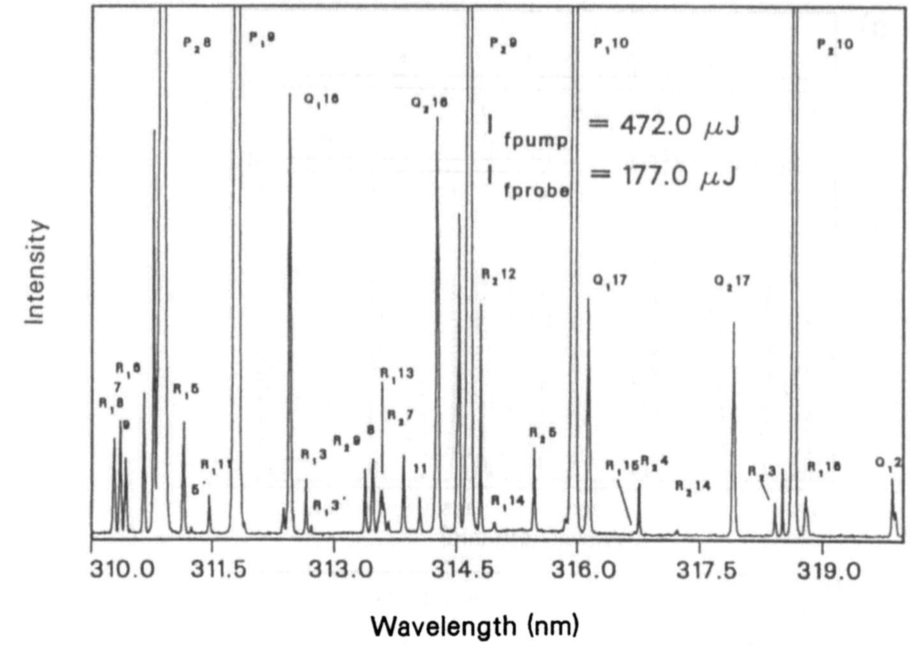

Figure 4. Degenerate four wave mixing spectrum of OH(v"=1) within the A-X (1-1) band

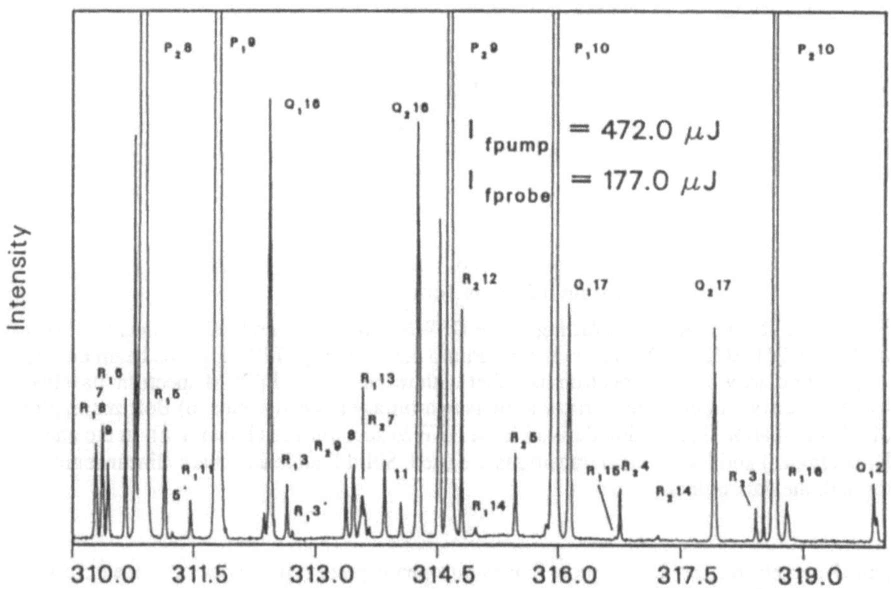

Analogous measurements have been made on the NH radical that is formed in an ammonia/oxygen/nitrogen flame [30]. At atmospheric pressure NH is formed within the flame front very close to the burner surface where DFWM spectra and comparative nitrogen CARS Q-branch spectra were taken. A DFWM spectrum in the Q-branch manifold of the $A(^3\Pi)$ - $X(^3\Sigma)$ transition at 334 nm is shown in Fig. 5a. It should be noted that the DFWM signals in this flame were exceptionally strong and spectra with good signal-to-noise ratio

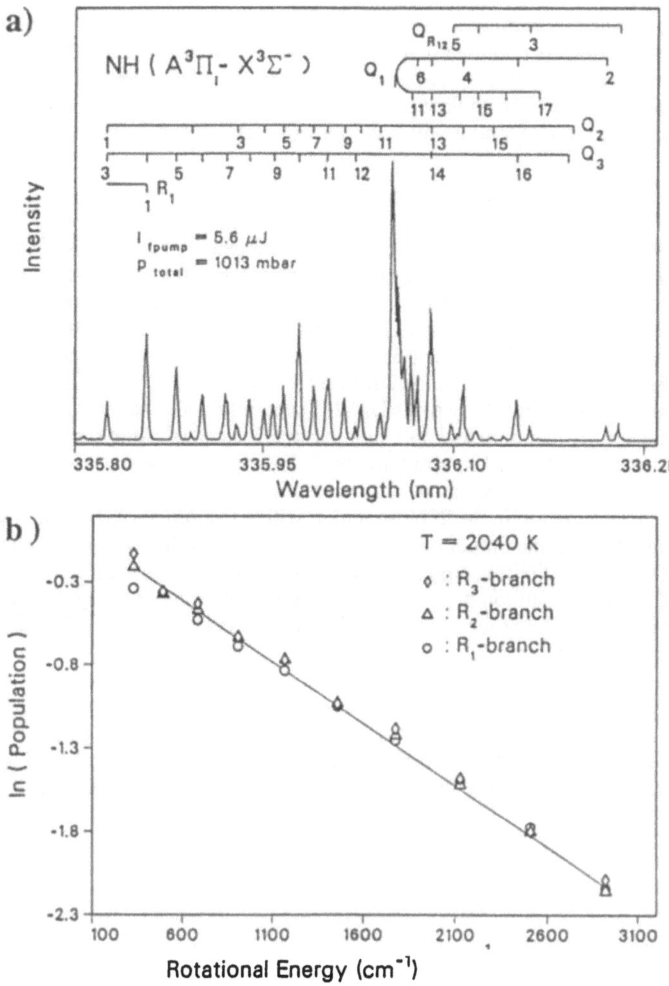

Figure 5. a) Experimental DFWM spectrum of the Q-branch region of the NH ($A^3\Pi$ - $X^3\Sigma$) transition taken in a 1 atm. pressure $NH_3/O_2/N_2$ flat flame 1.6 mm above the burner surface. The transitions are labeled by the ground state total angular momentum quantum number N. b) Boltzmann plot of rotational population distribution in the R-branches of the NH A-X (0,0) band. The straight line is a least squares regression through all data points, giving a temperature of 2042 K.

could be obtained using a total laser intensity of 10 kW/(cm^2 wavenumber). It was not possible, however, to collect LIF spectra at the same point because of too much light scattered off the burner surface.

Due to the spectral congestion in the Q-branch region temperatures have been obtained from Boltzmann plots of rotational populations deduced from the nearby well-separated lines of the three R-branches. Again, as Fig. 5b shows, all points fall on a reasonably straight line. Mainly due to temperature gradients within the thin flame front, the error bars of the DFWM and CARS measurement results (Table 1) are larger than those for the OH molecule taken in the more homogeneous post-flame gases. Additionally, at the low nitrogen content of this flame uncertainty in CARS thermometry also arises from the interference of the resonant signal with the spectrally flat nonresonant signal simultaneously generated by dilution gases [31]. Again emphasis should be put on the fact that for the determination of population fractions from integrated DFWM line intensities, no quenching corrections for the interrogated rotational levels have been made. This is in contrast to LIF temperature measurements where the neglect of rotational level as well as composition dependent quenching can lead to temperature misreadings of several 100 K [32].

4.3 NO detection and thermometry

The production and emission of NO in everyday combustion processes is of serious environmental concern. Consequently much attention has been given to the development of optical diagnostics in laboratory flames [33] and, in particular, considerable work on the spectroscopy of the $A^2\Sigma^+$ - $X^2\Pi$ band near 226 nm has been reported [34]. DFWM may offer an attractive alternative to LIF for measurements in hostile and turbulent environments of restricted access. Contrary to OH and NH discussed above, the NO radical as a stable molecule facilitates a systematic study of its DFWM spectrum under varying experimental conditions of pressure, temperature and collision partner. Instead of using flames, NO was contained in a stainless steel cell, 130 mm long, with three quartz windows to allow for entrance and exit of laser beams as well as the observation of fluorescence radiation.

Tunable uv-laser radiation in the 226 nm region with a resolution of 0.05 cm^{-1} was produced by frequency tripling the output of a frequency doubled Nd:YAG laser pumped dye laser (Lambda Physik FL 2002E) running with LDS-698 dye (Exciton). Up to 4 mJ per pulse is obtained at 226 nm, although no more than 0.6 mJ pump pulse energy is employed in the actual experiments. The probe beam intensity is recorded behind the NO cell to ensure that absorption by NO is below 2%.

A promising spectral region for rotational thermometry is the O_{12}-branch of the $\Pi_{3/2}$-band, which consists of transitions free of overlap with other branches. An experimental spectrum of this region, including a section of the P_2+P_{12} branch, is shown in Fig. 6. Line assignments are based on the rotational analysis of ref. [34]. Only $67 \cdot 10^{-3}$ mbar of NO pressure were necessary to produce these data with a signal to noise ratio greater than 300.With

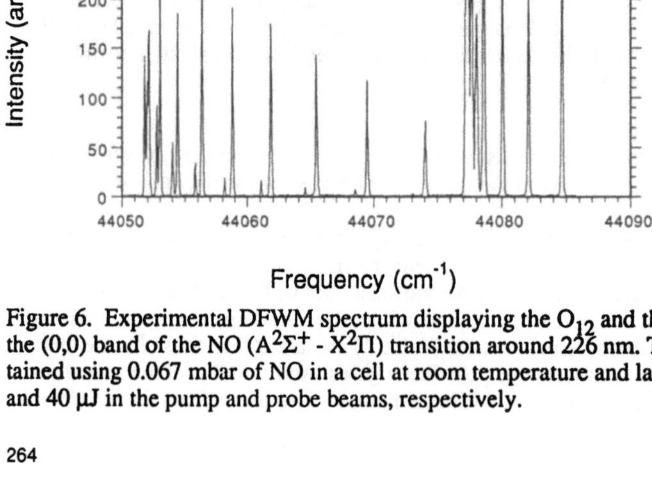

Figure 6. Experimental DFWM spectrum displaying the O_{12} and the P_2+P_{12} branches of the (0,0) band of the NO ($A^2\Sigma^+$ - $X^2\Pi$) transition around 226 nm. The spectrum was obtained using 0.067 mbar of NO in a cell at room temperature and laser energies of 120 µJ and 40 µJ in the pump and probe beams, respectively.

the laser energies employed (650 μJ pump, 200 μJ probe pulse) evidence of saturation in the DFWM signals was observed (see section 4.4.). This saturation will not necessarily be deleterious for thermometry if the relative rotational line intensities are unperturbed as was shown already in the analysis of OH and NH spectra. Therefore a Boltzmann analysis was attempted by fitting O_{12}-lines to Gaussian functions for integrated line intensity measurements. Using known line strengths of the O_{12}-branch [35] a straight line fit to a Boltzmann plot of several data sets gave a temperature of (296 ± 6) K in good agreement with the measured cell temperature of (295 ± 1) K.

Interestingly temperatures deduced from branches with overlapping lines (for example the Q_1+Q_{21} region) were lower by about 60 K relative to the ambient value of 296 K. Although the cause of this discrepancy still is not known with certainty, one possible explanation might be the way the known line strengths are added in the population fraction expression. As a result of saturation and/or the use of crossed pump polarizations, we observe lower than expected intensity ratios between branches having greatly different line strengths. Thus individual line strengths of unresolved lines may not be combined with their correct weights. Also since DFWM is a $\chi^{(3)}$-process similar to CARS, spectral interference of overlapped lines causes a more complicated intensity dependence than a simple isolated line model (as anticipated in the analysis) would suggest.

4.4 Saturation effects

The power densities employed in most of the experiments described here all were above the saturation limit as given by equation (6). The expression (2) for the intensities of the phase conjugated signal at line center becomes expression (8) and (9) for weak and strong pump powers, respectively. If one laser is used for all three input beams, eq. (8) states that the unsaturated signal will vary as the cube of the laser intensity. On the other hand in the case of strong saturation (eq. (9)) the signal will be independent of the laser intensity. We note that the signal will depend differently on transition line strengths in the two saturation regimes since I_s is inversely proportional to the line strength. As can be seen from the intensity ratios of main and satellite lines in the spectra of Fig. 3 the degree of saturation is different in the LIF and DFWM techniques: Whereas the LIF intensity ratio is far off the calculated values the square root dependence of the DFWM intensity gives ratios that are in good agreement with theoretical line strength results (see Table 2). This shows that the degree of saturation at the one photon level does not seem to perturb this squared line strength dependence of the DFWM intensities.

In the case of the NO molecule preliminary results have been obtained in the low pressure limit, where collisional effects of the pumped levels may be neglected. With laser intensities of approximately 1 MW/cm^2, as employed in the generation of the spectrum represented in Fig. 6, evidence of saturation in the DFWM signals was observed, including moderate line broadening (0.122 cm^{-1} FWHM compared to 0.051 cm^{-1} observed with lower powers) and a reduced dependence of the signal intensity on the laser intensity. Estimated values for the collision free saturation intensities for selected transitions for the three radicals investigated are given in Table 3.

A comparison of DFWM and LIF excitation spectra of the O_{12} bandhead recorded at low pump beam energies is shown in Fig. 7. The DFWM line intensity ratios are seen to vary approximately as the square of the corresponding LIF ratios, which are proportional to absorption. Contrary to the LIF lines the DFWM lines are sub-Doppler, averaging to 0.06 cm^{-1} which is close to the measured laser bandwidth since collisional or natural broadening mechanisms can be neglected (p_{tot}=0.067 mbar).

Both the DFWM and LIF linewidths were observed to narrow with decreasing intensity of the uv-laser pulse. The laser induced absorption rate [36], W_{12}, for the OH(R_1(7)) transition at pump energies of 500 μJ (7 MW/(cm2 wavenumber)) is $5\cdot10^{10}$ s^{-1}, which is much larger than the decay rate of the excited level due to spontaneous emission and rotational relaxation of $1.5\cdot10^{9}$ s^{-1}. Therefore the stimulated pumping rate rather than collision-

TABLE 3. Calculated or estimated parameters of the DFWM experiments for the species investigated under different experimental conditions.

Parameter	NO (cell exp.)	OH (H_2/O_2-) flame	OH (C_2H_8/air-) flame	NH ($NH_3/O_2/N_2$-flame)
T [K]	298	1850	1730	2150
p [mbar]	0.067	1013	1013	66.7
Doppler width [cm^{-1}]	0.1	0.25	0.24	0.3
Collisional width [cm^{-1}]	0.0001	0.11	0.08	
Laser linewidth [cm^{-1}]	0.05	0.002	0.15	0.15
DFWM width [cm^{-1}]	0.06	0.08	0.16	0.16
LIF width [cm^{-1}]	0.14		0.30	
Pump beam [kW/cm^2]	184	60		10
Probe beam [kW/cm^2]	57	50		

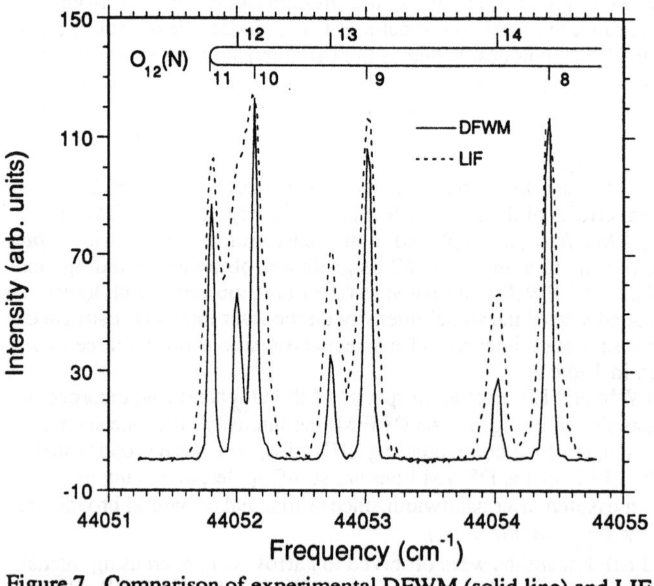

Figure 7. Comparison of experimental DFWM (solid line) and LIF (dashed line) excitation spectra of the O_{12} branch of NO. The spectra were obtained using an average of 10 laser pulses per frequency step. The DFWM line intensity ratios are seen to vary approximately as the square of the corresponding LIF ratios, which are proportional to absorption.

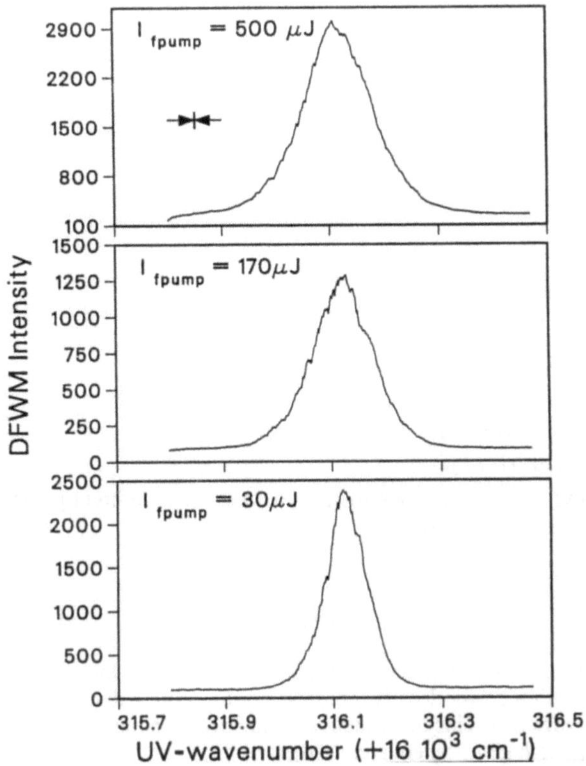

Figure 8. DFWM spectral line shape of the OH ($R_1(9)$) line taken at three different pump beam energies (probe beam energy kept low at 17 μJ (<50 kW/cm^2)) in an atmospheric pressure stoichiometric H_2/O_2 flame. Limiting spectral resolution of the laser is shown by arrows.

al removal dictates the lifetime of the electronic excited OH. Saturation for the DFWM transitions of the NH radical is achieved even more easily than for OH. We partly ascribe this to the larger transition line strengths for NH.

The saturation power dependence of the DFWM line shape of OH in a H_2-O_2 flame at atmospheric pressure is shown in Fig. 8. In this experiment the $R_1(9)$ line was scanned with a high resolution pulse amplified and frequency doubled ring dye laser (Coherent Inc. Mod. 699) where the uv-light had a bandwidth of 0.002 cm^{-1} (as indicated in Fig. 8). At the crossing point of the beams 5 mm above the flat flame burner we expect the flame temperature to be about 1460 K [37] with H_2O being the main collision partner of OH. The probe beam was maintained at less than 17 μJ (≈ 50 kW/cm^2) while the pump beam was varied from 14 μJ (≈ 40kW/cm^2) to 500 μJ (<1.4 MW/cm^2). The line center saturation power density for this transition is approximately 250 kW/cm^2. The OH transition is observed to have a Voigt line shape, with about equal Lorentzian and Gaussian contributions for all of the pump pow ers explored. Using the expression for the saturation intensity

$$I_s = I_s^0(1+\delta^2) \tag{12}$$

in the reflectivity term of Abrams/Lind (eq. (2)) - with $\delta = \Delta/\gamma_{12}$ - one arrives at

$$R^2 \equiv \frac{I_c^2}{I_p^2} = \alpha_0^2 \frac{4\gamma^6 (I/I_s^0)^2}{(\Delta^2 + \Gamma_{eff}^2)^3} \tag{13}$$

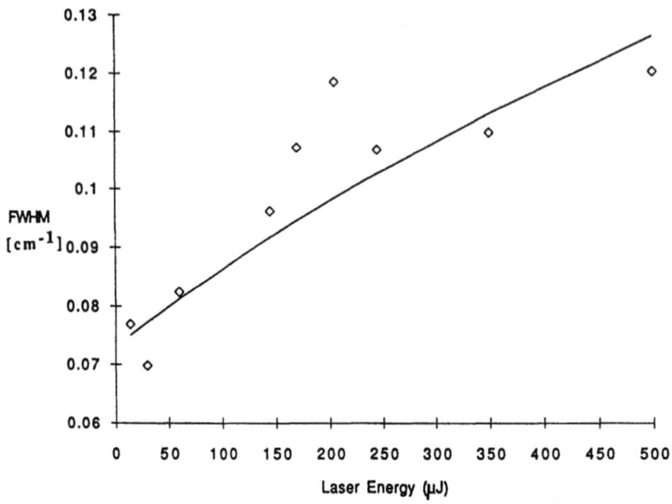

Figure 9. Values of Γ_{eff} (see text) from fitting DFWM spectra acquired at different pump energies. The solid line is a fit to Equation (13).

where

$$\Gamma_{eff}^{2} = \gamma^{2}(1 + 4I / I_{s}^{0}) \tag{14}$$

In these equations γ is the transition dephasing rate, I is the intensity of one of the pump beams, I_S^0 is the line center saturation power density (compare eq. 6), Δ is the detuning, and α_0 is the field absorption coefficient. This line shape is essentially the cube of a Lorentzian having a power broadened width given by equation (14).

The values obtained for Γ_{eff} from the data in Fig. 8 and other scans are shown in Fig. 9. The error bars shown are estimated from the quality of the fits. The solid line is a fit to equation (14) yielding γ=0.171 cm^{-1} and I_S^0= 956 μJ (2.64 MW/cm^2). Both of these values are larger than expected. The dephasing of the OH transition is due to rotationally inelastic collisions with water molecules and has been measured [37] to occur at a rate less than half of the fit value for γ. Thus the simple two level model for parallel polarizations inadequately predicts the saturation behaviour of the OH DFWM spectrum. Multilevel systems with Doppler broadening have been considered (e.g. ref [19]). Such theories often predict more complex line shapes, including saturation dips and dispersive behaviour. Using pump beam energies of 3 mJ a dip in the $R_1(9)$ line of OH in a 1 atm pressure methane/air flame has been observed.

4.5 Collisional effects

Degenerate Four Wave Mixing is a nonlinear optical process that is resonantly enhanced when atoms or molecules undergo a dipole allowed transition initiated by the incoming laser radiation. Therefore, like many other excitation phenomena involving transitions between discrete energy levels, collisions of radiator with perturber molecules will change the material excitation and radiation behaviour of these laser coupled states incoherently. This, in turn, influences the efficiency of the four wave mixing process by shifting molecules in and out of resonance with the exciting radiation. The DFWM lineshape of isolated lines will reflect the ensemble averaged collisional effects on the molecule depending on the pressure regime of the sample. Several investigations have appeared that deal with pressure induced effects in four wave mixing [20, 38, 39].

As was mentioned earlier and shown in Table 3 the DFWM linewidths are sub-Doppler so that collisional broadening/narrowing effects are more obvious in line shape studies. We find that the DFWM signals vary significantly with pressure, although not nearly to the ex-

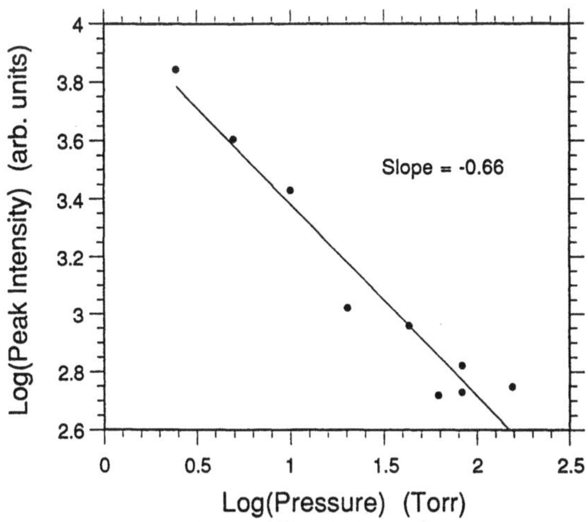

Figure 10. Plot of the logarithm of the integrated DFWM line intensity of the $O_{12}(13)$ transition of NO versus the logarithm of N_2 pressure.

tent predicted by Eq. (8). This was investigated in a cell experiment where at a NO partial pressure of 67 mbar nitrogen was added to the cell at pressures from 2.7 to 206 mbar. Figure 10 shows a plot of the logarithm of the buffer gas pressure, p_{buff}, versus the logarithm of the integrated intensity of a nonoverlapped $O_{12}(13)$ transition. The data are described by a straight line of slope -0.66.

The averaged slopes of similar fitted data for transitions with rotational quantum numbers between 9 and 13 suggest that the DFWM line intensities vary as $I \cdot p_{buff}^{-0.67}$ over this pressure range (a significant dependence on rotational excitation was not observed). We note that the two level model approach for DFWM intensities contains a pressure dependence of p_{buff}^{-6}, resulting from the factor α_0^2/I_s^2 when T_1 and T_2 are assumed to be determined by collisions with buffer molecules (a good approximation above 3 mbar). For nonmonochromatic laser excitation, however, this relation will not be fully obeyed since the laser spectral width is broader than the absorption linewidth at all pressures. Thus, I_s will increase less rapidly with increasing p_{buff}.

So far in the higher pressure range (>1 bar) no investigations of the pressure dependence of the DFWM signal intensity or the line shape for combustion relevant molecules are available. Preliminary results obtained by us from DFWM measurements on OH inside the combustion chamber of a spark ignition engine suggest comparatively low signal intensities at pressures up to 7 bar in the post-flame gases, although OH concentration is high so that only weakly absorbing transitions in the $R_{1,2}$ branches could be selected to reduce self-absorption of the DFWM beams. High pressure DFWM therefore constitutes an interesting and expanding area of active research in combustion diagnostics.

4.6 Planar DFWM imaging

Two-dimensional imaging within a single laser pulse is a very important means for the study of turbulent mixing of reacting and non-reacting flows. The spatial distribution of stable and unstable species [40], velocities [41] and the location of flame zones [42] can be determined. Using Rayleigh scattering or laser induced fluorescence the structure of the turbulent flame front and the spatial distribution of burned and unburned regions within the illuminated sheet can be visualized. These data validate comparisons to theoretical models for turbulent combustion and mixing so that refinements in both measurement and theory will help in a deeper understanding of technical combustion phenomena.The high signal strength

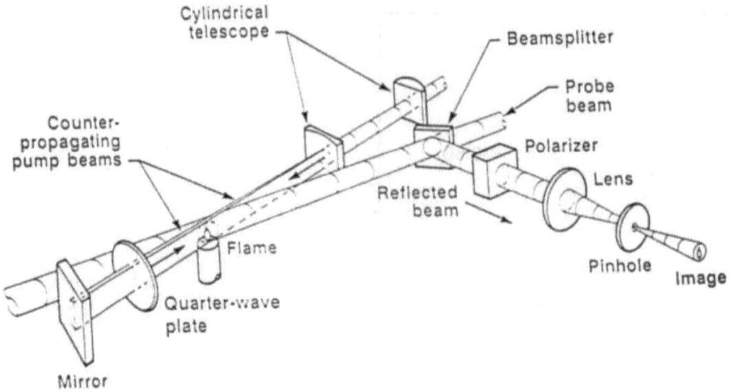

Figure 11. Schematic diagram of a DFWM imaging experiment.

and the coherence properties of the DFWM process make it an ideal tool for two-dimensional imaging applications. It is obvious from the previous sections that DFWM provides excellent selectivity and a sensitivity approaching that of LIF [43] in the detection of selected species.

Figure 11 depicts an experimental realization of a two-dimensional DFWM imaging geometry [44]. The only difference compared to the point measurement setup in Fig. 2 is that the two counterpropagating pump beams are formed into a thin (8 mm high, 0.3 mm thick) light sheet by a beam expanding telescope followed by a cylindrical telescope and that a 380 x 512 pixel CCD camera is used as a detector. Polarization discrimination in the signal beam path against scattered light originating from the forward pump and probe beams is accomplished by the insertion of a quarter wave plate into the retroreflected pump beam. The angle between forward pump and probe beams is 20° to give a reasonable spatial resolution without too much signal loss at large pump/probe angles [45]. The uv-output beam from a frequency doubled single mode ring dye laser (10 mJ per pulse) that was pulse amplified by a single mode frequency doubled Nd:YAG laser was used.

A principle objective of the experiment was to determine whether sufficient signal strength was available in this DFWM imaging geometry to permit spatial mapping of an intermediate combustion species like OH. Figure 12 shows, in a three-dimensional contour plot, a single shot image of OH obtainedby overlapping the beams in the flame tip of a methane/air diffusion flame. The observed intensity distribution reflects the shape of the reaction and post combustion zone where OH is present predominantly. No attempt was undertaken to deduce concentration information from these distributions since temperature has to be known for each image point. Also a correction for the intensity variation of the sheet and probe laser beams in the overlap region has to be performed to extract concentration values from calibrated signal intensities. Based on the number of pixel elements comprising the image (2×10^4) and the detector quantum efficiency (23%), we estimate that a total of 5×10^6 signal photons were incident on the CCD detector. The importance of the phase conjugated nature of the signal beam has long been recognized [46] and becomes particularly useful for remote detection in hostile environments. This has been demonstrated in the signal beam restoration capability of DFWM where the probe beam was distorted by passing it through a structured mask and through the hot flame gases. The sharp edges of this mask were well restored in the OH image of the probe beam when the two pump beams were unblocked. We also generated single shot high quality images of OH distributions from the bright luminous zone of an acetylene/oxygen welding torch flame burning under fuel rich conditions. The strong broadband emission of this flame make LIF measurements of OH difficult if not impossible. On the other hand the directionality of the DFWM signal beam together with the background light suppression techniques described above makes DFWM superior in these diagnostics applications.

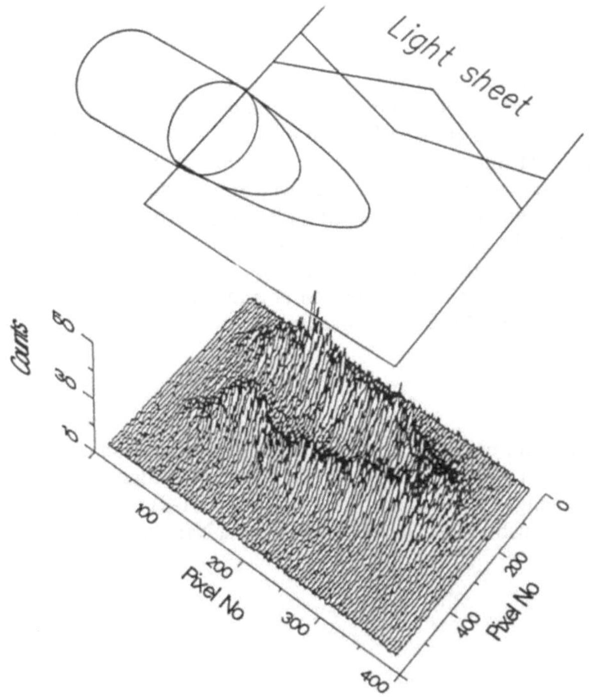

Figure 12. 2D-image of OH intensities (vertical axis) produced by DFWM in an axisymmetric methane/air diffusion flame. The flame burns from the upper left to lower right in the figure as indicated by the schematic drawing. The frequency doubled dye laser was tuned to 311.88 nm, corresponding to the $P_2(8)$ transition. Thickness of the laser sheet: 300 µm.

5. CONCLUSIONS

While DFWM is still in its infancy as a combustion diagnostics tool, experience so far suggest it has the potential to become an important detection method for trace species in hostile combustion environments.

Judged from its properties DFWM seems to offer desirable features from both CARS and LIF diagnostics: the CARS-like directionality of the signal beam generated in the nonlinear medium together with the high sensitivity of a resonantly enhanced process. Contrary to resonance CARS where three laser beams of different frequency are tuned in the vicinity of electronic transitions the difference of two of them simultaneously matching a Raman allowed rovibrational transition, DFWM only needs one single frequency, which drastically reduces experimental complexity. The sensitivity of DFWM is generally lower than that of LIF but sufficient to detect trace species at absolute concentration levels as low as $8 \cdot 10^{11}$ cm^{-3} at room temperature [47] so that detection limits in the 1 ppm range at atmospheric pressure seem feasible. Using phase sensitive detection techniques these limits can be reached even with low intensity cw-laser sources [48]. Currently, as a combustion diagnostics technique DFWM has only been demonstrated using electronic resonance enhancement. However, it is also possible to detect molecules via vibrationally allowed dipole transitions in the infrared. In fact, many DFWM experiments in the area of optical phase conjugation have been done using infrared transitions of molecules such as SF_6 [49] or NH_3 [50]. The low laser intensity requirements for DFWM makes the method - similar to LIF - suitable for two-dimensional imaging. Whereas in LIF large aperture imaging optics close to the measurement point is necessary to collect enough signal photons the coherence of the

DFWM signal beam permits small aperture optical access and remote detection of the probe volume which also significantly reduces stray light interferences and background flame emissions.

Certainly collisional quenching of the ground and excited electronic states coupled by the interacting laser fields has been of great concern in LIF experiments. This complication in the evaluation of DFWM signals is less well understood at this time. As has been shown experimentally in Sect. 4.5 the DFWM signal of NO decreases with increasing pressure of added nitrogen at low total pressures. On the other hand, for OH and NH, measurements at one atmospheric pressure and above produce high signal strengths at approximately the same low laser pulse energies used in the low pressure experiments. This discrepancy in experimental findings still has to be resolved theoretically. In this context, the application of picosecond laser sources to tackle the quenching problem at 1 atm and higher pressures seems to be an attractive extension of DFWM in flame diagnostics.

Acknowledgements

The authors thank Larry Rahn for performing the experiments and data evaluation on the high resolution DFWM work of OH in the H_2/O_2 flame. T. Dreier acknowledges the financial support of the Deutsche Forschungsgemeinschaft (grant No. Dr 195/1-2) for his stay in the US. Work performed at the Combustion Research Facility, Sandia National Laboratories, Livermore.

REFERENCES

[1] A. C. Eckbreth, P. A. Bonczyk and J. F. Verdieck: *Laser Raman and fluorescence techniques for practical combustion diagnostics*: Appl. Spectr. Rev. **13**, 15 (1977).

[2] J. A. Shirley and L. R. Boedeker: AIAA/ASME 24th Joint propulsion Conference, Boston, MA, paper No. 88-3038 (1988).

[3] B. Attal, D. Débarre, K. Müller-Dethlefs and J. P. E. Taran: Rev. Phys. Appl. **18**, 39 (1983).

[4] B. Attal-Trétout and P. Bouchardy: La Réch. Aerosp. **5**, 19 (1987).

[5] K. Kohse-Höinghaus, U. Meier and B. Attal-Trétout: Appl.Opt. **29**, 1560 (1990).

[6] T. Dreier and J. Wolfrum: J. Chem. Phys. **80**, 957 (1984).

[7] B. Attal-Trétout, O. O. Schnepp and J. P. E. Taran: Opt. Commun. **24**, 77 (1978).

[8] H, J, Eichler, D. Pohl and P. Gunter: *Laser enhanced dynamical gratings,* Springer Ser. in Optical Sciences, Vol. 50, Berlin (1986).

[9] T. S. Rose, W. L. Wilson, G. Wackerle and M. D. Fayer: J. Chem. Phys. **86**, 5370 (1987).

[10] D. H. Pepper: IEEE J. Quant. Electr. **QE-25**, 312 (1989).

[11] Y. Prior and E. Yarkoni: Phys. Rev. A **28**, 3689 (1983).

[12] J. Pender and L. Hesselink; Opt. Lett. **10**, 264 (1985).

[13] P. Ewart and S. V. O'Leary: J. Phys. B **15**, 3669 (1982).

[14] P. Ewart and S. V. O'Leary: Opt. Lett. **11**, 279 (1986).

[15] *Optical phase conjugation* (R. A. Fisher, Ed.), Acad. Press, N. Y., (1983).

[16] D. G. Steel, R. C. Lind, J. F. Lam and C. R. Giuliano: Appl. Phys. Lett. **35**, 376 (1979).

[17] R. L. Abrams and R. C. Lind: Opt. Lett. **2**, 94 (1978).

[18] S. M. Wandzura: Opt. Lett. **4**, 208 (1979).

[19] M. Ducloy, F. A. M. de Oliveira and D. Bloch: Phys. Rev. A **32**, 1614 (1985).

[20] D. G. Steel and R. A. McFarlane: Phys. Rev. A **27**, 1687 (1983).

[21] J. Cooper, A. Charlton, D. R. Meacher, P. Ewart, G. Alber: Phys. Rev. A **40**, 5705 (1989).

[22] see Ref. 15 p. 211

[23] G. H. Diecke and H. M. Crosswhite: J. Quant. Spectrosc. Radiat. Transf. **2**, 97 (1962).

[24] T. Dreier and D. J. Rakestraw: Opt. Lett. **15**, 72 (1990).

[25] D. A. Greenhalgh: *Quantitative CARS spectroscopy* in: *Advances in nonlinear spectroscopy* (R. J. H. Clarke, R. E. Hester, Eds.), J. Wiley & Sons Ltd., N. Y., (1987).

[26] F. Y. Yueh and E. J. Beiting: Computer Phys. Commun. **42**, 65 (1986).

[27] M. Péalat, F. Bouchardy, M. Levebvre and J. P. Taran: Appl. Opt. **24**, 1012 (1985).

[28] S. Druet and J. P. Taran: *Coherent anti-Stokes Raman spectroscopy* in: *Chemical and biochemical applications of lasers* (C. B. Moore, Ed.), Acad. Press, Cambridge (1979).

[29] A. Lawitzki, R. Tirgrath, U. Meier, K. Kohse-Höinghaus, A. Jörg and Th. Just: Joint Meeting of the German + Italian Sections of the Combustion Institute, Naplas (Italy), (1989).

[30] T. Dreier and D. J. Rakestraw: Appl. Phys. B **50**, 479 (1990).

[31] R. J. Hall and L. R. Boedeker: Appl. Opt. **23**, 1340 (1984).

[32] R. Cattolica and T. Mataga: to be published.

[33] M. S. Chou, A. M. Dean and D. Stern: J. Chem. Phys. **78**, 5962 (1983).

[34] I. Deézsi: Acta Physica (Hungarian) **9**, 125 (1958).

[35] See, for example, M. Nicolet: J. Geophys. Res. **70**, 679 (1965).

[36] R. P. Lucht in *Laser spectroscopy and its applications* (L. J. Radziemski, R. W. Solarz, J. A. Paisner, Eds.), Marcel Dekker Inc., N. Y. (1987).

[37] J. E. M. Goldsmith and L. Rahn: J. Opt. Soc. Am. B **5**, 749 (1988).

[38] R. K. Ray, D. Bloch, J. J. Snyder, G. Camy and M. Ducloy: Phys. Ref. Lett. **44**, 1251 (1980).

[39] W. H. Richardson, L. Maleki and E. Garmire: IEEE J. Quant. Electr. **QE-25**, 382 (1989).

[40] R. K. Hanson: 21st. Symp. (Int.) on Combustion, (The Combustion Institute, Pittsburgh, 1986), p. 1677.

[41] P. H. Paul, M. P. Lee and R. K. Hanson: Opt. Lett. **14**, 417 (1989).

[42] R. Suntz, H. Becker, P. Monkhouse and J. Wolfrum: Appl. Phys. B **47**, 287 (1988).

[43] P. Ewart, P. Snowdon and I. Magnusson: Opt. Lett. **14**, 563 (1989).

[44] D. J. Rakestraw, R. L. Farrow and T. Dreier: Opt. Lett. **15**, 709 (1990).

[45] L. M. Humphrey, J. P. Gordon and P. F. Liao.: Opt. Lett. **5**, 56 (1980).

[46] J. Feinberg: Opt. Lett. **7**, 486 (1982).

[47] D. J. Rakestraw: priv. commun.

[48] W. G. Tong, J. M. Andrews and Z. Wu: Anal. Chem. **59**, 896 (1987).

[49] J. W. R. Tabosa, C. L. César, M. Ducloy and J. R. R. Rios Leite: Opt. Commun. **40**, 77 (1981).

[50] A. Elci, D. Rogovin, D. Depatie and D. Haueisen: J. Opt. Soc. Am. **70**, 990 (1980).

Spatially Resolved CARS in the Study of Local Mixing of Two Liquids in a Reactor

H.P. Kraus and F.W. Schneider

Institute of Physical Chemistry, University of Würzburg,
Marcusstr. 9/11, W-8700 Würzburg, Fed. Rep. of Germany

We have used the CARS signal of the breathing vibration of toluene to follow the mixing process of liquid toluene and liquid p-xylene as a function of position inside a small (1.5 ml) continuous flow stirred tank reactor (CSTR).

The CARS measurements provide direct evidence for the presence of local concentration variations called fluctuations during the mixing process. This information is particularly important in the study of nonlinear chemical reactions which display chemical chaos, for example. The CARS signal is recorded over a time interval at various points inside the reactor. The arithmetic mean of the toluene concentration and the root mean square variations are then calculated. CARS has the advantage of being non-invasive. It is relatively sensitive in principle, since the intensity of the CARS signal depends on the square of the concentration in the interaction volume of the two laser beams. The pump and the Stokes beam have to be crossed in liquids in order to achieve phase matching. Thus one obtains spatial information about local concentrations in the reactor. For example, for a focal length of 20 mm and a crossing angle of ~1.3° the interaction volume is a narrow (~15 µm) and long (1.2 mm) volume of intense illumination in which the CARS signal is generated.

Another method, laser Doppler anemometry (LDA) [1], measures the velocity fields of streaming elements, where any information on local concentrations is indirect.

1. EXPERIMENTAL

A frequency doubled Nd-YAG laser (DCR 2A, Quanta Ray) pumps two dye lasers synchronously with a repetition rate of 10 Hz (Fig.1). One dye laser (PDL1) has a pulse duration of ~10 ns and a line width of 0.25 cm^{-1} at 570.8 nm (Rh6G). The second dye laser uses the dye Rh 101 at 605.5 nm with a line width of 4 cm^{-1} [2]. The difference in frequency is equal to the totally symmetric breathing vibration of toluene (1004 cm^{-1}).

A second CARS experiment is used as a reference in order to reduce the large (25%) pulse to pulse variations in the CARS signal to about 8 %, which represents the present experimental scatter.

1.1 The Reactor (CSTR)

In our experiments the CSTR is an ordinary spectrophotometric (1 cm) cell (volume 1.5 ml) [3] which has been shortened to a height of 13 mm. A Teflon stopper contains an inside dome to allow the removal of any air bubbles. The two input liquids enter through two thin Teflon hoses with an inside diameter of 0.86 mm. They terminate at the zero point of the x-axis (Fig.2). There are

Eds.: G. Marowsky and V.V. Smirnov

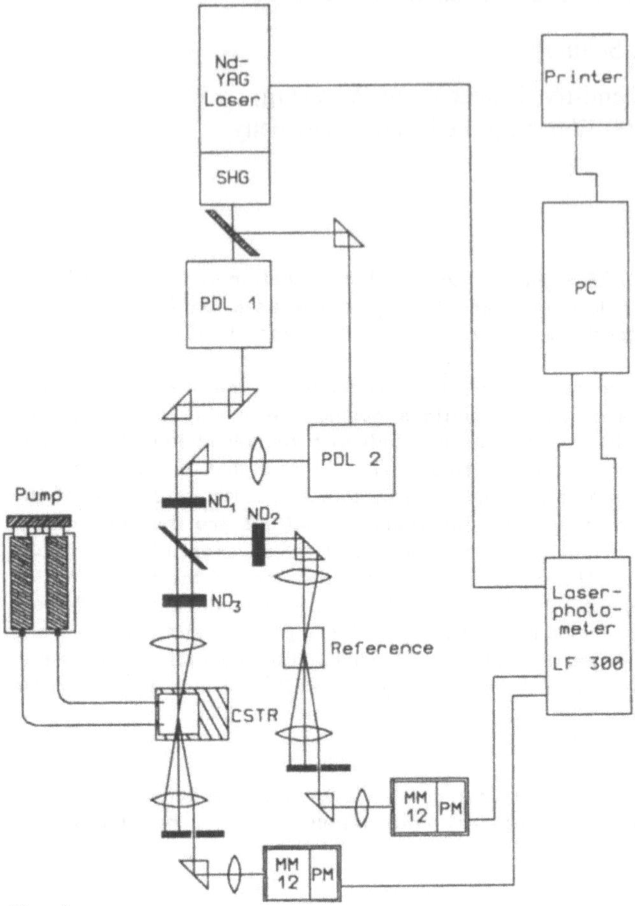

Fig. 1:

Experimental set-up. PDL 1 and PDL 2 are the two dye lasers, the ND_x are neutral density filters and the MM12 are Zeiss prism monochromators. The CSTR (see text) is mounted on a movable stage.

5 grid-like cross sections (x = 0 to 4) where 90 equally spaced points are measured in each grid. The magnetic stirrer is tooth shaped for high energy dissipation; its diameter is 8 mm. The reactor is placed on a stage that is moved by micrometers with great precision. The magnetic stirring motor is placed under the reactor. A precise piston pump (Infors Precidor, Basel) is driven by a stepping motor with a constant frequency of 20 Hz which leads to a flow rate of 0.64 ml/min per syringe. The two liquids are delivered by two identical 50 ml Hamilton syringes. The average residence time is $\tau_{res} = V/u$ where V is the volume of the reactor and u is the flow rate.

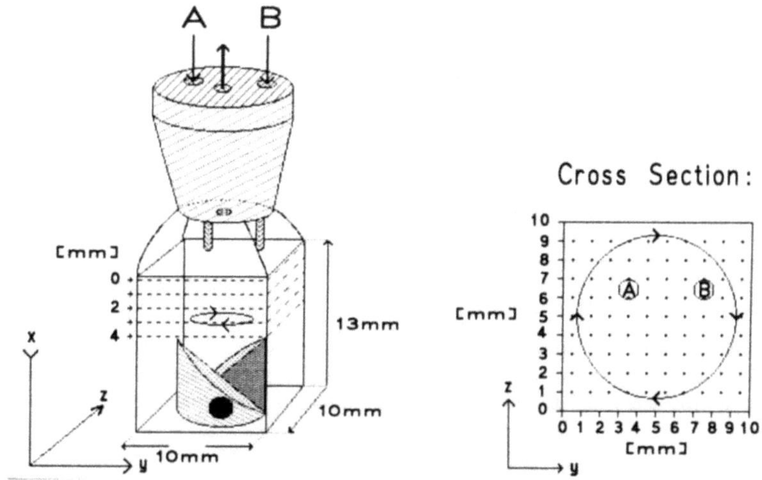

Fig. 2:
Continuous Flow Stirred Tank Reactor (CSTR) as a shortened spectrophotometric cell of 1.5 ml volume, Teflon stopper with inside dome, two entry tubes as shown in the cross section, one exit port, 5 horizontal layers where x=0 to 4 mm counting downwards, tooth shaped magnetic stirrer (Teflon), with arrows indicating direction of stirrer rotation.

1.2 Toluene/p-Xylene Mixtures

For non-reactive mixing we chose toluene and p-xylene since their mixing ratio is unlimited, their indices of refraction are practically identical and their CARS bands do not overlap. If pure toluene and p-xylene are used, large schlieren effects are observed during mixing. In order to eliminate any schlieren effects we used two premixed liquids A and B: one syringe contained a 80/20 mixture (liquid A) and the other syringe a 20/80 mixture (liquid B) of toluene/p-xylene. Equal amounts of both liquids are fed into the reactor. A final 50/50 mixture is obtained. The reference cell contained a 50/50 mixture without flow.

2. RESULTS

A plot of the square root of the CARS signal as a function of the percent toluene gives a straight line (Fig. 3). From the straight line the value of the percentage of liquid A in the bulk (% A) can be calculated for each point according to % A = 5/3(% T) - 33.3 %.

As an example, two time series are shown which show the percentage of liquid A with flow (Fig.4a) and without flow (Fig.4b). Measurements across the width of the reactor show the constancy of the CARS signal from x=1-9 mm in a cell without flow (Fig.5).

In the CSTR for each flow point (there are 90 points in a grid) the arithmetic mean of the CARS signal (1000 pulses each) is plotted (Fig.6). Each square corresponds to one measurement point which has the dimensions of an approximate cylinder with a diameter of ~15 µm and a length of ~1.2 mm. The standard deviation σ of a given point inside the reactor is

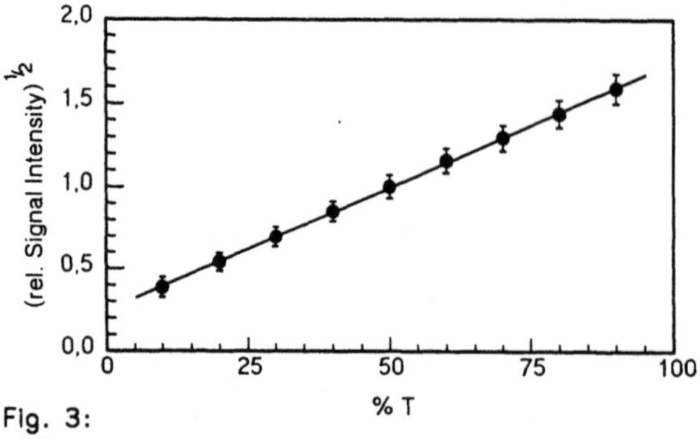

Fig. 3:
$I^{1/2}$ vs volume percent of toluene, where I is the intensity of the CARS signal due to the 1004 cm^{-1} vibration of toluene measured in a cell without flow; the line has a finite intercept due to the presence of some nonresonant CARS background.

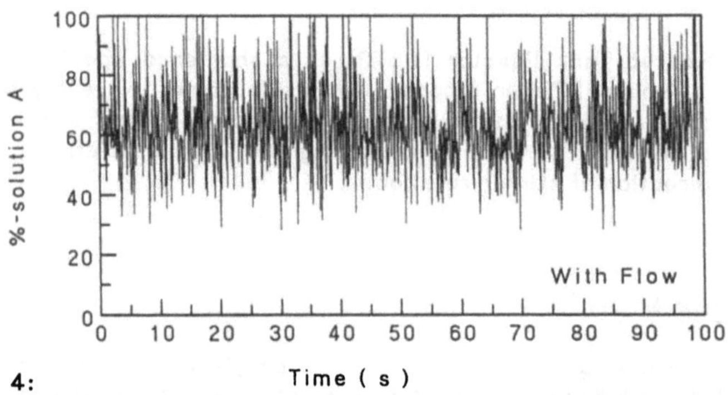

Fig. 4:

a) Time series of the percentage of liquid A in the CSTR with flow (τ_{res}= 70 s, 500 rpm) at the point x=0, y=3.5 nm and z= 7 mm. Notice that pure segregated solution A is measured in about 3-4% of all cases close to the entry port. The time average corresponds to 62.4 % solution A and $\sigma_{rel}= \pm 1.6$, where $\sigma_{rel} = \sigma_{flow}/\sigma_{no\ flow}$.

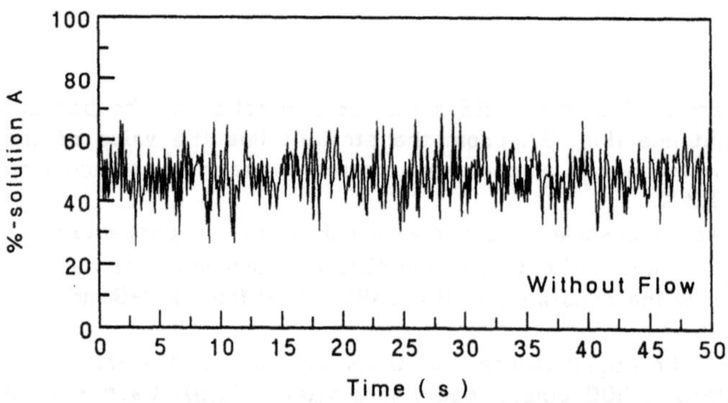

b) Time series of the percentage of liquid A in the CSTR without flow at the same spatial point as in Fig.4a. The time average corresponds to 50 % solution A and the average scatter is ±8 %.

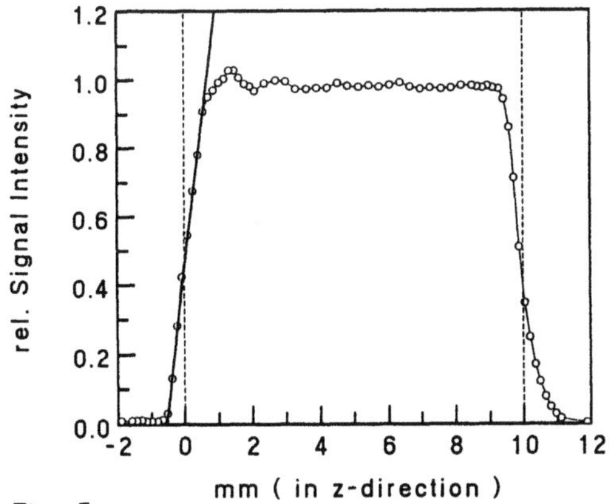

Fig. 5:
Rel. CARS intensity in the z direction in the CSTR. The CARS signal is practically constant over the width (1 to 9 mm) of the reactor for complete mixing without flow. The dotted lines represent the inside cell walls.

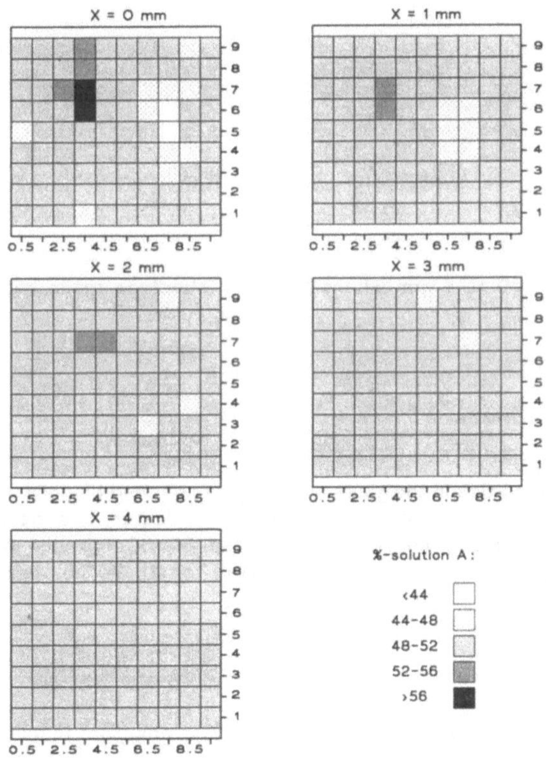

Fig. 6:
Average concentration $\bar{x}$ of solution A (in % as indicated) as a function of position, where $\bar{x}$ is the arithmetic mean of a given time series. The dark shadings indicate large values of $\bar{x}$ as indicated. Each CARS measurement is made in the middle of a square where the width and the length of the CARS interaction volume (cylinder) are ~15 µm and 1.2 mm, respectively.

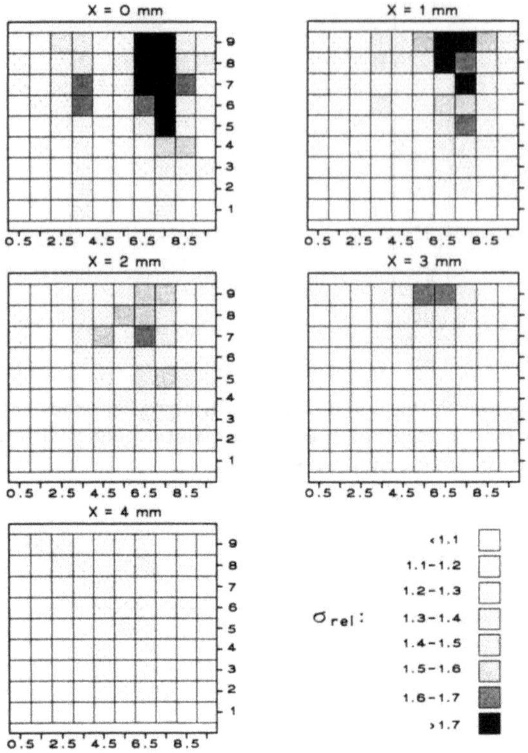

Fig. 7:
σ_{rel} in various cross sections x=0 to 4 mm where the dark shadings mean high values of σ_{rel} as indicated.

$$\sigma = \left[\frac{\Sigma_{i=o,n} (x_i - \bar{x})^2}{n-1} \right]^{1/2} .$$

The effect of local inhomogeneities in the glass wall may be reduced by normalizing all CARS measurements σ_{flow} by $\sigma_{no\ flow}$ where the latter has been measured in the reference cell (Fig.1) containing a stirred 50/50 mixture of toluene/p-xylene. The ratio $\sigma_{flow}/\sigma_{no\ flow} = \sigma_{rel}$ is plotted for all 90 points in each cross section (Fig.7).

3. DISCUSSION

Two processes describe the nonideal mixing of liquids in a reactor: macromixing and micromixing [4-6]. The macromixing process describes the convection of one type of liquid through the other on a macroscopic scale, which leads to relatively large segregated filaments of liquid. The micromixing process is usually viewed as a stretch and fold process of the liquid filaments which thereby become smaller and smaller until they disappear by molecular diffusion to produce a perfectly mixed liquid [4,7]. The transition from macromixing to micromixing is gradual.

It is concluded that macromixing occurs mainly in the general vicinity of the ports of entry of liquids A and B in layers 0 and 1 where the concentration map (Fig.6) shows an excess of liquid A (dark squares) around one tube and an excess of liquid B (light squares) around the other. Liquid B is spread out horizontally about 4 mm above the magnetic stirrer which is 8 mm in diameter. Pure liquid A and pure liquid B are occasionally segregated as seen from the time series (Fig.4a, with flow) at a point (x=0, y=3.5 mm, z=7 mm) close to the inlet tube A. The rotatory motion of the stirrer transports the macromixed liquids into the right hand corner of the reactor due to the excentric location of the inlet tubes. Here the fluctuation maps (Fig.7) display large σ_{rel} values (>1.7) indicating that the composition in the laser observation window changes dramatically from pulse to pulse in portions of layers x=0 and 1.

Roughly speaking, micromixing takes place in the lower layers (x=2 to 4) as σ_{rel} progressively diminishes until it reaches unity at the lowest layer. This means that the micromixing process and the ensuing molecular diffusion have produced a uniform composition approaching the bulk liquid. The liquid around the stirrer is probably completely mixed.

It is assumed that the mixing process is qualitatively similar when dilute aqueous solutions are used since their viscosity is similar to that of the toluene/ p-xylene mixture. The future use of lenses of various focal lengths will allow the calculation of the actual size of the liquid filaments. Furthermore, the optimization of the geometry of a reactor including baffles for optimal mixing is expected to be the result of future measurements.

Thus it has been demonstrated for the first time by non-invasive CARS measurements that mixing of liquids in a reactor is rather heterogeneous around the entry ports. In fact, a CSTR may be classified as a "noise generator" under the conditions of flow rate and geometry used in this work.

Acknowledgements

We thank the Deutsche Forschungsgemeinschaft and the Fonds der Chemischen Industrie for partial financial support.

References

[1] F. Durst, A. Melling and J.H. Whitelaw , Eds., Theorie und Praxis der Laser-Doppler-Anemometrie, Braun, Karlsruhe 1987.

[2] F.W. Schneider, R. Brakel and H. Spiegel; Ber. Bunsenges. Phys. Chem., 93, (1989) 304.

[3] F.W. Schneider and A.F. Münster, J. Phys. Chem. 95, (1991) 2130.

[4] J. Villermaux, in Spatial inhomogeneities and transient behaviour in chemical kinetics, (Eds. P. Gray, G. Nicolis, F. Baras, P. Borckmans and S.K. Scott), Manchester University Press, Manchester, (1990) p. 119.

[5] R. Aris, Introduction to the Analysis of Chemical Reactors, Prentice Hall, Englewood Cliffs, NJ,(1965).

[6] L. Lapidus, N.R. Amundson, Eds., Chemical Reactor Theory. A Review, Prentice Hall, Englewood Cliffs, NJ, (1977).

[7] J.M. Ottino, C.W. Leony, H. Rising and P.D. Swanson, Nature 333, (1988) 419.

Pure Rotational CARS for Temperature Measurement in Turbulent Gas Flows

V.V. Moiseenko, S.A. Novopashin, and A.B. Pakhtusov

Institute of Thermophysics, SU-630090 Novosibirsk, USSR

Abstract. The broadband polarization pure rotational CARS is presented. The ratio of the two spectrum parts is used to determine the gas temperature. The CARS spectrometer was used to study the temperature distribution in a choked, underexpanded air jet.

1. Introduction

CARS (coherent anti-Stokes Raman spectroscopy) is a well-established technique for the nonintrusive investigations of the temperature fields in gas flows [1-4].

The broadband polarization pure rotational CARS with folded BOXCARS geometry is presented. This method has a number of advantages for investigations of gas flows at low temperatures. Firstly, this process has a large cross section [5]. Secondly, broadband variant CARS can be realized easily, because rotational quanta have energy values of some cm^{-1} and the spectral widths of the dye lasers may reach 200-300 cm^{-1}. Moreover, it allows measurements on molecular species in gas mixtures. An additional advantage is that rotational lines may be much more easily spectrally resolved than the piled-up rotational lines in a vibrational Q-branch. Since pure rotational CARS lines are arranged near the exciting line it is necessary to use a planar or folded BOXCARS scheme, often combined with polarization techniques, to isolate the CARS beam sufficiently.

2. Experimental

For rotational CARS generation a doubled Nd:YAG laser at 532 nm, operating at the power 70 mJ and the bandwidth 0.25 cm^{-1} are used. Rotational energies are excited with two photons of different frequencies from a broadband dye laser with an amplifier. The bandwidth of dye laser spectrum is 200 cm^{-1}, the power of each dye beam is 10 mJ. The dye laser is pumped by a fraction of green radiation (60 mJ). The green beam and two dye laser beams are aligned in a folded BOXCARS configuration, focused, and crossed with a crossing angle of 4°. The spatial crossing region has a dimension $25 \times 25 \times 400\ \mu m^3$.

The polarization of one dye beam is orthogonal to the second one and the green beam. With the polarizations so arranged, the CARS spectrum is generated through the third-order non-linear susceptibility component χ_{1212} and is orthogonally polarized to the pump beam at 532 nm. Behind the sample region the beams are recollimated and the CARS signal beam is selected by means of the diaphragm and polarizer. The spectrum of the CARS signal is dispersed by a single monochromator with a dispersion of 2 nm/mm.

Springer Proceedings in Physics, Volume 63 **Coherent Raman Spectroscopy**
Eds.: G. Marowsky and V.V. Smirnov

Fig. 1 Calibration

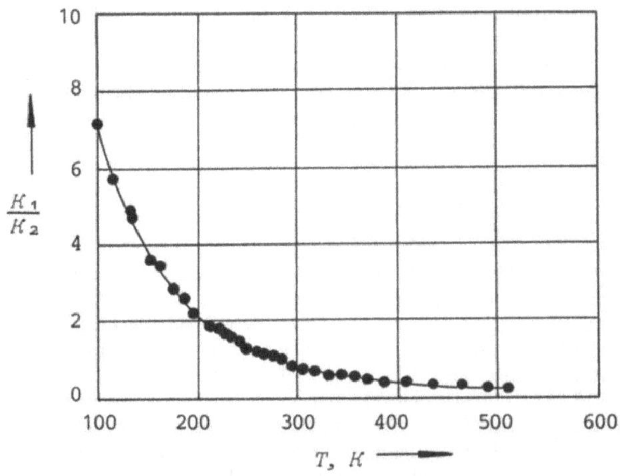

Detection systems of two types are used. In the first case the ratio of the two spectrum parts is used to determine the gas temperature. The two-channel fibre-optic light conduit and two photomultipliers are used to measure the intensities of two spectral parts. The both bandwidthes are 30 cm^{-1}. For calibration a subsonic gas stream was formed. The stream was exhausted from the tube with inner diameter 10 mm. The gas was heated or cooled in a heat exchanger, through which gas was fed. The temperature in the stream was measured by means of the thermocouple, the junction of which was placed at a distance of 0.5 mm from the focal point. Fig. 1 shows the ratio of intensities as a function of temperature for air. Experiments show that the error of a single measurement fot the whole temperature range is no more than 6 %. It is necessary to note that this technique permits measurement the temperature in a constant gas composition.

In the second case an optical multichannel analyser camera with an image intensifier is used to record CARS spectra. The number of elements in the analyser matrix is 1024. Each 25 μm detector element covers about 1.5 cm^{-1}, and a spectral resolution of about 3 cm^{-1} resulted. Figure 2 shows only the anti-Stokes part of the spectrum for air at 295 K, averaged over 20 lasers shots. The mathematical algorithm is needed to calculate the concentrations of components and temperature.

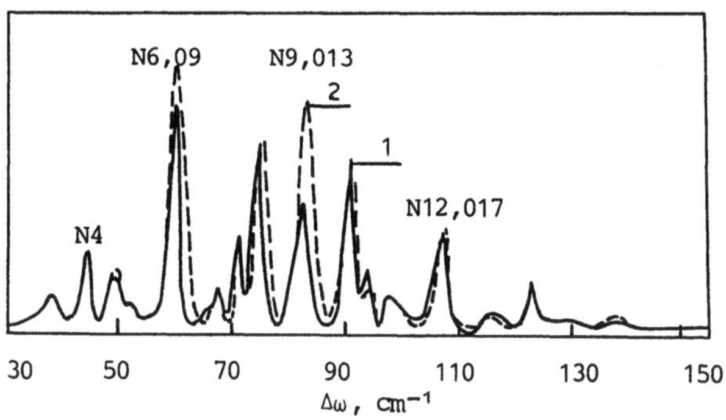

Fig. 2. CARS spectrum for air at 295 K. 1 - experiment, 2 - theory

There are a number of causes of the error of the measurement by means of CARS method. In probing techniques presented, some factors do not play a role. These are: 1) fluctuations of pulse duration ; 2) fluctuations of spatial intensity distribution at the focal point; 3) fluctuations of pulse energies. The main sources of the measurement error for pure rotational CARS are the fluctuations of the spectra of the two lasers [6].

3. Results

A two-channel variant of the CARS spectrometer was used to study the temperature distribution in a choked, underexpanded air jet. The jet was exhausted from a circular sonic nozzle of diameter 2 mm into the air at room conditions. The pressure in the prechamber was $5 \cdot 10^5$ Pa, the temperature was 290 K. The well-known cell structure of shocks is realized at these conditions. Figure 3 shows the jet axis temperature distribution. The ambient gas does not influence the flow in the core of the first shock-cell. A comparison with known data [7] is shown for this region. The difference at a certain distance from the nozzle exit is connected with the cell structure of the jet and the finite value of the spatial resolution. All experimental points were obtained by averaging 100 pulses. Simultaneously the value of RMS temperature fluctuations was calculated. The transverse temperature distribution for the distance 5 mm from the nozzle exit is shown in Fig. 4. The RMS temperature distribution is shown in this figure as well. The increase of temperature fluctuations in the mixing layer characterizes the flow regime as turbulent.

The first multi-channel experiments have been performed. The spectra of nitrogen and air in the temperature range 100 - 300 K have been recorded. To calculate species concentrations and temperature from the spectra we compare the model spectra with the experimental one. For example, the comparison for air at normal conditions is shown in Fig. 2. The lasers'bandwidth, the broadening of the rotational lines [8] and apparatus broadening were taken into account. As can be seen, the agreement is good, but dif-

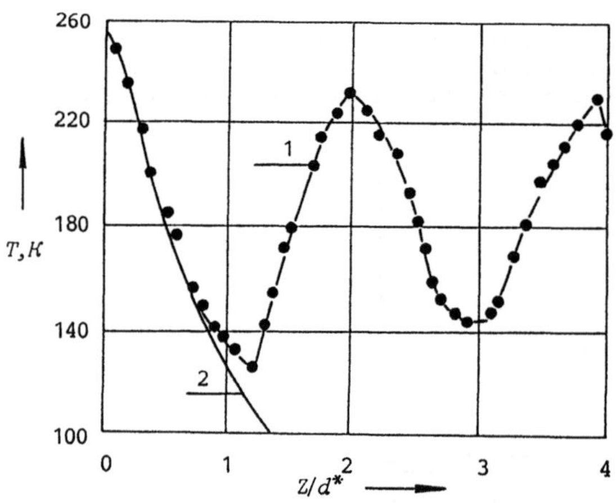

Fig. 3. Axial temperature distribution
1 - experiment
2 - theory

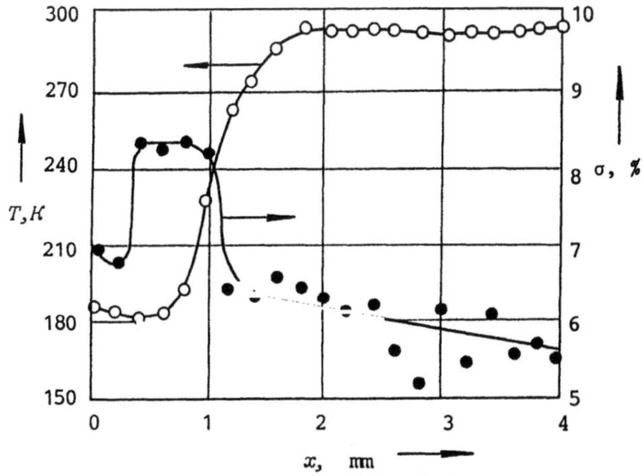

Fig. 4. Transverse temperature distribution and RMS temperature distribution

ficultes arise concerning the interference of lines of nitrogen and oxygen.

4. Conclusions

Let us point out the characteristics of the two-channel probing technique:

1. Temperature range is 100-500 K.
2. Localization is (25x25x400) μm^3.
3. Temporal resolution is 20 ns.
4. Number density $> 10^{18}$ molecules/cm^3.

References

1. Moya F., Druet S. A. J., Taran J. P. E.: Gas spectroscopy and temperature measurements by coherent Raman anti-Stokes scattering. Opt. Comm. 13, 169-174 (1975).
2. Smirnov V.V., Fabelinskii V.I.: Temperature measurements and spectroscopy of vibrational-rotational levels of nitrogen excited in a discharge by CARS. JETP Lett. 28, 427 (1978).
3. Eckbreth A. C., Dobbs G. M., Stafflebeam J. H., Tellex P. A.: CARS temperature and species measurements in augmented jet engine exhausts. Appl. Opt. 23, 1328-1333 (1984).
4. Dick B., Gierulski A.: Multiplex rotational CARS of N_2, O_2, and CO with excimer pumped dye lasers: species identification and thermometry in the intermediate temperature range with high temporal and spatial resolution. Appl. Phys. B 40, 1-7 (1986).
5. Inaba H.: Detection of atoms and molecules by Raman scattering and resonance fluorescence, in Laser Monitoring of the Atmosphere, ed. Hinkley E.D. (Springer, Berlin, Heidelberg 1976).
6. Kroll S., Sandell D.: Influence of laser-mode statistics on noise in nonlinear-optical processes - application to single-shot broadband coherent anti-Stokes Raman scattering thermometry. J. Opt. Soc. Am. B 5, 1910-1926 (1988).
7. Zhohov V.A., Homutskii A.A.: The Atlas of Supersonic Free-Expansion Flows of an Ideal Gas from Axisymmetric Nozzle, Trudi of TsAGI, 1224 (1970).
8. Jammu K. S., John G. E. St., Welsh H. L.: Pressure broadening of the rotational Raman lines of some simple gases. Can. J. Phys. 44, 797-814 (1966).

Coherent Raman Scattering in High-Pressure/High-Temperature Fluids: An Overview*

S.C. Schmidt and D.S. Moore

Los Alamos National Laboratory, Los Alamos, NM 87545, USA

The present understanding of high-pressure/high-temperature dense-fluid behavior is derived almost exclusively from hydrodynamic and thermodynamic measurements. Such results average over the microscopic aspects of the materials and are, therefore, insufficient for a complete understanding of fluid behavior. At present, dense-fluid models can be verified only to the extent that they agree with the macroscopic measurements. Recently, using stimulated Raman scattering, Raman induced Kerr effect scattering, and coherent anti-Stokes Raman scattering, we have been able to probe some of the microscopic phenomenology of these dense fluids. In this paper, we discuss primarily the use of CARS in conjunction with a two-stage light-gas gun to obtain vibrational spectra of shock-compressed liquid N_2, O_2, CO, their mixtures, CH_3NO_2, and N_2O. These experimental spectra are compared to synthetic spectra calculated using a semiclassical model for CARS intensities and best fit vibrational frequencies, peak Raman susceptibilities, and Raman linewidths. For O_2, the possibility of resonance enhancement from collision-induced absorption is addressed. Shifts in the vibrational frequencies reflect the influence of increased density and temperature on the intramolecular motion. The derived parameters suggest thermal equilibrium of the vibrational levels is established less than a few nanoseconds after shock passage. Vibrational temperatures are obtained that agree with those derived from equation-of-state calculations. Measured linewidths suggest that vibrational dephasing times have decreased to subpicosecond values at the highest shock pressures.

1. INTRODUCTION

Presently most models of explosive and shock-induced chemical behavior treat the medium as a continuum[1,2] that chemically reacts according to either a pressure dependent or Arrhenius kinetics rate law. One or more parameters are used to incorporate the global chemical behavior, hydrodynamic phenomenology and effects of material heterogeneity. In the past few years, several studies[3-10] have been started that attempt to improve the methodology by defining the continuum, not as a single component, but as one that incorporates ideas such as hot spots, voids or multicomponents. However, in all of these studies essentially no effort is made to incorporate any of the microscopic details of the shock-compression/energy transfer and release phenomenology that constitutes the detonation or reactive process.

The objective of our work has been two fold; (1) to determine the molecular structure and identify chemical species in unreacting and reacting shock-

* This work was supported by the US Department of Energy.

Springer Proceedings in Physics, Volume 63 Coherent Raman Spectroscopy
Eds.: G. Marowsky and V.V. Smirnov

compressed molecular systems and (2) to study the effect of pressure and temperature on condensed phase energy transfer. Also, we would like to identify the unique features of a shock wave that contribute to the energy transfer processes. Achievement of these goals would contribute significantly to understanding the initial mechanisms governing shock-induced chemical reactions and possibly to the steps controlling product formation. Gas guns are being used to dynamically shock compress molecular liquids to pressures where chemical reaction occurs. The high-density/high-temperature fluid is then probed using coherent Raman scattering techniques.

2. COHERENT RAMAN SCATTERING IN SHOCK-COMPRESSED LIQUIDS

Three coherent Raman scattering techniques have been attempted in shock compressed liquid samples. Advantages of these techniques, primarily because of large scattering intensities and the beam-like nature of the scattered signal, are increased detection sensitivity, temporal resolution limits approaching laser pulse lengths, and possible spatial resolution approaching the diffraction limit of the optical components. As with all optical methods in shock-wave applications, optical accessibility because of material opacity or particulate scattering remains a major difficulty with coherent Raman scattering.

Stimulated Raman scattering (SRS) has been observed in shock-compressed benzene up to pressures of 1.2 GPa.[11] Stimulated Raman scattering[12,13] (Fig. 1) occurs when the incident laser intensity, ω_1, in a medium exceeds a threshold level and generates a strong, stimulated Stokes beam, ω_3. The threshold level is determined by the Raman cross-section and linewidth of the transition and by the focusing parameters of the incident

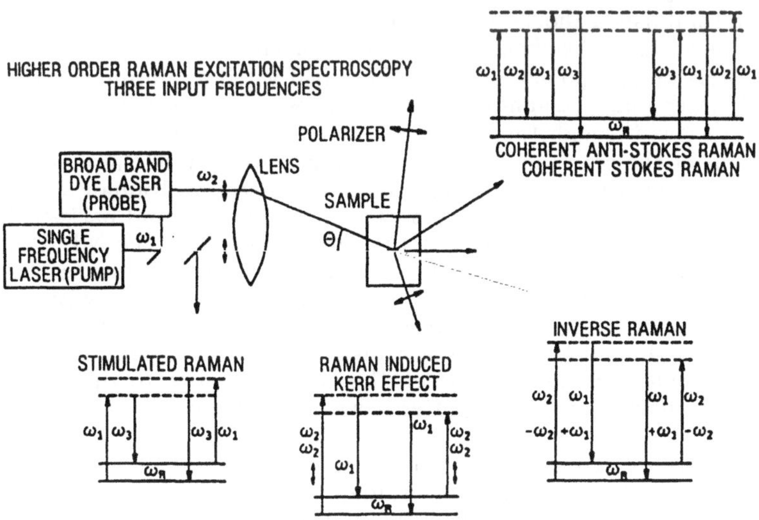

Fig. 1. Coherent Raman scattering techniques.

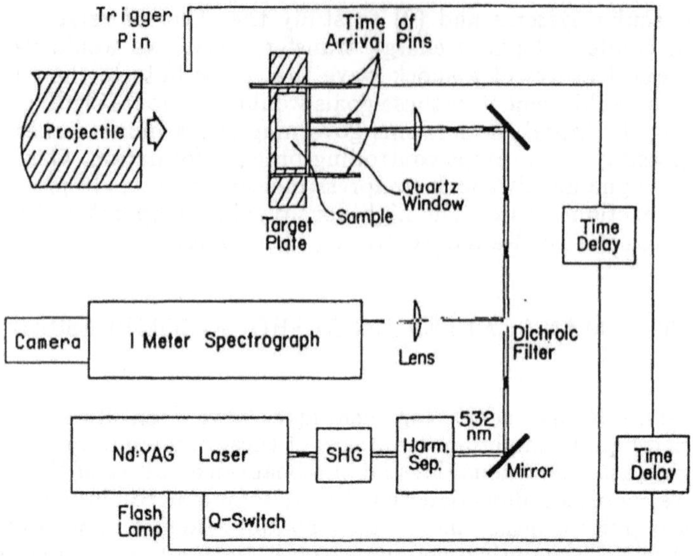

Fig. 2. Schematic representation of backward stimulated Raman-scattering experiment. SHG, second harmonic generator; Harm. Sep., harmonic separator. Sample; liquid benzene.

beam. Typical threshold intensities are ~10–100 GW/cm^2. Figure 2 illustrates the arrangement used for the backward stimulated Raman scattering experiment.

An aluminum projectile of known velocity from a 51-mm-diameter, 3.3-m-long gas gun impacted a 6061 aluminum target plate producing a shock wave, which ran forward into ca. 8-mm-thick benzene sample. Standard data reduction techniques[14] using shock velocities determined from time-of-arrival pins and published shock-velocity/particle-velocity data[15] were used to determine the state of the shock-compressed benzene.

A single 6-ns-long frequency-doubled Nd-doped yttrium aluminum garnet (Nd:YAG) laser pulse (Quanta-Ray DCR1A) was focused using a 150-mm focal length lens through the quartz window to a point in the benzene 2–6 mm in front of the rear sample wall. The high intensity of the laser at the focus, coupled with the presence of a large cross-section Raman active vibrational mode in the sample, produces gain in the forward and backward directions along the beam at a frequency that is different from the Nd:YAG frequency by the frequency of the active mode. The timing sequence was determined by the incoming projectile. Trigger pins, in conjunction with an appropriate time delay, triggered the laser flash lamp approximately 300 μs prior to impact. A time-of-arrival pin activated just before impact and the appropriate time delay served to Q-switch the laser just prior to the shock wave striking the quartz window and after it was past the focal point of the incident laser light.

In liquid benzene, the ν_1 symmetric stretching mode[16] at 992 cm^{-1} has the lowest threshold for stimulated Raman scattering induced by 532-nm light, and was the transition observed in these experiments. As depicted in Fig. 2, the backward stimulated Raman beam was separated from the incident laser by means of a dichroic filter and was then focused onto

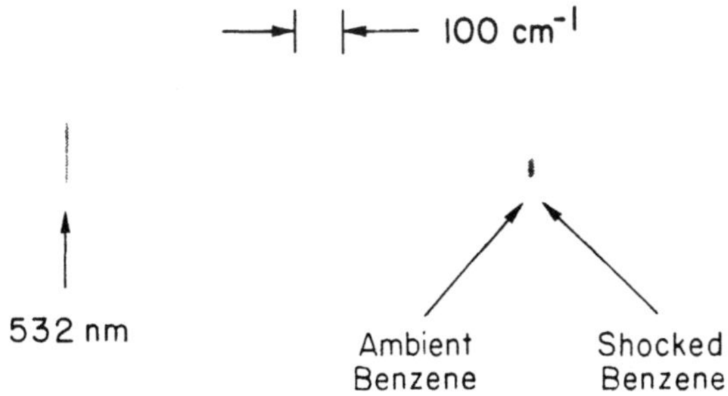

Fig. 3. Scattered light spectrogram for shock-compressed benzene.

the 10-μm-wide entrance slit of a 1-m Czerny-Turner spectrograph equipped with a 1200-grooves/mm grating blazed at 500 nm and used in first order. Figure 3 shows the resulting spectrogram for benzene shock-compressed to 0.92 GPa. The reflected incident laser line and the backward stimulated Brillouin-scattering line at 532 nm are observable, as is the backward stimulated Raman-scattering line from ambient benzene. The latter feature resulted as a consequence of the shock wave having passed only about two-thirds of the way through the sample, and hence a stimulated Raman signal was also obtained from the unshocked liquid.

The frequency shift of the Raman line has small contributions of approximately 0.1 cm^{-1} because the light crosses the moving interface between two media of different refractive indices and because of the material motion behind the shock wave.[17] Since these errors are considerably less than the experimental uncertainty of ± 0.5 cm^{-1} for the measured frequency shifts and are a small fraction of the shift due to compression, no attempt was made to correct for these effects.

The measured shift of the ν_1 ring-stretching mode vibrational wavenumber versus pressure of the shocked benzene has been published elsewhere.[11] Observation of the ring-stretching mode at 1.2 GPa strongly suggests that benzene molecules still exist several millimeters behind the shock wave at this pressure, but does not, however, exclude some decomposition.[18,19,20]

Beam intensities using SRS are sufficiently large that film can be used as a detector. The large incident intensities required, however, can cause damage to optical components near focal points. Spatial and temporal resolution are determined by the confocal parameter of the focusing lens and the incident laser pulse duration. The SRS technique also suffers because only certain molecules produce stimulated Raman scattering and of those molecules only the lowest threshold transition can be observed. Because of these limitations other coherent Raman scattering processes affording more experimental flexibility were attempted.

Raman-induced Kerr effect spectroscopy (RIKES)[21] has been discussed as a diagnostic technique[22,23,24] for performing measurements in shock-compressed systems which may have a large non-resonant background. RIKES requires a single frequency pump beam, ω_1, a broad-band probe source, ω_2, no phase matching and lower incident power levels than stimulated Raman scattering (Fig. 1).

RIKES occurs when a linearly polarized probe laser beam is passed through the rotating electric field of a circularly polarized pump beam. The

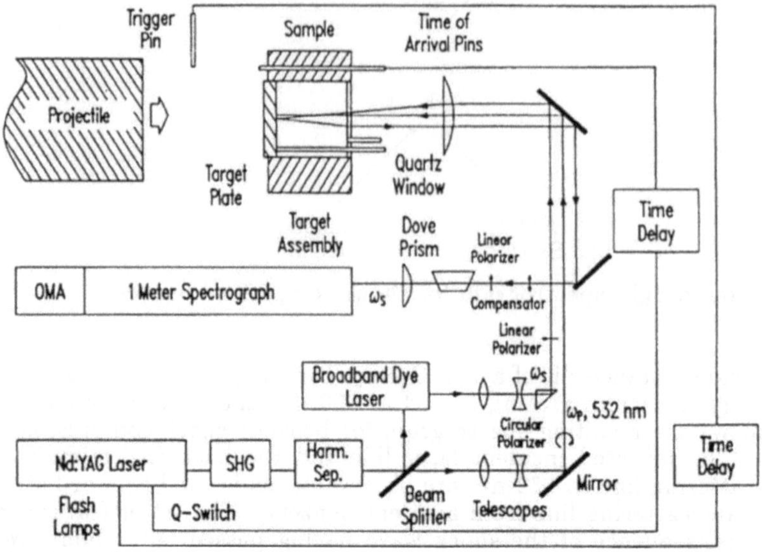

Fig. 4. Schematic representation of the RIKES experiment. SHG, second harmonic generator; Harm. Sep., harmonic separator; OMA, optical multichannel analyzer; sample, benzene.

four-wave parametric process described induces an ellipticity on the probe beam whenever the frequency difference between the two lasers equals that of a Raman active transition in the sample.[25,26]

A schematic of the experimental apparatus used to perform RIKES in shock-compressed benzene is shown in Fig. 4. The shock-compressed benzene sample was obtained using the technique described previously for the SRS experiments. Timing of the 6-ns-long frequency-doubled Nd:YAG laser pulse used for the circularly polarized pump laser beam was also accomplished as described for the SRS experiment. The probe laser beam was obtained by using a portion of the frequency-doubled Nd:YAG laser passed through a Fresnel rhomb to produce a beam of >99% circular polarization. The dye laser beam (Stokes frequencies) is passed through a high-quality Glan-Taylor (air-gap) prism to produce a beam of ~1 part in 10^6 linear polarization. The two beams are focused and crossed in the sample using a 150-mm focal length, 50-mm-diameter lens. The crossing angle is near 6 degrees, giving an overlap length of ~150 μm at the focus. The Stokes beam is then reflected by the highly polished front surface of a 304 stainless steel target plate back through the sample and along a path parallel to the incoming beams. A series of reflectivity measurements showed that polished steel shocked to pressures in excess of 1 Mb and then released, would retain ca. 10 to 40 percent of its original reflectivity. A mirror separated the reflected dye laser beam from the other beams and directed it first through a Babinet-Soleil polarization compensator and then through a Glan-Taylor polarization analyzer. The compensator was found to be necessary to remove the ellipticity introduced into the linearly-polarized Stokes laser beam by the birefringence inherent in the optical components located between the polarizers, including the ambient sample. When the two Glan-Taylor prisms are crossed, the dye laser beam is blocked except at frequencies corresponding to Raman resonances, where

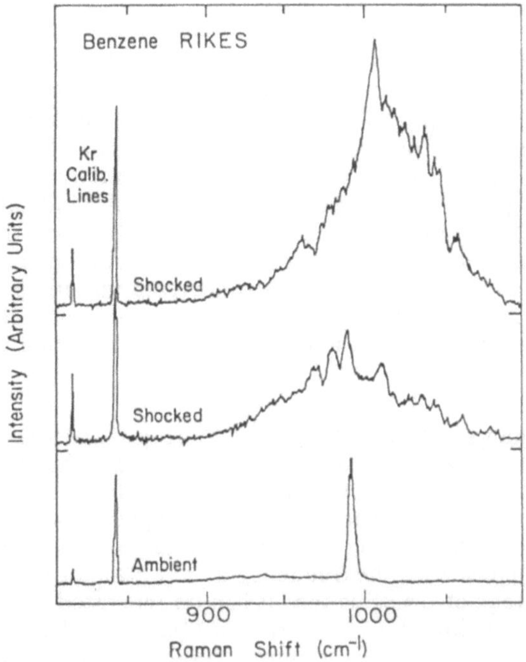

Fig. 5. RIKES of an ambient and two shock-compressed liquid benzene samples. The shock pressure was 1.17 GPa and the 557.03-nm and 556.22-nm Kr calibration lines are shown. All spectra are obtained at the same power levels.

the RIKES signals are passed. These signals are directed through a dove prism and focused into the entrance slit of a 1-m spectrometer equipped with a 1200-grooves/mm grating blazed at 500 nm and used in first order. The signals were detected at the exit of the spectrometer using an intensified diode array (Tracor Northern Model TN-6133) coupled to an optical multichannel analyzer (OMA) (Tracer Northern Model TN-1710). The instrument spectral resolution was approximately 4.2 cm^{-1}.

Figure 5 shows RIKES spectra near 992 cm^{-1} of an ambient and two shock-compressed liquid benzene samples. The shock pressure was 1.17 GPa, and the 557.03-nm and 556.22-nm Kr calibration lines are shown. All spectra are obtained at the same power levels. Both traces have spectral features, however, they are not consistent and do not exhibit the pressure-induced frequency shift expected for the benzene ring stretching mode based on previous SRS and coherent anti-Stokes Raman scattering (CARS) experiments. In a polarization sensitive coherent Raman experiment, such as RIKES, the possibility exists that shock-induced changes in a material could perturb the probe laser polarization sufficiently to obscure the desired signals. Therefore, the sensitivity of the RIKES apparatus to minor rotations of the dye laser polarization was investigated. The figure of merit used was the polarization analyzer rotation angle necessary to saturate the detector with unblocked dye laser. It was found that the detector could be driven from zero signal to saturation with a polarization rotation angle of ~20-arc minutes (20′), using

50-μm slits and ~50-μJ dye laser energy. The RIKES signal found for the ring-stretching mode of ambient liquid benzene nearly saturated the detector through 25-μm slits (using ~200-μJ pump laser energy and 6° beam crossing angle). These data suggest that, if the shock-compressed sample induced a rotation of the probe laser polarization $\geq 20'$, the signal would be masked by the broad-band dye laser background passed by the analyzer. The RIKES spectra (Fig. 5) obtained in shocked samples show only broad-band dye laser which has been passed by the polarization analyzer. These results indicate that the shock-compressed sample induces a rotation of at least 20′ on the dye laser polarization. They also lead to the conclusion that, while it may be possible to perform RIKES experiments in shock-compressed materials in spite of our failure, the experiment is considerably more difficult than techniques not sensitive to the absolute polarization of the laser beam (such as SRS and CARS).

Coherent anti-Stokes Raman scattering (CARS)[25–31] (Fig. 1) occurs as a four-wave parametric process in which three waves, two at a pump frequency, ω_p or ω_1, and one at a Stokes frequency, ω_s or ω_2, are mixed in a sample to produce a coherent beam at the anti-Stokes frequency, ω_{as} or $\omega_3 = 2\omega_p - \omega_s$. The efficiency of this mixing is greatly enhanced if the frequency difference, $\omega_p - \omega_s$, coincides with the frequency ω_j of a Raman active mode of the sample. An advantage of CARS is that it can be generated at incident power levels considerably below those required for stimulated Raman scattering. However, since phase matching is required, possible geometrical arrangements are limited.

A schematic of the experimental apparatus used to perform CARS[32,33,34,35] in shock-compressed benzene, nitromethane, liquid nitrogen, liquid oxygen, liquid carbon monoxide both neat and in mixtures with nitrogen and oxygen, and nitrous oxide, is shown in Fig. 6. For pressures greater than 2 GPa, a two-stage light gas gun was used to accelerate a polycarbonate projectile with 4-mm-thick AZ31B magnesium, 2024 aluminum, or 304 stainless steel impactors to a desired velocity. The target assemblies were of two types, one for ambient liquids (as described previously) and one for cryogenic liquids.[33,36] The cryogenic target assembly was used to condense and hold the liquid N_2, O_2, CO, their mixtures, and N_2O. It consisted of a toroidal cooling chamber, through which liquid N_2 or cold N_2 gas was passed, surrounding a cylindrical sample chamber. The sample chamber included a highly polished 304 stainless steel target plate at the front and a 6.3-mm-diam quartz or lithium fluoride window at the rear. Lithium fluoride was used at higher pressures because of better optical transmission. Impactor and target plate thicknesses were chosen, and time-of-arrival pin assemblies were installed in the ca. 1.5-mm-long liquid sample, so as to insure that rarefaction waves would not compromise the one-dimensional character of the compression in the region observed optically. Sample gases were condensed into the target from standard stainless steel sample cylinders. High purity gases were used, and the mixtures were homogenized at least 24 hours using convection.

A pair of HeNe laser beams interrupted by the projectile and detected by fast photomultiplier tubes provided redundant trigger signals for the flashlamps of an injection-seeded Nd:YAG laser, after appropriate time delays based on the expected projectile velocity. Time-of-arrival pins located in the target were used to trigger the Q-switch on the Nd:YAG laser after an appropriate time delay, as well as to measure the shock velocity in the liquid. The

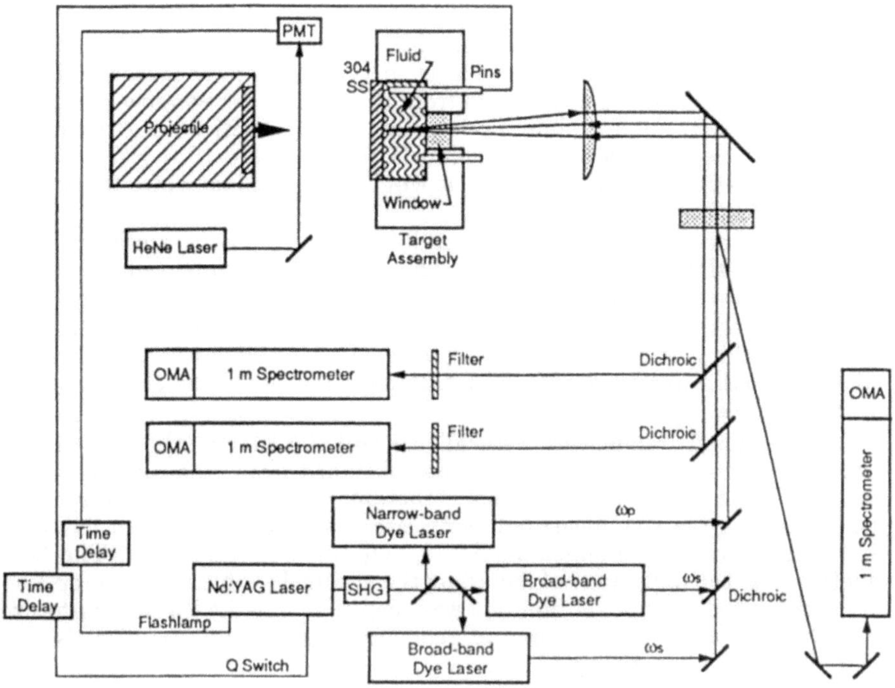

Fig. 6. CARS experiments for shock-compressed materials.

Q-switch time was usually set so that the laser pulses arrived at the target at the time the shock arrived at the window. Because of imprecision in the actual projectile velocity and projectile tilt, a thin layer of either unshocked (ambient) sample or double-shocked sample was present when the lasers arrived at the sample. These situations will be discussed more completely below.

The pump frequency in the CARS process was obtained by using approximately 40% of the 6-ns-long frequency-doubled output of the Nd:YAG laser (Quanta-Ray DCR-1A or Quanta-Ray DCR-11) to pump a narrow-band dye laser (Quanta-Ray PDL-1) at near 557 nm (Rhodamine 590, Exciton) for the nitrogen data, near 582 nm (Kiton Red, Exciton) for the oxygen data, and near 560 nm (Rhodamine 590, Exciton) for the CO and CO/N_2 mixtures. Two broad ranges of Stokes frequencies were produced using homebuilt broad-band dye lasers utilizing the laser dye DCM (Exciton, lasing region 630 to 650 nm) in one and Rhodamine 640 (Exciton) in the other, each pumped by half the remaining Nd:YAG output. The second broad-band dye laser was used for measurements of the O_2 vibration in the mixtures. For benzene and nitromethane samples, the frequency-doubled Nd:YAG laser pulse was used directly as the pump beam.[32] The CARS signals produced in the sample were reflected by the stainless steel target plate back out the window, collimated by the lens, picked off by a dichroic beam splitter, directed through a 6-nm-band-width filter monochromator and then dispersed by a 1-m spectrometer. Multichannel detection of the CARS signals was done using an intensified photodiode array (Tracor Northern 6132) and analyzer (Tracor Northern 6500). If two spectral regions were being studied, a second dichroic beamsplitter was used along with another 1-m spectrometer and an intensified photodiode array (Princeton Instruments IRY-512G) and analyzer (Princeton Instruments ST-120). In addition, because of significant shot-to-shot variation, the broad-band dye laser spectral profile was measured in each experiment by directing a reflection from the target tank window into a third 1-m spectrometer and a photodiode array (EG&G Reticon 512S) and analyzer (DSP Technologies 2012S/LeCroy CAMAC crate/IBM PC-XT). This measured spectral profile was used, as will be described later, for the analysis of each CARS spectrum.

Phase matching was experimentally optimized in the ambient sample for the focusing conditions used. The dispersion in the sample was assumed to linearly scale with the increase in refractive index due to volume compression.[33,34,35] Linear scaling of the dispersion results in the same phase-matching angle at all compressions. The focusing was also chosen so that the spatial region producing CARS signals covered the entire thickness of the sample.

Vibrational frequencies were all calibrated ($\pm 1\ cm^{-1}$) using vacuum wave numbers of atomic emission lines obtained from standard calibration lamps. The spectral instrument function of the CARS spectrometer/ photodiode array was measured using either an atomic emission line or by extraction from the ambient liquid N_2, O_2, or CO transition. The latter method has the advantage of including the spectral profile of the pump laser. This measured instrument function was then convoluted with the synthesized CARS spectra discussed below, to give spectra that could be directly compared with the experimental data.

The intensity of the beam at ω_{as} is given by the semiclassical description:

$$I_{as} \propto \sum_i \frac{\omega_{as}^2 I_p^2 I_s (N_i L_i)^2 |\chi_i^3|^2}{n_p^2 n_s n_{as}} \times \left(\frac{n_{as}^2+2}{3}\right)\left(\frac{n_s^2+2}{3}\right)\left(\frac{n_p^2+2}{3}\right)^4 , \quad (1)$$

where I_p and I_s are the incident intensities of the pump and Stokes beams, respectively, and n_{as}, n_s, and n_p are the refractive indices at ω_{as}, ω_s, and ω_p, respectively.[33,34,35] $N_i L_i$ corresponds to the Lagrangian density of the ith layer and the sum is over noninterfering layers. χ_i^3, the third order susceptibility of the ith layer is given by

$$|\chi_i^3|^2 \propto \left(\sum_j \frac{\Gamma_j \chi_j^{pk}(\omega_j - \omega_p + \omega_s)}{(\omega_j - \omega_p + \omega_s)^2 + \Gamma_j^2} + \chi^{nr}\right)^2 + \left(\sum_j \frac{\Gamma_j^2 \chi_j^{pk}}{(\omega_j - \omega_p + \omega_s)^2 + \Gamma_j^2}\right)^2 , \quad (2)$$

where χ_j^{pk} is the peak third-order susceptibility, χ^{nr} is the nonresonant susceptibility, Γ_j is the HWHM linewidth and the sum on j is over vibrational transitions. This equation only holds in the case of no electronic resonance enhancement.[30] The resultant spectra are then convoluted with the ca. 3-cm^{-1} slit function.

Figure 7 shows a series of CARS spectra of shock-compressed liquid N_2. These spectra are illustrative of the kinds of spectra obtained in these exper-

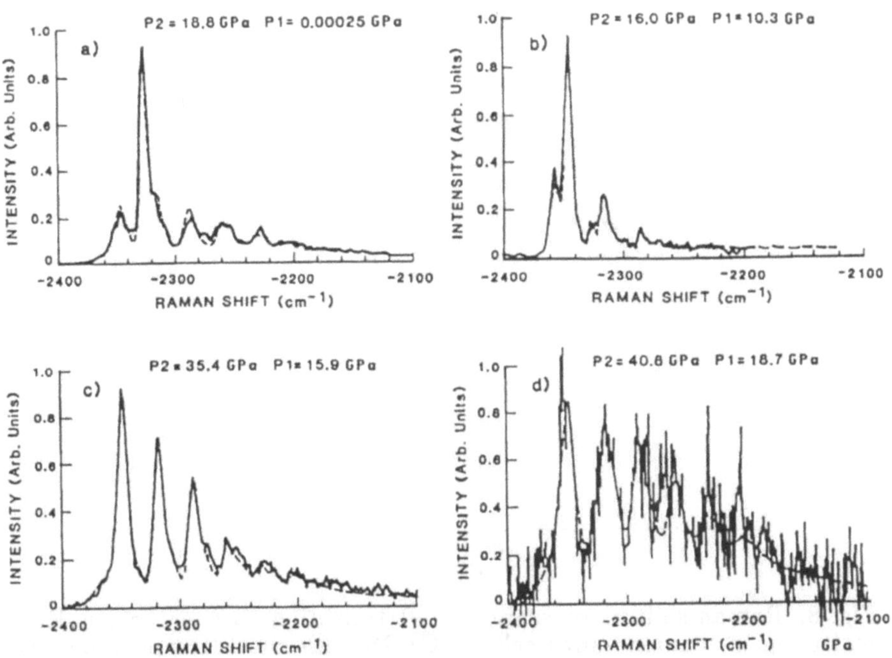

Fig. 7. Representative CARS spectra of shock-compressed N_2. (a) is representative of CARS spectra recorded when the shock has not reached the window and contains contributions from ambient (pressure P_1) and single-shocked (pressure P_2) N_2. (b), (c), and (d) are CARS spectra recorded after the shock has hit the window, and contain contributions from single (pressure P_1) and double-shocked (pressure P_2) N_2. The solid curves are the experimental data and the dashed curves are computed synthetic spectra.

iments. For experiments where the shock wave had not reached the window, spectra similar to Fig. 7a were obtained. The large peak at 2327 cm^{-1} is the CARS signal from unshocked nitrogen and the remaining progression of lines are the fundamental and hot band transitions from the single-shocked fluid. Because the unshocked liquid has such a narrow linewidth[37] compared to the width of several wave numbers for the shock-compressed fluid, two difficulties were found. At the laser intensities used to produce CARS in the shock-compressed fluid, the CARS process could either be easily saturated leading to an increase in the apparent linewidth of the ambient liquid, or could result in large enough signals from the ambient liquid to locally saturate the detector.

If the shock wave reached and reflected from the rear window, both the single and double-shocked regions in the sample were interrogated by the incident laser beams. The resultant spectra, similar to those depicted in Figs. 7b–7d, consisted of two partially overlapped progressions of transitions arising from the two interrogated regions. Figure 7b illustrates the case for which the lines have not broadened sufficiently to obscure the individual peaks of the two progressions. At higher shock pressures and temperatures, the lines broaden considerably (Fig. 7c) and it is difficult to distinguish the two progressions without comparison to computed spectra. Figure 7d is sim-

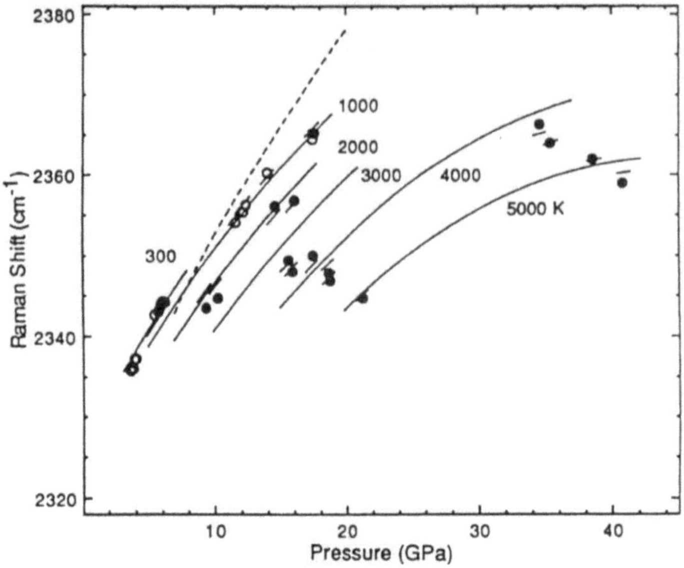

Fig. 8. Raman shift vs pressure for 0–1 transition. o–spontaneous Raman scattering in diamond-anvil cell. •–CARS with shock compression. Solid lines are isotherms obtained using an empirical fit to the data.[38,39] Dotted line is the 2000-K isotherm from Belak et al.[41]

ilar to 7c except with much poorer signal-to-noise ratio. Also shown in Fig. 7 as the dashed lines are the computed spectra using Eqs. (1) and (2). The measured broad-band dye laser spectral profiles noted above were used for I_s in these computations. Figure 8 shows for the 0–1 fundamental transition of N_2 the measured frequencies and by a small line segment the frequencies calculated using an empirical fit[38,39] at the corresponding experimental pressures and temperatures. The values at the lower pressures and temperatures were obtained from spontaneous Raman scattering measurements in a high-pressure diamond anvil cell.[40] The CARS data in the 10 to 20-GPa pressure range and the 2340 to 2360-cm^{-1} frequency region are from single-shock experiments, while the remaining CARS data were obtained in double-shocked liquid N_2. In the single-shocked material, there is a monotonic increase of the vibrational frequency with increasing pressure up to a pressure of ca. 17.5 GPa. Above this pressure the frequency no longer increases, and appears to begin to decrease. The vibrational frequency in the double-shocked material (whose temperature is lower than single-shocked material at the same pressure) shows similar behavior, but the reversal occurs at higher pressures. It is interesting to note the effect of temperature in these data. When the fluid is single or double-shocked to the same density, the difference in measured Raman shift is due to the effects of temperature on the potential and on the portion of the potential sampled on average. Within the precision of the data presented, the anharmonicity of the intramolecular potential appears to be constant for all pressures and is the same as that expected from gas-phase data.[34]

Also shown for comparison in Fig. 8 are the Raman frequency shifts along a 2000-K isotherm from an empirical approximation to Monte Carlo

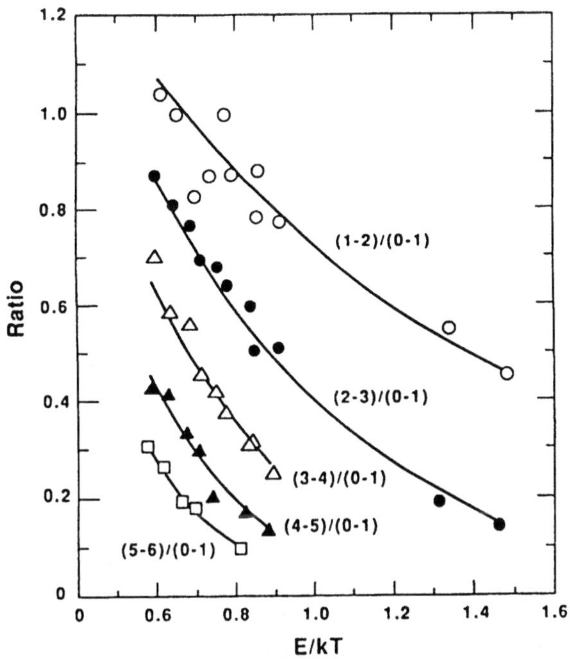

Fig. 9. The ratios of the left-hand side of Eq. (3) for the observed hot band transitions to that of the fundamental transition (plotted as the symbols) vs E/kT_{eos}, where E is the energy of the transition. Also shown as the lines in the figure are the expected ratios assuming Boltzmann population distributions (using T_{vib}) and the harmonic-oscillator-approximation dependence of Raman cross section on vibrational level.

calculations by Belak et al.[41] As pointed out by the authors, the Monte Carlo calculations give frequencies that are slightly larger than experimental results, with the discrepancy becoming greater for increasing pressure.

The experimentally determined linewidths and peak Raman susceptibilities of the fundamental and hot band transitions were used to estimate the vibrational relaxation time and to determine a vibrational temperature. The relationship between CARS intensities and the spontaneous Raman cross section and vibrational-level populations can be approximated by[30]

$$\Gamma_j \chi_j^{pk} \frac{h}{2\pi c^4} \omega_p \omega_s^3 = \left(\frac{d\sigma}{d\Omega}\right)_j (\rho_j - \rho_k) \quad , \tag{3}$$

where h is Planck's constant, c is the speed of light, $(d\sigma/d\Omega)_j$ is the spontaneous Raman cross section of the j to k vibrational transition, and ρ_j and ρ_k are the number densities in vibrational levels j and k, respectively. Ratios of Eq. (3) for the excited-state transitions to the fundamental transition were used to explore the possibility of a non-Boltzmann population distribution and a decreased vibrational relaxation time at the pressures and temperatures shown in Fig. 8. The right-hand side ratios were calculated using the harmonic oscillator approximation for the dependence of the Raman cross section on vibrational level, $(d\sigma/d\Omega)_j \propto (j+1)$, and assuming Boltzmann

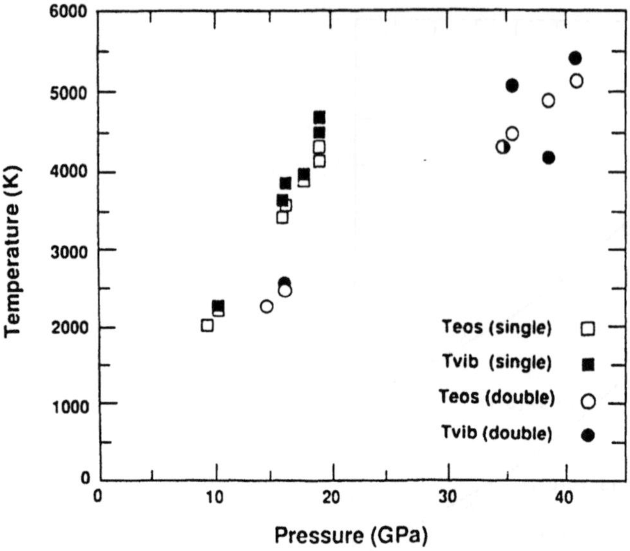

Fig. 10. Equation-of-state temperatures and vibrational temperatures (extracted from the computed synthetic spectra) vs shock pressure. The lowest pressure single and double shock points have no T_{vib} because those data have insufficient signal/noise to observe excited state transitions.

population distributions, with vibrational temperature T_{vib}, for ρ_j and ρ_k. In Fig. 9, these results are presented vs E/kT_{vib} for several transitions. The energy, E, of the vibrational transition is slowly varying, hence the abscissa depends primarily on the temperature of the shocked fluid. Similar ratios of the left-hand side of Eq. 3 using the susceptibilities and linewidths used to generate the synthetic spectra[34] are also depicted in Fig. 9. Within the previously stated errors it is evident that the experimental parameters, and consequently the spectra, are adequately represented by a simple model based on Boltzmann equilibrium of the vibrational levels. This result implies that energy has been transferred from the bulk translational motion into the vibrational energy levels in a time less than or comparable to the characteristic time of the shock-compression experiment. For the highest pressure experiments, this time is 10 ns or less, a change of ten orders of magnitude from the ambient vibrational relaxation time of ca. 60 s.[42] Because the vibrational populations appear to be in equilibrium, they can be used as a measure of the vibrational temperature. Figure 10 shows the equation-of-state temperature and vibrational temperature vs shock pressure for both single and double-shocked N_2. The good agreement between the measured and calculated temperatures lends strong support to the model used to obtain the equation-of-state temperatures.[43]

In Fig. 11, the Raman linewidths[34] used to obtain the spectral fits of the shock-compressed fluid nitrogen spectra are presented vs temperature. Note that Eqs. (1) and (2) used to calculate the CARS spectra use Lorentzian profiles to represent the spectral lines shapes. While this choice provides an adequate representation of our measured CARS spectra, it does not preclude the possibility of inhomogeneous broadening (i.e., Gaussian contribution to the line shapes). This effect would appear predominately in the line wings where the accuracy of our data is insufficient to distinguish between shapes.

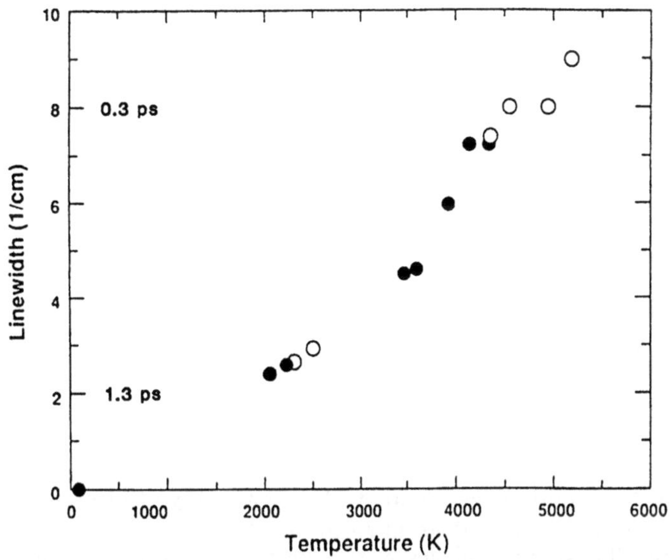

Fig. 11. Raman linewidths extracted from the synthetic spectral fits to the fluid N_2 CARS spectra plotted against shock temperature. The solid and open circles represent single and double-shock data, respectively. The dephasing times noted on the ordinate assume homogeneous broading.

In addition, the limited signal/noise does not permit the observation of line shape differences between the fundamental transition and the observed hot bands. Lorentzian profiles were chosen because they are generally a good first approximation far from the critical region.[44,45]

Two statements can be made about the spectral linewidth data. First, the vibrational dephasing time T_2 (where $T_2^{-1} = 2\pi c\Gamma_j$) has decreased from tens of ps at liquid and near-critical densities[45,46] to about 0.3 ps near 40 GPa and 5000 K. Collapse of the Q branch from motional narrowing[47,48] is probably complete at the pressures and temperatures investigated here, and the line broadening probably results only from pure dephasing. (However, it must be noted that this notion may not be entirely correct because other lower density studies[49] have shown that vibrational dephasing may not be separable from energy relaxation processes.)

The second statement to be made about Fig. 11 is that as the shock pressure increases along the Hugoniot, the spectral linewidth appears to almost linearly increase with temperature.[34] A dependence (within experimental error) on density is not readily apparent. For example, the single and double-shock data appear to fall on the same line. At the shock densities investigated here, the steep repulsive core of the intermolecular potential is being sampled. Consequently, the change in density with increasing shock pressure is much less than the change in temperature. This fact, coupled with the observed weak density dependence of the linewidth, produces the result observed in Fig. 11.

Visible absorption bands have been observed in both gaseous and liquid oxygen by several investigators.[35,50–58] The presence of these proximity (collision) induced absorption bands in the spectral region where CARS experi-

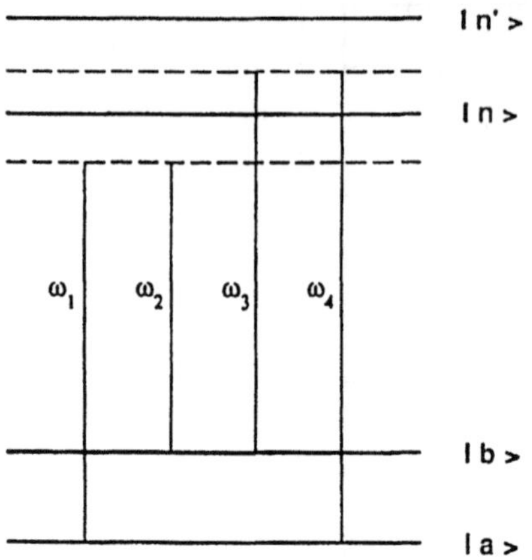

Fig. 12. Electronic, vibrational, and laser translation frequencies.

ments were performed for liquid oxygen leads to the possibility of electronic resonantly enhanced CARS,[59] as well as absorption of the pump, Stokes, and anti-Stokes beams. These possibilities were investigated by incorporating them into the CARS calculations of synthetic spectra using Eq. (1) and an empirical model for the absorption. The empirical model included the following simultaneous transitions: (a) $2\ {}^3\sum_g^-(\nu''=0) \rightarrow 2\ {}^1\Delta_g(\nu'=0)$; (b) $2\ {}^3\sum_g^-(\nu''=0) \rightarrow {}^1\Delta_g(\nu'=1) + {}^1\Delta_g(\nu'=0)$; and (c) ${}^3\sum_g^-(\nu''=1) + {}^3\sum_g^-(\nu''=0) \rightarrow 2\ {}^1\Delta_g(\nu'=0)$ as well as the contribution from the high pressure UV bands: (d) ${}^3\sum_g^- \rightarrow {}^3\Delta_u$.

For electronic resonantly enhanced CARS, the third-order susceptibility for a single-vibrational transition ω_{ba} of the ith layer can be written[30,59–61]

$$\chi_i^3 \propto \chi^{pk} \sum_{a,b,n,n'} (A - B_1 + B_2 - B_3) + \chi^{nr}$$

with

$$\begin{aligned}
A &= \rho_{aa}^{(0)}\left[(\omega_{ba} - \omega_1 + \omega_2 - i\Gamma_{ba})(\omega_{na} - \omega_1 - i\Gamma_{na})(\omega_{n'a} - \omega_4 - i\Gamma_{n',a})\right]^{-1} \\
B_1 &= \rho_{bb}^{(0)}\left[(\omega_{ba} - \omega_1 + \omega_2 - i\Gamma_{ba})(\omega_{nb} - \omega_2 + i\Gamma_{nb})(\omega_{n'a} - \omega_4 - i\Gamma_{n',a})\right]^{-1} \\
B_2 &= \rho_{bb}^{(0)}\left[(\omega_{n'n} - \omega_3 + \omega_2 - i\Gamma_{n'n})(\omega_{n'b} - \omega_3 - i\Gamma_{n'b})(\omega_{n'a} - \omega_4 - i\Gamma_{n'a})\right]^{-1} \\
B_3 &= \rho_{bb}^{(0)}\left[(\omega_{n'n} - \omega_3 + \omega_2 - i\Gamma_{n'n})(\omega_{nb} - \omega_2 + i\Gamma_{nb})(\omega_{n'a} - \omega_4 - i\Gamma_{n'a})\right]^{-1}
\end{aligned} \tag{4}$$

where χ^{pk} is the peak susceptibility, and $\rho_{aa}^{(0)}$ and $\rho_{bb}^{(0)}$ are the initial number densities in the states a and b, respectively. The transition frequencies are as shown in Fig. 12, Γ_{ba} and $\Gamma_{n'n}$ are the vibrational half-width, at half-maximum (HWHM) linewidths, and the remaining Γ's are the electronic transition linewidths. Druet et al.[59] have suggested that for a broad absorption continuum such as described above, the resonantly enhanced third-order susceptibility can be described by Eq. (4) using a broad Lorentzian for the excited electronic state. For our experiment $\omega_1 = \omega_3 = \omega_p$, $\omega_2 = \omega_s$, $\omega_4 = \omega_{as}$. Synthetic spectra were calculated using Eqs. (1) and (4) in conjunction with

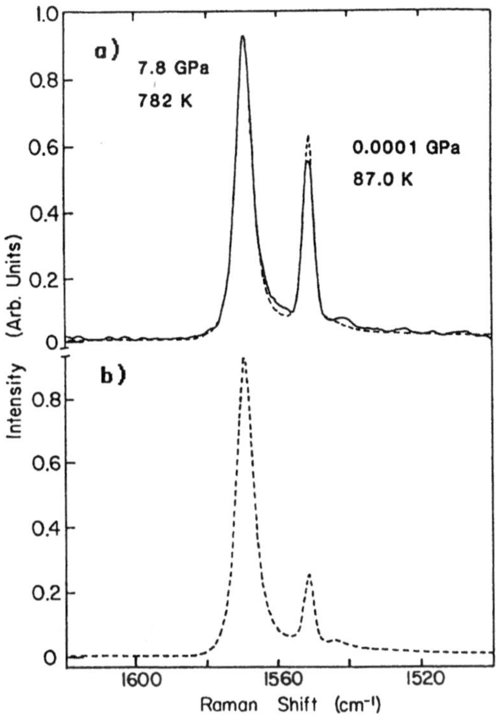

Fig. 13. Representative CARS spectrum: (a) — experimental data at 7.8 GPa and 782 K, --- synthetic spectrum including effects of absorption, (b) synthetic spectrum neglecting effects of absorption.

a broad 1400-cm^{-1} wide absorption band centered at 17,700 cm^{-1} to simulate the proximity induced absorption, taking into account absorption of ω_p, ω_s, and ω_{as} for a range of values for χ^{pk}/χ^{nr}, transition frequencies, and linewidths.[35] Attempts to favorably compare these results with experimental spectra were unsuccessful. The calculated spectra were always distorted giving too large a scattering intensity at Raman shifts greater than the vibrational transition frequency and insufficient intensity at lower energies. A qualitative examination of the intensities expected based on the relative locations of the absorption bands and the laser frequencies indicates the correct intensity distribution was calculated if resonances were occurring. From this lack of agreement, we concluded that even though ω_p, ω_s, and ω_{as} are being attenuated by proximity induced dipole transitions, an electronic resonance enhancement of the CARS signal is not occurring.[35] Based on this finding, we calculated synthetic CARS spectra for O_2 as was done previously for N_2 using Eqs. (1) and (2). A sample spectrum and fit[35] are shown in Fig. 13a. The prominent peak results from the 0–1 transition of the shock-compressed oxygen and the smaller feature to the right is the 0–1 transition from a thin layer of unshocked material. Because the temperatures obtained in these experiments were too low to produce hot-band transitions of sufficient intensity to be readily visible, estimates of vibrational temperatures were not possible. Future work using CARS data from shock-compressed mixtures of O_2 with N_2 and CO will, because of increased temperatures, allow estimates of vibrational temperatures. Also shown in Fig. 13b is a spectrum calcu-

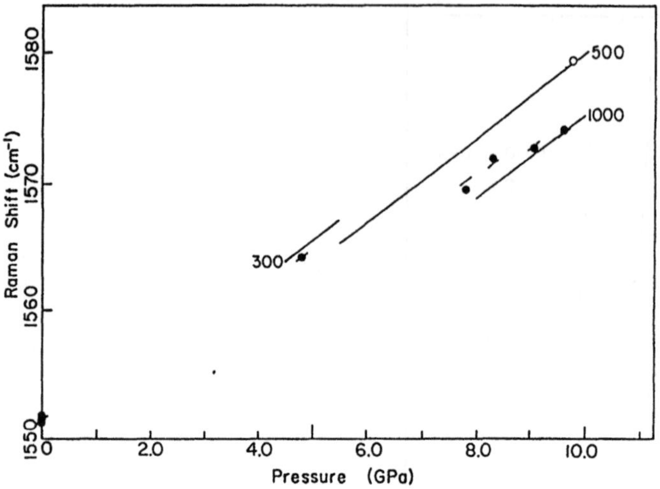

Fig. 14. Raman shift vs pressure. • O_2, single shock; ∘ O_2, double shock; — isotherms.

lated using the same parameters as for the synthetic spectrum depicted by Fig. 13a, but leaving out the effects of absorption. The large change in the ratio of peak intensities demonstrates the importance of including absorption for good simulation of the experimental data.

Figure 14 shows the measured Raman shifts versus pressure for shock states of liquid O_2. There is a monotonic increase of the vibrational frequency with increasing pressure. This is in contrast to the behavior observed in nitrogen,[34,38,39] where the frequency first increases with pressure, then reverses and begins to decrease with further increases in pressure. However, the maximum pressure of these experiments is considerably less than the pressure, ~ 17.5 GPa, where the frequency reversal occurred for nitrogen. When the fluid is single or double-shocked to the same pressure, the difference in the measured Raman shift results from the effects of temperature on the potential and on the portion of the potential sampled on average.[34,35] An empirical fit was obtained using the measured Raman frequency shifts for ambient, single-shocked and double-shocked oxygen.[35] The short segment near each data point in Fig. 14 gives the calculated frequency value from the empirical fit at the measured pressure and temperature and the long curves show the positions of the 300, 500, and 1000-K isotherms. Again it should be noted that Eq. (2) used to calculate the CARS spectra utilizes Lorentzian profiles to represent the spectral line shapes. While this choice provides an adequate representation of our measured CARS spectra, it does not preclude the possibility of inhomogeneous broadening (i.e., Gaussian contribution to the line shapes). This effect would appear predominately in the line wings where the accuracy of our data is insufficient to distinguish between shapes. At lower pressures the liquid oxygen line shape may be influenced by motional narrowing and fluctuations near the critical point.[62,63] Assuming that the broadening results from pure dephasing, the vibrational dephasing time $T_2/2$ (where $T_2^{-1} = 2\pi c\Gamma_j$) decreased from a few tens of ps at ambient conditions to about 1 ps at the highest pressure shock conditions.

CARS spectra obtained from shock-compressed liquid CO appear to show a small vibrational frequency increase with pressure compared with

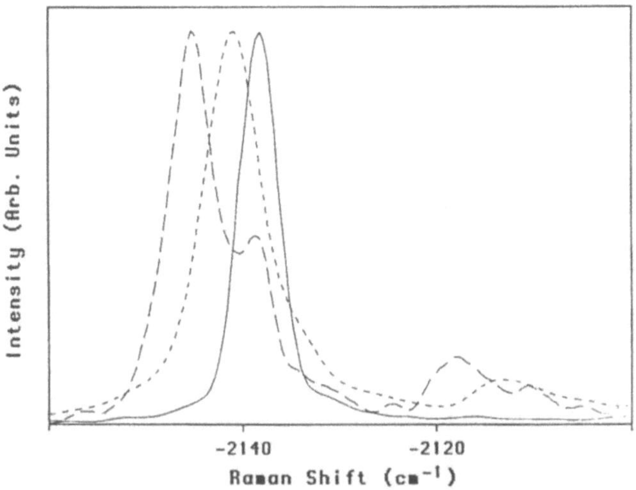

Fig. 15. CARS spectra of CO in neat CO (dotted line) and in a 1/4:CO/N_2 mixture (dashed line), shocked to near 7 GPa. The feature in the mixture at ca. 2137 cm^{-1} is from a small amount of unshocked material. The features between 2110 and 2120 cm^{-1} result from the first hot-band transition. The unshocked CO spectrum (solid line) is shown for reference.

N_2, which is iso-electronic with CO. N_2 has very similar density (0.793 vs 0.808) and essentially the same Hugoniot (locus of shock states) up to 10 GPa. Yet, the vibrational frequency shift with shock pressure along the Hugoniot in the two materials appears to be quite different. We have also obtained CARS spectra of mixtures of CO in N_2 and CO in O_2. Figure 15 shows CARS spectra of unshocked CO, neat CO single-shocked to about 7 GPa, and a 20% mixture of CO in N_2 shock compressed to a similar pressure. There is a clear difference in vibrational frequency shift of the CO in the shocked mixture versus that in the shocked neat material. Similar measurements with N_2 show that its vibrational frequency appears to show no such variation from the neat material to the mixture.

These observations suggest a fundamental difference in the intermolecular interactions in CO versus N_2. One possibility is dipolar interaction, which has been used to explain the larger Raman linewidth of pure liquid CO.[64] Differences in the CO vibrational frequency shift in neat CO versus CO/O_2 mixtures were also observed. However, the temperature is also substantially different in these mixtures. Double-shock experiments, not shown here, in neat CO imply that there is a temperature dependence of the CO vibrational frequency similar to what was found in N_2. Complete analyses of the CARS spectra of the mixtures will be used to help elucidate this phenomenology, estimate relaxation times, and extract vibrational temperatures for comparison to equation-of-state calculations. The vibrational frequency information, along with the vibrational temperatures, will be used to help refine mixture equation-of-state theory, particularly by providing molecular-level details of the effects of the cross-potentials.

For a polyatomic molecule such as nitromethane, the vibrational spectroscopy is considerably more complicated than that discussed above for diatomic molecules. At low temperatures, only the vibrational fundamentals

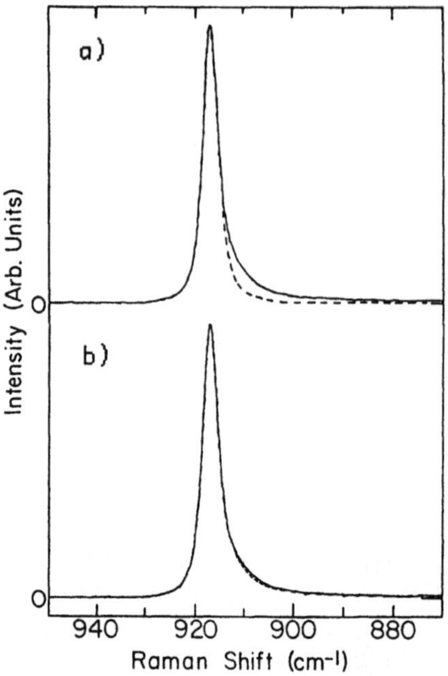

Fig. 16. CARS experimental spectrum of ambient nitromethane CN-stretching mode. a) Synthetic spectrum (dashed) includes no hot bands; b) synthetic spectrum (dashed) includes contribution from hot bands.

need be considered, leading to much simplification in the spectra. In the case of nitromethane, there are 15 possible vibrational normal modes,[65,66] one of which is the nearly free rotation of the methyl group around the CN bond. As the temperature is increased, even just to room temperature, the low frequency modes start to become thermally populated to a measurable extent. The measured band contour is not just a collisionally-broadened single vibrational frequency, but contains contributions from hot bands due to thermal excitation of the level being probed, as well as thermal excitation of all other modes. This situation would still be completely tractable if all the anharmonicity constants were known. In the case of nitromethane, however, we could find only one reliable anharmonicity constant involving the vibration of interest in this work, namely the CN stretching mode, ν_4.[65]

In order to interpret the CARS spectra of shock-compressed nitromethane correctly, a model[67,68] of the vibrational band contour was developed that includes the contributions of the hot bands. The model was used to calculate the positions and intensities of the hot bands of the CN stretching mode for ambient nitromethane and several shock-compressed pressure and temperature states of nitromethane using Eqs. (1) and (2). Figure 16 shows the improved fit of the synthetic CARS spectrum when including the hot bands for ambient pressure and temperature nitromethane. The low frequency side of the peak is not satisfactorily fit without inclusion of the available hot bands.

CARS spectra of the ν_4 mode at several pressures in shock-compressed nitromethane have been previously reported,[69–71] as has a preliminary estimate of the ν_4 vibrational frequency versus shock pressure.[71] The spectral

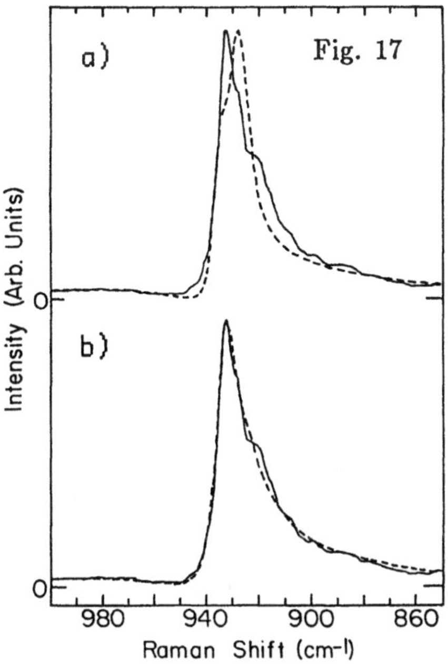

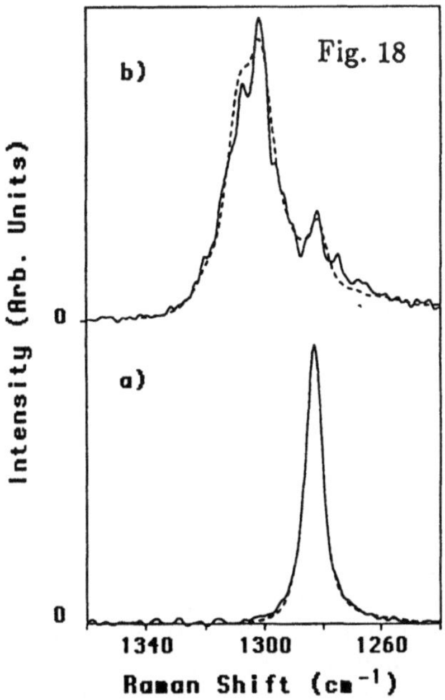

Fig. 17. CARS experimental spectrum of the CN-stretching mode of liquid nitromethane shock compressed to ca. 33 GPa. a) Synthetic spectrum (dashed) with all hot bands and using model; b) synthetic spectrum (dashed) with combination hot bands set to zero.

Fig. 18. CARS spectra of a) unshocked N_2O and b) shocked N_2O. The solid lines are experimental data and the dashed lines are spectral simulations.

fit of the shock-compressed nitromethane CARS spectrum in Fig. 17a is that given by the model with the assumption that the anharmonic coefficients do not change with pressure. The linewidth of the transition was the only variable adjusted for the fit shown. There is something obviously wrong with this model when it is applied to the shock-compressed material, even at modest pressures and temperatures.

The discrepancy probably arises because of changes in the anharmonicity coefficients with pressure, contrary to assumption. Also, because the ν_4 vibrational frequency increases with pressure, indicating a stiffening of the potential for that motion, it might become more resilient to the presence of other thermally populated vibrational motions. The consequence of this picture would be that the off-diagonal anharmonic coefficients would be comparatively smaller. Figure 17b shows a synthetic CARS spectrum assuming the off-diagonal anharmonicity coefficients are zero and only the diagonal terms and Γ are varied for the best fit. The quality of the fit in Fig. 17b is clearly better than that of 17a. There is obviously more work to do to understand these spectra.

Figure 18 shows spectra of the ν_1 mode of unshocked (0.0003 GPa and 191 K) N_2O and N_2O shocked to approximately 4.3-GPa pressure and 735-K temperature. Because the temperature of the unshocked liquid is low, the data can be fit with a synthetic spectrum comprised of only the fundamental transition. Inclusion of hot bands as was done for the ν_4 mode of CH_3NO_2 is not necessary. However, a preliminary fit of the spectra for shock-compressed N_2O suggests inclusion of both the first excited state transition of the ν_1 mode and also transitions originating from the excited states of the ν_2 bending mode. The energy level separations used in this initial fit were the gas phase values given by Herzberg.[72] The spectral structure resulting from this scheme is readily visible in Fig. 18b. The shoulder to the left of the highest peak is thought to be the shock shifted 0–1 transition. The large peak is the result of the fundamental 1–2 transition and the hot band originating from the first excited state of the ν_2 bending mode. Analyses of this and other high-pressure/high-temperature liquid N_2O spectra are in progress and will be reported in the near future.

3. SUMMARY

Fundamental understanding of the detailed microscopic phenomenology of shock-induced chemical reaction and detonation waves is being sought by using pulsed coherent optical scattering experiments to determine the molecular structure, constituents, and energy transfer mechanisms in shock-compressed, high-pressure/high-temperature fluids. To date we have confirmed that N_2, O_2, CO, N_2O, CH_3NO_2, and C_6H_6 still exist as molecules on the nanosecond time scale behind the shock front, but that energy transfer is occurring from the translational degrees of freedom into the vibrational modes. For N_2, O_2, and CO, the vibrational relaxation times have decreased to less than a few nanoseconds in the shock-compressed state. For nitrogen, this is about ten orders of magnitude rate increase from ambient conditions. For N_2 (work is in progress with O_2 and CO), the vibrational population distribution appears Boltzmann-like and gives temperatures that agree satisfactorily with those obtained from equation-of-state calculations. We have experimentally shown the effect of the intermolecular potential (reflecting the thermodynamic environment of the shock-compressed state) on the intramolecular potential through frequency shift measurements. These results show differences for N_2 and CO behavior previously not elucidated by shock Hugoniot data. Linewidth data have been used, assuming homogeneous line broadening, to provide an estimate of dephasing times. For N_2 and O_2 these have decreased to less than one picosecond at the highest pressures.

Also we have shown that the dynamic shock-compression technique can be used to produce a high-pressure/high-temperature dense fluid, which can then be interrogated at the molecular level using short pulsed coherent Raman scattering techniques.

REFERENCES

[1] C. A. Forest, "Burning and Detonation," LA-7245 (Los Alamos National Laboratory Report, Los Alamos, New Mexico 1978).

[2] C. L. Mader, *Numerical Modeling of Detonation* (University of California Press, Berkeley, California 1979).

[3] E. L. Lee and C. M. Tarver, Phys. Fluids **23**, 2362 (1980).

[4] J. Wackerle, R. L. Rabie, M. J. Ginsberg and A. B. Anderson in *Proceedings of the Symposium on High Dynamic Pressures* (Commissariat à l'Energie Atomique, Paris, France 1978) p. 127.

[5] M. Cowperthwaite in *Proceedings of the Symposium on High Dynamic Pressures* (Commissariat à l'Energie Atomique, Paris, France 1978), p. 201.

[6] J. W. Nunizato in *Shock Waves in Condensed Matter—1983*, J. R. Asay, R. A. Graham, and G. K. Straub, eds. (Elsevier Science Publishers, Amsterdam, 1984) p. 293.

[7] J. W. Nunizato and E. K. Walsh, Arch. Rational Mech. Anal. **73**, 285 (1980).

[8] J. N. Johnson, P. K. Tang and C. A. Forest, J. Appl. Phys. **57**, 4323 (1985).

[9] P. K. Tang, J. N. Johnson, and C. A. Forest in *Proc. 8th Symp. Detonation* (Albuquerque, New Mexico 1985), p. 375.

[10] C. Mader and J. Kerschner in *Proc. 8th Symp. Detonation* (Albuquerque, New Mexico, 1985) p. 366.

[11] S. C. Schmidt, D. S. Moore, D. Schiferl, and J. W. Shaner, Phys. Rev. Lett. **50**, 661 (1983).

[12] M. Maier, W. Kaiser, and J. A. Giordmaine, Phys. Rev. **177**, 580 (1969).

[13] D.V.J. Linde, M. Maier, and W. Kaiser, Phys. Rev. **178**, 178 (1969).

[14] M. H. Rice, R. G. McQueen, and J. M. Walsh, *Solid State Physics 6* (Academic Press, New York 1958) p. 1.

[15] R. D. Dick, J. Chem. Phys. **57**, 6021 (1970).

[16] W. D. Ellenson and M. Nicol, J. Chem. Phys. **61**, 1380 (1974), this mode is called ν_2 in G. Herzberg, *Infrared and Raman Spectra* (Van Nostrand Reinhold, New York 1968).

[17] R. N. Keeler, G. H. Bloom and A. C. Mitchell, Phys. Rev. Lett. **17**, 852 (1966).

[18] A. N. Dremin and V. Yu. Klimenko, "On the Role of the Shock Wave Front in Organic Substances Decomposition," Gas Dynamics of Explosions and Reactive Systems, Minsk, USSR, 1981.

[19] A. N. Dremin and L. V. Barbare in *Shock Waves in Condensed Matter—1981*, Am. Inst. Phys. Proc. 78, W. S. Nellis, L. Seaman, and R. A. Graham eds. (New York, 1983), p. 270.

[20] L. V. Barbare, A. N. Dremin, S. V. Pershin, and V. V. Yakovlev, Fiz. Gor. i Var **5**, No. 4, 528 (1969).

[21] D. Heiman, R. W. Hellworth, M. D. Levenson, and G. Martin, Phys. Rev. Lett. **36**, 189 (1976).

[22] S. C. Schmidt, D. S. Moore, and J. W. Shaner in *Shock Waves in Condensed Matter—1983*, J. R. Asay, R. A. Graham, and G. K. Straub, eds. (Elsevier Science Publishers, Amsterdam, 1984) p. 293.

[23] D. S. Moore, S. C. Schmidt, D. Schiferl, and J. W. Shaner in *High Pressure in Science and Technology, Part II*, C. Homan, R. K. MacCrone and E. Whalley, eds. (North-Holland Publishing, New York, 1984) p. 87.

[24] W. G. VonHolle and R. A. McWilliams in *Laser Probes for Combustion Chemistry (American Chemical Society Symposium Series 134)*, D. R. Crosley, ed. (American Chemical Society, Washington, DC 1983), p. 319.

[25] G. L. Eesley, *Coherent Raman Spectroscopy* (Pergamon Press, Oxford 1981).

[26] M. D. Levenson in: *Chemical Applications of Nonlinear Raman Spectroscopy*, A. B. Harvey, ed. (Academic Press, New York 1981) pp. 214–222.

[27] P. D. Maker and R. W. Terhune, Phys. Rev. **137**, A801 (1965).

[28] W. M. Tolles, J. W. Nibler, J. R. McDonald, and A. B. Harvey, Appl. Spectrosc. **31**, 253 (1977).

[29] N. Bloembergen, *Nonlinear Optics* (Benjamin, Reading, MA, 1965).

[30] S.A.J. Druet and J.-P.E. Taran, Prog. Quantum Electron **7**, 1 (1981).

[31] W. B. Roh, P. W. Schreiber, and J.-P.E. Taran, Appl. Phys. Lett. **29**, 174 (1976).

[32] D. S. Moore, S. C. Schmidt, and J. W. Shaner, Phys. Rev. Lett. **50**, 1819 (1983).

[33] S. C. Schmidt, D. S. Moore, and M. S. Shaw, Phys. Rev. **B35**, 493 (1987).

[34] D. S. Moore, S. C. Schmidt, M. S. Shaw, and J. D. Johnson, J. Chem. Phys. **90**, 1368 (1989).

[35] S. C. Schmidt, D. S. Moore, M. S. Shaw, and J. D. Johnson, J. Chem. Phys. **91**, 6765 (1989).

[36] W. J. Nellis and A. C. Mitchell, J. Chem. Phys. **73**, 6137 (1980).

[37] S. A. Akhmanov, F. N. Gadjiev, N. I. Koroteev, R. Yu. Orlov, and I. L. Shumay, Appl. Opt. **19**, 859 (1980).

[38] S. C. Schmidt, D. Schiferl, A. S. Zinn, D. D. Ragan, and D. S. Moore, High Pressure Science and Technology **4**, 577 (1990).

[39] S. C. Schmidt, D. Schiferl, A. S. Zinn, D. D. Ragan, and D. S. Moore, submitted to J. Appl. Phys.

[40] A. S. Zinn, D. Schiferl, and M. F. Nicol, J. Chem. Phys. **87**, 1267 (1986).

[41] J. Belak, R. D. Etters, and R. LeSar, J. Chem. Phys. **89**, 1625 (1988).

[42] D. W. Chandler and G. E. Ewing, J. Chem. Phys. **73**, 4904 (1980).

[43] M. S. Shaw, J. D. Johnson, and B. L. Holian, Phys. Rev. Lett. **50**, 1141 (1983); J. D. Johnson, M. S. Shaw, and B. L. Holian, J. Chem. Phys. **80**, 1279 (1984); M. S. Shaw, J. D. Johnson and J. D. Ramshaw, J. Chem. Phys. **84**, 3479 (1986).

[44] S. A. Akhmanov, F. N. Gadzhiev, N. I. Koroteev, R. Yu. Orlov, and I. L. Shumai, JETP Lett. **27**, 243 (1978).

[45] J. Chesnoy, Chem. Phys. Lett. **125**, 267 (1986).

[46] J. Chesnoy and J.-J. Weis, J. Chem. Phys. **84**, 5378 (1986).

[47] S. I. Temkin and A. I. Burstein, JETP Lett. **24**, 86 (1976).

[48] S.R.J. Brueck, Chem. Phys. Lett. **50**, 516 (1977).

[49] D. W. Oxtoby, Annu. Rev. Phys. Chem. **32**, 77 (1981).

[50] J. W. Ellis and H. O. Kneser, Z. Phys. **86**, 583 (1933).

[51] R. P. Blickensderfer and G. E. Ewing, J. Chem. Phys. **51**, 5284 (1969).

[52] P. H. Krupenie, J. Phys. Chem. Ref. Data **1**, 423 (1972).

[53] V. I. Dianov-Klokov, Opt. Spectrosc. **6**, 290 (1959).

[54] V. I. Dianov-Klokov, Opt. Spectrosc. **13**, 109 (1962).

[55] V. I. Dianov-Klokov, Opt. Spectrosc. **21**, 233 (1966).

[56] C. W. Cho, E. J. Allin, and H. L. Welsh, Can. J. Phys. **41**, 1991 (1963).

[57] K. Syassen and M. Nicol, in *Physics of Solids Under High Pressure*, edited by J. S. Schilling and R. N. Shelton (North-Holland, Amsterdam, 1981), p. 33.

[58] M. Nicol and K. Syassen, Phys. Rev. B **28**, 1201 (1983).

[59] S.A.J. Druet, B. Attal, T. K. Gustafson, and J.-P.E. Taran, Phys. Rev. A **18**, 1529 (1978).

[60] N. Bloembergen, H. Lotem, and R. T. Lynch, Jr., Indian J. Pure and Appl. Phys. **16**, 151 (1978).

[61] B. Attal-Trétout, P. Berlemont, and J.-P.E. Taran, Indian J. Pure Appl. Phys. **26**, 159 (1988).

[62] H. Kiefte, M. J. Clouter, N. H. Rich, and S. F. Ahmad, Chem. Phys. Lett. **70**, 425 (1980).

[63] M. J. Clouter and H. Kiefte, J. Chem. Phys. **66**, 1736 (1977).

[64] S.R.J. Brueck, Chem. Phys. Lett. **53**, 273 (1978).

[65] D. C. McKean and R. A. Watt, *J. Mol. Spectrosc.*, Vol. 61, 184 (1976).

[66] G. Malewski, M. Pfeiffer, and P. Reich, *J. Mol. Structure*, **3**, 419 (1969).

[67] D. S. Moore and S. C. Schmidt, in *Proc. 9th Sym. Detonation*, preprint (Portland, Oregon, 1989), p. 80.

[68] J. R. Hill, D. S. Moore, S. C. Schmidt, and C. B. Storm, "Infrared, Raman, and Coherent Anti-Stokes Raman Spectroscopy of the Hydrogen Deuterium Isotopomers of Nitromethane," submitted to J. Phys. Chem.

[69] S. C. Schmidt, D. S. Moore, J. W. Shaner, D. L. Shampine, and W. T. Holt, Physica **139** & **140B**, 587. (1986).

[70] D. S. Moore, S. C. Schmidt, J. W. Shaner, D. L. Shampine, and W. T. Holt, in *Shock Waves in Condensed Matter—1985*, Y. M. Gupta, Ed. (Plenum Publishing, NY, 1986) p. 207.

[71] S. C. Schmidt, D. S. Moore, D. Schiferl, M. Châtelet, T. P. Turner, J. W. Shaner, D. L. Shampine, and W. T. Holt, in *Advances in Chemical Reaction Dynamics*, R. M. Rentzepis and C. Capellos, Eds. (D. Reidel Publishing, NY, 1986) p. 425.

[72] G. Herzberg, *Infrared and Raman Spectra* (Van Nostrand Reinhold, NY, 1945).

Index of Contributors

Springer Proceedings in Physics

Managing Editor: H. K. V. Lotsch

MIX
Papier aus verantwortungsvollen Quellen
Paper from responsible sources
FSC® C105338
FSC
www.fsc.org

If you have any concerns about our products,
you can contact us on
ProductSafety@springernature.com

In case Publisher is established outside the EU,
the EU authorized representative is:
Springer Nature Customer Service Center GmbH
Europaplatz 3, 69115 Heidelberg, Germany

Printed by Libri Plureos GmbH
in Hamburg, Germany